Environmental Friendly Catalysts for Energy and Pollution Control Applications

Environmental Friendly Catalysts for Energy and Pollution Control Applications

Editors

José Ignacio Lombraña
Héctor Valdés
Cristian Ferreiro

MDPI • Basel • Beijing • Wuhan • Barcelona • Belgrade • Manchester • Tokyo • Cluj • Tianjin

Editors

José Ignacio Lombraña
Chemical Engineering
University of the Basque
Country (UPV/EHU)
Leioa
Spain

Héctor Valdés
Clean Technologies
Laboratory
Catholic University of
the Most Holy Conception
Concepción
Chile

Cristian Ferreiro
Chemical Engineering
University of the Basque
Country (UPV/EHU)
Leioa
Spain

Editorial Office
MDPI
St. Alban-Anlage 66
4052 Basel, Switzerland

This is a reprint of articles from the Special Issue published online in the open access journal *Catalysts* (ISSN 2073-4344) (available at: www.mdpi.com/journal/catalysts/special_issues/environment_energy_pollution).

For citation purposes, cite each article independently as indicated on the article page online and as indicated below:

LastName, A.A.; LastName, B.B.; LastName, C.C. Article Title. *Journal Name* **Year**, *Volume Number*, Page Range.

ISBN 978-3-0365-4312-3 (Hbk)
ISBN 978-3-0365-4311-6 (PDF)

Contents

About the Editors

José Ignacio Lombraña

José Ignacio Lombraña received his PhD in Industrial Chemistry from the University of the Basque Country (UPV/EHU), where, since 1997, he has been a professor in the area of Chemical Engineering. Currently, he teaches Fluid Mechanics to chemical engineers and biotechnologists and, within the Master in Chemical Engineering, Water Treatment. He has guided a large number of pre-graduate and Master's studies at a rate of about 2 per year. His research activity deals with advanced technologies for the removal of water contaminants and applications of microwave drying processes, in which he has led different projects of his research group, funded by national (CICyT, MINECO) and international (CTP, FEDER) institutions. He has authored more than 80 papers that have appeared in internationally highly recognized journals. As an instructor for young post-graduate researchers, he was supervisor and director for 17 doctoral theses (13 defended). In the last ten years, he has been a committee member and has been the coordinator responsible for the PhD Chemical Engineering program for the UPV/EHU.

Héctor Valdés

Héctor Valdés has a Doctorate in Chemical Engineering from the Universidad de Concepción (2003). He was awarded with Postdoctoral Fellowship by the German Academic Exchange Service (DAAD) and conducted a postdoctoral research at the Technical University of Clausthal (2003-2004). In November 2017, he was appointed Professor (Full) at Universidad Católica de la Santísima Concepción. He teaches a wide variety of courses at the School of Engineering-UCSC. He has guided 39 pre-graduate reports, 7 Master's theses and two Doctoral theses. He is currently the Director of the Clean Technologies Laboratory, Engineering Faculty, UCSC. His main research areas are related to the development of new technologies for the control of air pollution and unconventional technologies for wastewater treatment based on advanced oxidation processes. Currently, he is the author of 75 Scopus articles (66 indexed in WoS), with 1652 Scopus citations and an H-index 19 (Scopus, WoS). He has two invention patents. He has been awarded with 10 FONDECYT project grants (6 as a Responsible Researcher, 1 as a Co-researcher and 3 as a Sponsoring Professor). He was the director of an international ECOS-CONICYT project. He contributes to bring closer the science and technology to high school students with the development of EXPLORA projects. He is a permanent reviewer of 15 prestigious journals in the WoS category.

Cristian Ferreiro

Cristian Ferreiro is a Postdoctoral Researcher in the Chemical Engineering Department at the University of the Basque Country (UPV/EHU). He completed his PhD degree in Chemical Engineering at UPV/EHU in 2021 and obtained a Master's Degree in Chemical Engineering at UPV/EHU in 2016. His main research interests are: i) advanced oxidation processes (AOPs), ii) tap water and industrial wastewater treatment, iii) the development of new catalytic materials to drive the circular economy in heterogeneous photocatalysis and ozonation processes, iv) water disinfection and v) water reuse through the integration of AOPs and membrane processes. He has been involved in four research projects and one enterprise project. He has published 12 papers in SCI journals, a book and more than 25 conference proceedings and communications with an H-index of 5 (WoS).

Preface to "Environmental Friendly Catalysts for Energy and Pollution Control Applications"

Catalysts are extensively used in various technologies, playing a fundamental role in the efficient generation of energy and in controlling industrial emissions. The use of catalysts has great importance in terms of its productive, economic and environmental implications. Catalysts not only improve production systems but also contribute controlling the emissions of pollutants. Catalysts are mainly applied in the gaseous phase (to control emissions from either fixed or mobile sources) and in the liquid phase (mostly for water and wastewater treatment). In the gas phase, catalysts are used to reduce the emissions of pollutants from power generation systems, chemical and manufacturing industries, and vehicle exhaust gases, whereas in the liquid phase, catalysts enable the catalytic oxidation of refractory organic pollutants, converting industrial wastewaters into admissible discharges.

Nevertheless, catalysts in some cases have a negative environmental impact due to their intensive production strategies. Therefore, the search for new materials such as environmentally friendly catalysts (EFCs) is urgently needed. Hence, the development of novel EFCs for sustainable energy production, to face climate change problems, and to abate industrial emissions has become a challenge in the current research fields. A great variety of catalytic materials, including single metals as well as mixed metals (and their oxides), are currently being used, either supported over alumina, silica, titania, platinum, ceria, strontium, cobalt, sodium citrate, activated carbons, and zeolites, or directly attached to the reactor itself, allowing their continuous use and avoiding waste emissions.

This Special Issue addresses applications of catalysts as an effective solution for the treatment of industrial emissions, focusing on catalytic ozonation, heterogeneous photocatalysis and Fenton catalysis, as well as the conversion into EFCs through optimized, heterogeneous Fenton-like variants and catalytic combustion. It also includes other applications involving biocatalysts for the design of bioanodes and new catalysts for energy production in fluid catalytic cracking (FCC) units. In addition to the above-mentioned catalysts, a significant part of the new research is focused on catalyst designs that allow recycling through a metal–organic framework (MOF). Thus, there is a marked tendency to develop materials that can be used in multiple operating cycles with high efficiency and selectivity.

This Special Issue of *Catalysts* comprises contributions from 57 authors distributed in 10 countries: Spain (11 authors), Chile (10 authors), Malaysia (10 authors), China (6 authors), Poland (6 authors), United States of America (6 authors), Ecuador (5 authors), Turkey (1 author), Bangladesh (1 author), and Nigeria (1 author). It is intended for academics and practitioners working in the energy and environmental sectors searching for environmentally sustainable catalysts. To gain a better insight into the essence of this Special Issue, summaries of the published papers are presented below. Please note that the first six contributions are about catalysts' design and their application and the others are related to the improvement of catalytic processes.

1. *Design of Co_3O_4@SiO_2 Nanorattles for Catalytic Toluene Combustion Based on Bottom-Up Strategy Involving Spherical Poly(styrene-co-acrylic Acid) Template.*

Co_3O_4@SiO_2 nanorattles were synthesized in this study, bearing in mind the need to develop optimal transition-metal-oxide-based catalysts for the combustion of volatile organic compounds (VOCs). The introduction of Co_3O_4 nanoparticles into empty SiO_2 spheres resulted in their loose distribution, facilitating the access of reagents to active sites and, on the other hand, promoting the involvement of lattice oxygen in the catalytic process. As a result, the catalysts obtained in this way showed a very high activity in the combustion of toluene, which significantly exceeded that achieved over a standard silica gel-supported Co_3O_4 catalyst.

2. *Catalytic Ozonation of Toluene over Acidic Surface Transformed Natural Zeolite: A Dual-Site Reaction Mechanism and Kinetic Approach.*

Volatile organic compounds (VOCs) can damage human health due to their carcinogenic effects. Catalytic ozonation using zeolite appears to be a valuable process to eliminate VOCs from industrial emissions at room temperature. Results obtained here provide a mechanistic approach during the initial stage of catalytic ozonation of toluene using an acidic surface on modified natural zeolite. Experimental evidence suggests that ozone is adsorbed and decomposed at Lewis acid sites, forming active atomic oxygen that leads to the oxidation of adsorbed toluene at Brønsted acid sites.

3. *Photocatalytic Study of Cyanide Oxidation Using Titanium Dioxide (TiO₂)-Activated Carbon Composites in a Continuous Flow Photo-Reactor.*

In this study, the photocatalytic oxidation of cyanide by titanium dioxide (TiO_2) supported on activated carbon (AC) was evaluated in a continuous flow UV photo-reactor. The continuous photo-reactor was made of glass and covered with a wood box to isolate the fluid of external conditions. These results showed that photocatalysis and the continuous photo-reactor's design enhanced the photocatalytic cyanide oxidation performance compared to an agitated batch system. Therefore, the use of TiO_2-AC composites in a continuous-flow photo-reactor is a promising process for the photocatalytic degradation of cyanide in aqueous solutions.

4. *Enhanced SO₂ Absorption Capacity of Sodium Citrate Using Sodium Humate.*

A novel method of improving the SO_2 absorption performance of sodium citrate (Ci-Na) using sodium humate (HA–Na) as an additive was put forward in this study. The influences of different Ci-Na concentration, inlet SO_2 concentration and gas flow rate on desulfurization performance were studied. The synergistic mechanism of SO_2 absorption by HA–Na and Ci-Na was also analyzed.

5. *Characterization of Anaerobic Biofilms Growing on Carbon Felt Bioanodes Exposed to Air.*

This research article elucidates the structure and performance of an electrogenic biofilm that develops on air-exposed, carbon-felt electrodes that are commonly used in bioelectrochemical systems. The research demonstrates the influence of a protective aerobic layer present in the biofilm (mainly formed by *Pseudomonas* genus bacteria) that prevents electrogenic bacteria (such as *Geobacter* sp.) from hazardous exposure to oxygen during its normal operation.

6. *Metal–Organic Frameworks (MOFs) and Materials Derived from MOFs as Catalysts for the Development of Green Processes.*

This review paper focused on the development of metal–organic frameworks (MOFs) serving as catalysts for the conversion of carbon dioxide into short-chain hydrocarbons and the generation of clean energies starting from biomass. The common patterns in the performance of the catalysts, such as the acidity of MOFs, metal nodes, and surface area and the dispersion of the active sites, are discussed in this review.

7. *Practical Approaches towards NOₓ Emission Mitigation from Fluid Catalytic Cracking (FCC) Units.*

This research article reviews options for cost-effective and emissions mitigation, using optimal amounts of precious metals while evaluating the potential benefits of current promoter dopant packages. Thus, the refinery is no longer forced to make a promoter selection based on preconceived notions regarding only precious metal activity but can rather make decisions based on the best financial strategy without measurable loss of CO/NO_x emission selectivity.

8. *Application of a Combined Adsorption-Ozonation Process for Phenolic Wastewater Treatment in a Continuous Fixed-Bed Reactor.*

This research article studies the removal of phenol from industrial effluents through catalytic ozonation using granular activated carbon in a continuous fixed-bed reactor. Based on the evolution of total organic carbon (TOC) and phenol concentration, a kinetic model was proposed to study the effect of the operational variables on the combined adsorption–oxidation (Ad/Ox) process. The interpretation of the constants allows for the study of the benefits and behaviour of the activated carbon during the ozonization process under different conditions affecting adsorption, oxidation, and mass transfer.

9. *Turbidity Changes during Carbamazepine Oxidation by Photo-Fenton.*

This research article studies the turbidity generated during the Fenton photoreaction applied to the oxidation of waters containing carbamazepine as a function of factors such as pH, H_2O_2 concentration and catalyst dosage. The results allow for establishing the degradation pathways and the main decomposition byproducts.

10. *Pragmatic Approach toward Catalytic CO Emission Mitigation in Fluid Catalytic Cracking (FCC) Units.*

This work reveals how CO promoter design strategies can afford a tangible and immediate CO conversion efficiency increase without the need for additional loading of precious metals. The key lies in the support material architecture that is essential to boost the CO conversion and reduce the NO_x generation in the FCC unit. It was demonstrated that the suppression of Pt sintering, as well as the enhancement of the oxygen mobility on the catalyst surface, leads to a lower amount and cost of Pt and a higher usage rate compared to current industry-standard designs.

11. *Optimization of Fenton Technology for Recalcitrant Compounds and Bacteria Inactivation.*

In this research article, Fenton technology was applied to decolorize methylene blue (MB) and to inactivate *E. coli* K12, used as recalcitrant compound and bacteria models, respectively, in order to provide an approach into single and combinative effects of the main process variables influencing the Fenton technology. Box–Behnken design (BBD) was applied to evaluate and optimize the individual and interactive effects of three process parameters, namely ferrous ion concentration, molar ratio between H_2O_2 and ferrous ion, and pH for Fenton technology.

12. *Photocatalysis for Organic Wastewater Treatment: From the Basis to Current Challenges for Society.*

This paper reviews details on the fundamentals, the common photocatalyst preparation for coupling heterojunction, the morphological effect and photocatalyst-characterization techniques. The important variables that potentially affect the process efficiency, namely catalyst dosage, pH, the initial concentration of sample pollution, irradiation time by light, temperature, durability and stability of the catalyst, are also discussed. Overall, this paper offers an in-depth perspective of photocatalytic degradation, according to pollution cases, and its future direction.

José Ignacio Lombraña, Héctor Valdés, and Cristian Ferreiro

Editors

 *catalysts*

Article

Design of Co₃O₄@SiO₂ Nanorattles for Catalytic Toluene Combustion Based on Bottom-Up Strategy Involving Spherical Poly(styrene-*co*-acrylic Acid) Template

Anna Rokicińska [1], **Magdalena Żurowska** [1], **Piotr Łątka** [1], **Marek Drozdek** [1], **Marek Michalik** [2] and **Piotr Kuśtrowski** [1,*]

[1] Faculty of Chemistry, Jagiellonian University, Gronostajowa 2, 30-387 Kraków, Poland; anna.rokicinska@uj.edu.pl (A.R.); magda.zurowska@student.uj.edu.pl (M.Ż.); piotr.latka@uj.edu.pl (P.Ł.); marek.drozdek@uj.edu.pl (M.D.)
[2] Faculty of Geography and Geology, Jagiellonian University, Gronostajowa 3a, 30-387 Kraków, Poland; marek.michalik@uj.edu.pl
* Correspondence: piotr.kustrowski@uj.edu.pl; Tel.: +48-12-686-2415

Abstract: Bearing in mind the need to develop optimal transition metal oxide-based catalysts for the combustion of volatile organic compounds (VOCs), yolk-shell materials were proposed. The constructed composites contained catalytically active Co_3O_4 nanoparticles, protected against aggregation and highly dispersed in a shell made of porous SiO_2, forming a specific type of nanoreactor. The bottom-up synthesis started with obtaining spherical poly(styrene-*co*-acrylic acid) copolymer (PS30) cores, which were then covered with the SiO_2 layer. The Co_3O_4 active phase was deposited by impregnation using the PS30@SiO₂ composite as well as hollow SiO_2 spheres with the removed copolymer core. Structure (XRD), morphology (SEM), chemical composition (XRF), state of the active phase (UV-Vis-DR and XPS) and reducibility (H_2-TPR) of the obtained catalysts were studied. It was proven that the introduction of Co_3O_4 nanoparticles into the empty SiO_2 spheres resulted in their loose distribution, which facilitated the access of reagents to active sites and, on the other hand, promoted the involvement of lattice oxygen in the catalytic process. As a result, the catalysts obtained in this way showed a very high activity in the combustion of toluene, which significantly exceeded that achieved over a standard silica gel supported Co_3O_4 catalyst.

Keywords: volatile organic compounds; toluene; core–shell structures; spherical polymer templates; Co_3O_4

Citation: Rokicińska, A.; Żurowska, M.; Łątka, P.; Drozdek, M.; Michalik, M.; Kuśtrowski, P. Design of Co₃O₄@SiO₂ Nanorattles for Catalytic Toluene Combustion Based on Bottom-Up Strategy Involving Spherical Poly(styrene-*co*-acrylic Acid) Template. *Catalysts* 2021, 11, 1097. https://doi.org/10.3390/catal11091097

Academic Editors: José Ignacio Lombraña, Héctor Valdés and Cristian Ferreiro

Received: 29 July 2021
Accepted: 10 September 2021
Published: 11 September 2021

Publisher's Note: MDPI stays neutral with regard to jurisdictional claims in published maps and institutional affiliations.

1. Introduction

Considering the necessity to implement efficient methods for the elimination of volatile organic compounds (VOCs) from contaminated air, active and selective catalysts have been extensively developed, which could allow the lowering of the reaction temperature of the total oxidation of VOCs. In general, systems tested so far can be divided into two main types due to the nature of the catalytically active phase, i.e., materials based on noble metals and transition metal oxides. The catalysts containing noble metals (e.g., Pt, Pd, Au or Ag) supported on various oxides (mainly SiO_2 or γ-Al_2O_3) have an obvious advantage due to their extraordinary activity at low temperatures [1,2]. However, high production costs and sensitivity to poisoning (by sulfur- and halogen-containing compounds) constitute a serious limitation in their application. A compromise solution is therefore to use transition metal oxides which combine the benefits of low manufacturing costs with relatively high efficiency and enhanced resistance to deactivation. The final performance of these catalysts is influenced by various factors, including properties of active components, dispersion of the active phase, and size and morphology of grains [1–5]. Co_3O_4, CuO, MnO_x, Fe_2O_3 and NiO are the most common oxide systems used in the combustion of VOCs. As in the case of noble metals, the involvement of a support also has a positive effect on dispersion and thermal stability of the catalytically active phase. The selection of the appropriate material forming the support depends on many factors (including its porosity, chemical activity and strength of interaction with the active phase), but a typical choice is γ-Al_2O_3, SiO_2, TiO_2, ZrO_2, CeO_2 and zeolites [1,5,6]. Mixed systems containing at least two transition or noble metals in their structure have also been increasingly studied. The effect of improved catalytic activity obtained in such cases is most often related to promoting the mobility of lattice oxygen and facilitating electron transfer [7].

Particularly promising catalysts for the total oxidation of VOCs could be core–shell composites, which contain a core located inside another domain forming an outer layer [8,9]. The structures belonging to the group of core–shell materials also include hollow core–shell and yolk–shell systems [10]. The advantages of using such composite materials include: (i) ability to protect the core against effects of environmental changes outside the material, (ii) limiting the possibility of increasing the volume and maintaining structural integrity, (iii) protecting the core particles against aggregation into larger clusters, and (iv) selective molecule penetration into the core. The core–shell materials combine properties of both the shell and the core. By covering nanoparticles with one or more layers of other materials, systems with intensified or completely changed physical or chemical properties in relation to the initial structure can be obtained [11,12]. Few core–shell structures for VOCs combustion, limited mainly to systems containing noble metal nanoparticles embedded in an oxide shell (e.g., Pd@CeO_2 [13]), have been described in the literature so far.

Due to availability and the relatively low price of substrates, as well as simplicity and short time of synthesis, polymer templates are often selected as structure-directing agents to obtain core–shell composites by the bottom-up approach [14,15]. Oxide layers, e.g., SiO_2, TiO_2, ZrO_2 or CeO_2, are deposited onto the polymer cores, and then the organic interior is easily removed by calcination or chemical etching with acidic, alkaline or organic agents [16–20]. In the synthesis of such core–shell systems, among other things, heterophase polymerization (leading to encapsulation of inorganic nanoparticles), sol-gel processes (leading to coating polymer colloids with an inorganic layer) and layer-by-layer deposition are used. Regardless of the choice of the synthesis method, it is necessary to facilitate interaction between a polymer and an inorganic component [18,20–22].

Amines are often used to modify a surface of polystyrene templates [23]. The amine layer adsorbed on the latex surface induces the formation of the silica layer through electrostatic interactions. Upon dissolution or calcination, the polymer is removed together with the amine, creating a hollow silica sphere. However, the attachment of the amine to the latex surface reduces the integrity of the resulting SiO_2 shell. This effect can be overcome by using silanes or silica nanoparticles as hollow shell precursors, but this drastically increases the cost of synthesis, and the process becomes more complicated.

One of the most frequently used techniques for the preparation of polymer cores is emulsion polymerization. It is a type of radical polymerization which is based on the use of an emulsion containing water, monomer and surfactant. Potassium or ammonium persulfate is widely used as an initiator in the emulsion polymerization [14]. The process often takes place in an o/w (oil in water) emulsion in which monomer droplets are dispersed in water by surfactants. Other types of polymerization, such as so-called reverse w/o emulsion (water in oil) as well as w/o/w and o/w/o microemulsion, are rarely chosen [14,16,20]. The mechanism of surfactant-free emulsion polymerization has also been described [24]. The amount of polymer contained in the aqueous phase influences the nucleation mechanism and the change in dispersion of the systems during the polymerization of molecules with different properties [25,26]. A common solution used during the polymerization of styrene is an addition of methacrylic or acrylic acid. Ge and co-workers [15] synthesized poly(styrene-*co*-methacrylic acid) spheres by the emulsion polymerization without emulsifiers. The addition of methacrylic acid reduced the size of the spheres formed (to ca. 370 nm) and the thickness of the SiO_2 layer (to ca. 50 nm) compared to the synthesis with the use of a purely polystyrene template. It was unequivocally found that the presence of methacrylic acid in the template has a beneficial effect on the formation of homogeneous silica layers on the surface of latex templates. Thus, in order to increase the degree of uniform bonding of SiO_2 with the surface of polystyrene spheres, they can be functionalized, for example, with acid groups, by copolymerization of styrene with methacrylic acid or sulfonation with sulfuric acid [14,15]. The deposition of the SiO_2 shell on the poly(styrene-*co*-methacrylic acid) spheres was carried out using various aging conditions—at 150 °C (obtaining materials with a mesoporous shell) or at room temperature (microporous spheres). Moreover, aging at room temperature resulted in a layer of SiO_2 with a thickness of ca. 30 nm, while the higher temperature of the thermal treatment reduced its thickness to ca. 15 nm.

In this paper a pioneering method of synthesis of core–shell catalysts with an active phase in the form of Co_3O_4 spinel nanoparticles located inside a porous SiO_2 shell is described. In the first step of preparation, a spherical copolymer template (namely poly(styrene-*co*-acrylic acid)) with a defined grain size within the range of 200–300 nm was obtained. The SiO_2 coating was deposited on the surface of the latex spheres by performing the hydrolysis and condensation of tetraethyl orthosilicate (TEOS) in a water-isopropanol solution of ammonia. The formed composite particles with a copolymer core retained or removed by calcination were modified with various amounts of cobalt-based active phase. The samples containing Co_3O_4 nanoparticles were characterized by a wide range of physicochemical techniques, and above all, their activity in the toluene combustion was determined. The role of selected parameters influencing the catalytic performance of the developed composite materials in the studied process is discussed. At the same time, a very significant influence of the proper path selection, resulting in various dispersion of the active component within the Co_3O_4@SiO_2 yolk–shell structure, on the achieved conversion of the organic pollutant is shown.

2. Results and Discussion

The main objective of this study was to obtain composite materials containing Co_3O_4 nanoparticles embedded and protected inside a spherical, porous SiO_2 shell, which would be highly active and selective in the catalytic combustion of VOCs. In the first step of the synthesis, a spherical poly(styrene-*co*-acrylic acid) latex containing 30 wt.% acrylic acid in relation to styrene was prepared. The formed PS30 emulsion was examined by dynamic light scattering (DLS), proving the high monodispersity of the particles with diameters within the narrow range of 240–260 nm and the average ζ potential of −42.8 mV measured in deionized water at pH = 7 (Figure 1a). The latter value is related to the high amount of carboxyl groups on the surface of the copolymer nanoparticles and confirms high colloidal stability of the formed emulsion [27].

The prepared latex spheres were further modified by the deposition of the silica shell using the Stöber strategy. The coverage of the PS30 surface with the silica coating slightly shifts the average ζ potential to −39.6 mV (Figure 1a). This change indicates a variation in the chemical composition of the surface. In this case, the negative zeta potential mainly comes from the ionization of silanol groups [28]. However, since the zeta potential value remains outside the range from −30 mV to +30 mV, it should be considered that the PS30@SiO_2 nanoparticles have sufficient repulsion force to maintain colloidal stability [29]. In the literature, the correlation between the SiO_2 particle size and the zeta potential value was discussed [30]. It was postulated that as the size of the spheres decreased, the average ζ potential raised and oscillated in the range from −37.7 mV to −45.5 mV.

The introduction of the inorganic part was additionally confirmed by thermal analysis (Figure 1b). The mass loss of 3.2% in the temperature range up to approx. 140 °C is associated with removal of residual solvents, unreacted monomers and physically adsorbed water. Desorption of these components is related to the heat consumption, which confirms the endothermic peak observed by differential thermal analysis (DTA). At higher temperatures, the parts of the chain of acrylic acid units slowly decompose and polystyrene depolymerizes. Rapid decomposition begins at ca. 320 °C with a maximum rate at 360 °C.

This exothermic step is mainly due to the combustion of the polymer. After the analysis, the copolymer-free SiO_2 remains, the content of which in the PS30_Si material is estimated to be 84.5 wt.%.

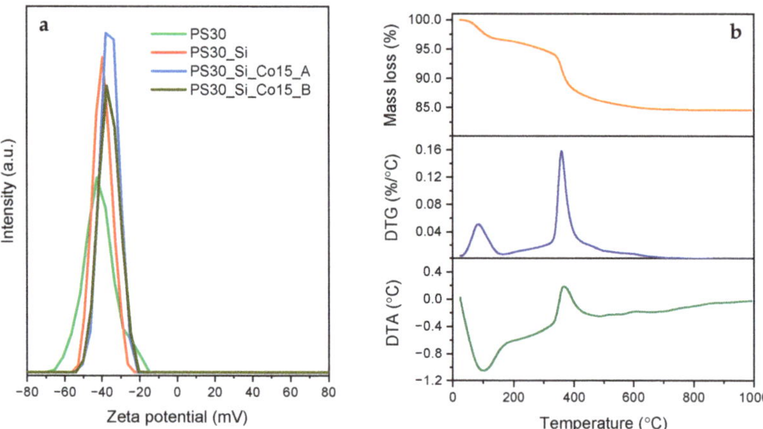

Figure 1. (**a**) Zeta potential distribution of samples dispersed in water, and (**b**) mass loss (TG), mass loss rate (DTG) as well as differential thermal analysis (DTA) curves for PS30_Si measured at air atmosphere.

The obtained PS30_Si composite was calcined at the oxidizing atmosphere to remove the copolymer core and subsequently studied using scanning electron microscopy (SEM). The SEM image shown in Figure 2a confirms a well-formed spherical structure of the grains. The particles are homogeneous, and their average size is close to ca. 300 nm. This means that, taking into account the size of the latex spheres determined earlier by DLS (240–260 nm), the thickness of the SiO_2 layer is approx. 20–30 nm.

Figure 2. SEM images of calcined PS30_Si (**a**), PS30_Si_Co15_A (**b**) and PS30_Si_Co15_B (**c**).

Simultaneously, textural properties of PS30_Si before and after removal of the copolymer core were analyzed. The measured isotherms of N_2 adsorption, shown in Figure 3, reveal a very interesting conclusion. The shape of the isotherm for the starting PS30@SiO$_2$ structure can be classified as type II, according to the IUPAC (International Union of Pure and Applied Chemistry, Research Triangle Park, NC, USA) nomenclature, which is characteristic of materials containing pores with diameters within the macroporosity range. It is manifested by a very intense adsorption effect observed at the relative pressure $p/p_0 > 0.9$. Interestingly, after removing the PS30 template, this effect largely disappears. It should therefore be assumed that the adsorption of N_2 prior to the core elimination corresponds to the condensation of the adsorbate in slits formed between the silica shell and the copolymer core, which during drying shrinks slightly, forming the free spaces. These observations are reflected in the determined textural parameters. For the PS30_Si filled by the copolymer core, the total pore volume (V_{total}, constituted mainly by wide meso- and macropores) is 0.319 cm^3/g, and consequently the surface area is determined according to the Brunauer–Emmett–Teller theory (S_{BET}) is 41 m^2/g (Table 1). The destruction of the copolymer scaffold results in a significant decrease in V_{total} and S_{BET} to 0.059 cm^3/g and 15 m^2/g, respectively, since the voids formed inside the hollow SiO_2 spheres are too large for the adsorbate to condense therein.

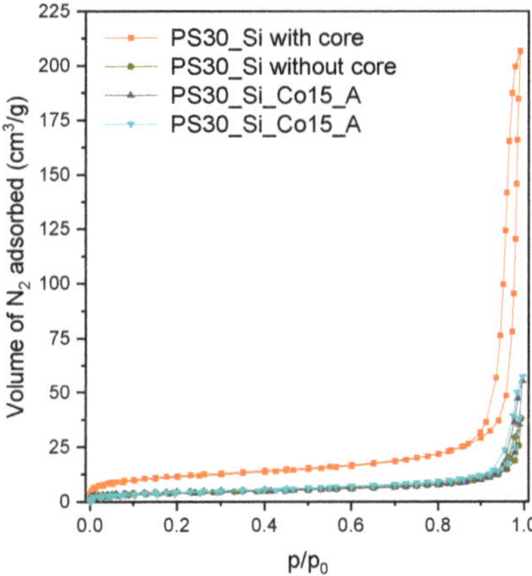

Figure 3. N$_2$ adsorption isotherms for PS30_Si before and after the PS30 core removal as well as the PS30_Si_Co15 catalysts synthesized in the paths A and B.

Table 1. Chemical composition, Co$_3$O$_4$ crystallite size and textural parameters of the studied PS30_Si-based samples.

Sample	Co Loading [wt.%]	Mean Co$_3$O$_4$ Crystallite Size [Å]	Textural Parameters	
			S_{BET} [m^2/g] [a]	V_{total} [cm^3/g] [b]
PS30_Si with PS30 core	-	-	41	0.319
PS30_Si without PS30 core	-	-	15	0.059
PS30_Si_Co5_A	5.9	225	18	0.063
PS30_Si_Co10_A	11.6	298	17	0.083
PS30_Si_Co15_A	13.6	338	16	0.086
PS30_Si_Co5_B	5.8	237	16	0.069
PS30_Si_Co10_B	11.1	282	19	0.089
PS30_Si_Co15_B	15.3	358	16	0.089

[a] specific surface area calculated using the Brunauer–Emmett–Teller (BET) model, [b] total pore volume measured as the amount of N$_2$ adsorbed at p/p$_0$ ~ 0.99.

X-ray fluorescence (XRF) analyses were carried out to confirm the introduction of the Co-containing active phase onto the silica support. The determined Co loadings are presented in Table 1. It should be stated that the chemical composition of the studied materials is close to the intended values.

The modification of PS30@SiO$_2$ with the Co$_3$O$_4$ phase causes a further change in average ζ potential to a value of ca. -37 mV which is very similar for the calcined PS30_Si_Co15 samples prepared both on path A and B (Figure 1a). Therefore, it should be stated that a part of the Co modifier was deposited on the outer surface of SiO$_2$ nanospheres, changing its charge a little. Figure 2b,c demonstrate the SEM images collected for the calcined PS30_Si_Co15 obtained in the paths A and B, respectively. For PS30_Si_Co15_A no significant differences in morphology are observed compared to the hollow SiO$_2$ spheres (Figure 2a). In the case of PS30_Si_Co15_B, a greater aggregation of nanoparticles is noted. They form distinct clusters, and consequently polydispersity increases significantly.

In Figure 3 two examples of the N$_2$ adsorption isotherms for the samples with the highest Co loadings are demonstrated, while the textural parameters for all synthesized

Co-loaded samples are summarized in Table 1. It can be concluded that the introduction of cobalt oxide results in a slight increase in the pore volume compared to the hollow SiO_2 spheres. This effect is also correlated with the amount of active phase deposited—the more it is, the greater value of V_{total}. Therefore, it should be assumed that the cobalt oxide nanoparticles formed as a result of calcination create secondary porosity. These nanograins are arranged partly on the outer surface of the SiO_2 nanospheres, partly inside their pores, but above all on the inner surface of the SiO_2 shells. In the case of the synthesis by the path B, the deposition of the active phase precursor takes place in the completely empty interior of the SiO_2 spheres compared to the path A, in which the Co(II) nitrate solution is introduced into the spaces between the shell and the copolymer core and only to a limited extent inside the PS30 template structure. This leads to different contents of the active phase precursor inside the spheres, and thus various arrangements of the formed oxide nanograins. For the PS30_Si_Co_B materials, they are a bit more loosely aggregated, and therefore in this case the V_{total} values are slightly higher than in the case of the analogous PS30_Si_Co_A catalysts.

The crystalline structure of the synthesized materials was studied by X-ray diffraction (XRD). Figure 4a,b present the diffractograms registered for the PS30_Si_Co_A and PS30_Si_Co_B catalysts, respectively.

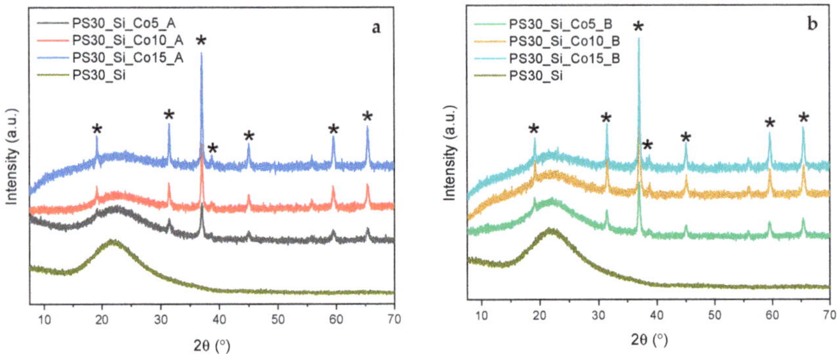

Figure 4. XRD patterns of PS30_Si-based catalysts synthesized in the path A (**a**) and B (**b**). Reflections characteristic for the Co_3O_4 phase are marked with an asterisk.

The diffraction patterns clearly display a broad, low, intense diffraction peak (101) at $2\theta{\sim}22°$, originating from the poorly ordered silica in the composite shells [31]. The presence of the transition metal oxide in the structure of the developed materials is confirmed by other characteristic reflections observed at 2θ equal to $19.0°$, $31.2°$, $37.0°$, $38.5°$, $44.8°$, $59.3°$ and $65.2°$ with intensities growing with the increasing Co content. They correspond to (111), (220), (311), (222), (400), (511) and (440) lattice planes in the crystal structure of Co_3O_4 spinel phase (space group $Fd3m$), respectively (PDF 00-009-0418). The calculated values of the parameter a of 8.06 Å and the unit cell volume of 523.2 Å are only slightly lower compared to the theoretical data of 8.084 Å and 528.30 Å, respectively [32]. The mean Co_3O_4 crystallite sizes determined on the basis of the Scherrer equation from the broadening of the most intense (311) diffraction are given in Table 1. A very clear correlation between the Co_3O_4 mean crystallite size and the Co content for the materials in both series is observed.

In order to determine the chemical environment of the active phase in the investigated catalysts, UV-Vis-DR measurements were performed. The collected spectra are shown in Figure 5a,b. The results for both series of the catalysts are very similar, and fully confirm the presence of the Co_3O_4 spinel phase. Around 260 nm, the band corresponding to the $O^{2-} \rightarrow Co^{3+}$ charge transfer is distinguished. However, the main bands occur at higher wavelengths. Their width is indicative of different ordering of spinel particles, but generally the absorption band in the range of 300–500 nm is attributed to octahedrally coordinated

Co^{3+} ions, whereas another wide band at 600–800 nm is assigned to the electron transitions characteristic for Co^{2+} ions in tetrahedral coordination in the Co_3O_4 spinel [33–35].

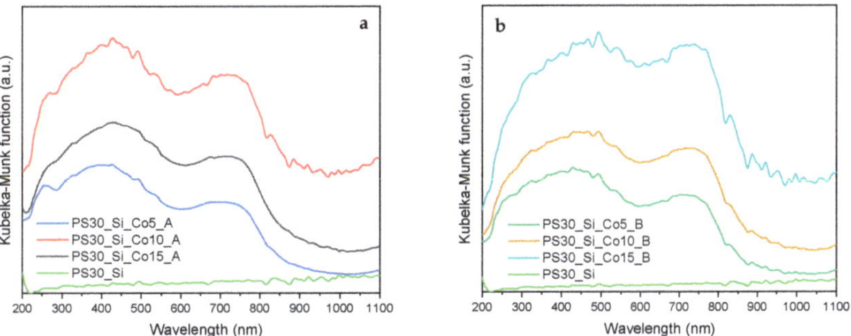

Figure 5. UV-Vis-DR spectra of PS30_Si-based catalysts synthesized in the path A (**a**) and B (**b**).

The surface composition of the Co-loaded materials synthesized in the paths A and B was more precisely characterized by X-ray photoelectron spectroscopy (XPS). The high-resolution XPS spectra recorded in the Co 2p region (Figure 6a,b) display a characteristic doublet of the Co $2p_{3/2}$ and Co $2p_{1/2}$ peaks, which is related to the spin-orbital splitting. The determined distance between the peaks in this doublet for all studied materials is $\Delta = 15.3$–15.4 eV, which is the typical value reported for the Co_3O_4 phase in the literature [36,37].

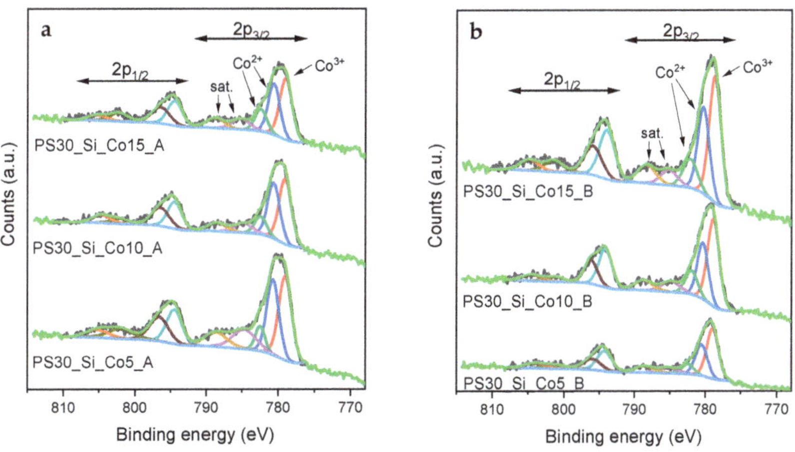

Figure 6. XPS Co 2p spectra of PS30_Si-based catalysts synthesized in the path A (**a**) and B (**b**).

The studied materials are characterized by the presence of Co^{3+} ions in octahedral coordination (779.0 ± 0.3 eV) and Co^{2+} in tetrahedral coordination (780.6 ± 0.2 eV) [38]. In the fitting model used, an additional component related to Co^{2+} multiplet splitting at 782.4 ± 0.3 eV was added [38–40]. In order to describe the surface composition in more detail, the Co^{2+}/Co^{3+} ratio was calculated based on the performed deconvolution. The values for the PS30_Si_Co_A catalysts are close to 1.0, while for PS30_Si_Co_B they are at the level of 0.8, which indicates that in the latter case the Co^{3+} ion content is higher. The presence of more Co^{3+} ions usually enhances the reducibility and increases the activity of the Co_3O_4 spinel phase in the catalytic combustion of VOCs [41,42]. The calculated ratio of the Co concentration on the surface (from the XPS results) to that in the total volume (from the XRF results) provides clear differences in the cobalt deposition on the surface of the

SiO_2 spheres (Figure 7). For the PS30_Si_Co_B catalysts, the enrichment of the surface with cobalt ranges from 40% to 90%. In the case of the catalysts obtained in path A, there is a clear tendency to change the distribution along with the Co content. The PS30_Si_Co5_B sample has almost four times more Co on the surface than in the volume, while for higher Co contents, the enrichment of the surface with the active phase significantly decreases (for PS30_Si_Co15_B, the Co content on the surface is only 10% higher than in the entire volume).

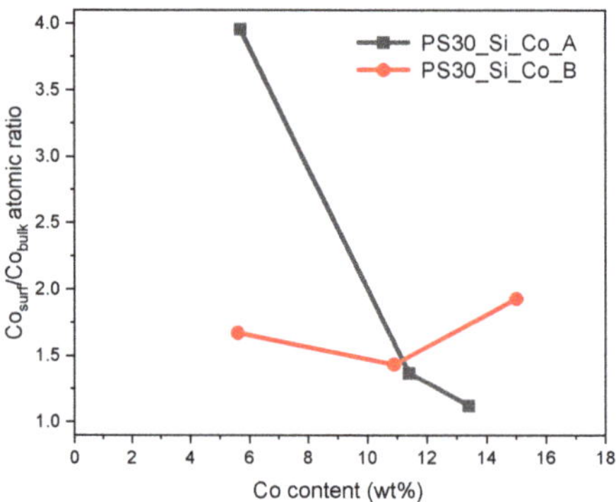

Figure 7. Relationship between the Co_{surf}/Co_{bulk} atomic ratio and the total cobalt content in the studied yolk–shell catalysts.

The activity of oxide catalysts in the total oxidation of VOCs is often directly related to their redox properties, as the reaction involves lattice oxygen from the subsurface layers and even the interior of the solid phase [43]. Susceptibility to reduction oxides more easily creates surface defects, which promotes catalytic activity [44]. Therefore, we decided to use the temperature-programmed reduction (H_2-TPR) technique to study the reducibility of the developed catalysts (Figure 8a,b).

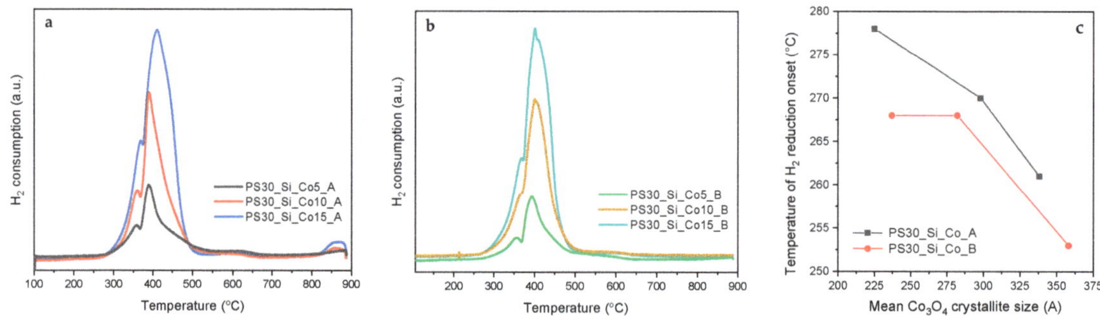

Figure 8. Temperature-programmed reduction profiles of PS30_Si-based catalysts synthesized in the path A (**a**) and B (**b**) as well as relationship between temperature of reduction onset and mean Co_3O_4 crystallite size (**c**).

Mandal et al. [45] postulated that large Co_3O_4 particles are typically reduced directly to metallic cobalt in one step, while the reduction of Co_3O_4 nanoparticles is often a two-step process. In the case of the tested materials, two main reduction peaks in the H_2-TPR profiles are observed, which confirm the nanometric dimensions of the active phase particles. The reduction of Co_3O_4 spinel proceeds through two steps described by the following reaction equations:

$$Co_3O_4 + H_2 \rightarrow 3CoO + H_2O$$

$$CoO + H_2 \rightarrow Co^0 + H_2O$$

The first reduction step begins in the temperature range of 250–280 °C and reaches its maximum rate at about 360–370 °C. A deeper reduction in the active phase occurs at higher temperatures (a maximum of greater intensity at approximately 390–410 °C). For both series, an increase in the cobalt content in the studied catalysts results in an increase in the intensity of the reduction maxima and their shift towards higher temperatures. These effects are rather expected, since the higher content of the component to be reduced requires the consumption of more hydrogen, and on the other hand, kinetic circumstances lead to the achievement of the maximum rate of the polythermal reduction process at increasingly higher temperatures.

However, very interesting conclusions can be deduced by analyzing the temperatures of reduction onset determined for the individual catalysts (Figure 8c). These values were determined based on an increase in a detector signal to a value exceeding by 1% a baseline value determined in the temperature range of 200–220 °C, in which no reaction resulting in the consumption of H_2 occurs. It is clearly visible that the reduction process begins at lower temperatures in the case of larger Co_3O_4 crystallites, the formation of which is favored by the higher Co content. Nevertheless, it is not the most important parameter responsible for easier reducibility. Clearly, the PS30_Si_Co_A catalysts require higher temperatures to initialize the reduction process than the analogous PS30_Si_Co_B materials, containing crystallites of similar sizes and comparable Co loadings. It should therefore be assumed that the observed differences result from a different distribution of Co_3O_4 nanoparticles, which in turn are varied by the synthesis procedure. In the case of PS30_Si_Co_A, the infiltration of the $Co(NO_3)_2$ solution leads to the aggregation of more active phase nanoparticles on the inner walls of the SiO_2 spheres, which closely adhere to the surface of the support stabilizing them. Differently, the introduction of the $Co(NO_3)_2$ solution into the hollow SiO_2 spheres provides a larger space for crystallization and the subsequent transformation of the active phase precursor to the oxide form. Hence, in this case, the Co_3O_4 crystallites are loosely bonded to the silica support and thus are easier to be reduced. The shift of the reduction onset towards lower temperatures with the activation of lattice oxygen should result in higher catalytic activity in the combustion of VOCs [38,46].

The Co_3O_4-containing catalysts synthesized on the basis of the PS30_Si composite template were tested for the catalytic activity in the total oxidation of toluene (Figure 9a). The catalysts of both series A and B showed high activity in this reaction, the main products of which were CO_2 and H_2O. Selectivity to CO_2 remained at an extremely high level. For each material, regardless of the reaction temperature, it was > 99.9%. In only selected cases, the formation of traces of benzene, as a by-product, was detected. It is observed that the PS30_Si_Co_A catalysts demonstrate very similar activity, while the materials calcined prior to the inclusion of the cobalt precursor (PS30_Si_Co_B) exhibit apparent differences in the catalytic performance. Nevertheless, it should be noted that for the PS30_Si_Co_B catalysts, which are characterized by easier reducibility as previously shown by the H_2-TPR measurements, significantly better catalytic results were achieved. The toluene conversion at the level of 50% for the most active catalysts of both series was attained at temperatures of 306 °C and 266 °C for PS30_Si_Co10_A and PS30_Si_Co15_B, respectively (Table 2). The extraordinary catalytic activity of the PS30_Si_Co_B catalysts can be found even better when the reaction rate (r) is compared:

$$r = \frac{F_{toluene} * C_{toluene}}{m_{Co}}$$

where $F_{toluene}$ represents the flow rate of toluene (mol/s), $C_{toluene}$ is the toluene conversion, and m_{Co} is the content of Co in the working catalyst. The results collected for the experimental points measured at 275 °C are shown in Figure 9b.

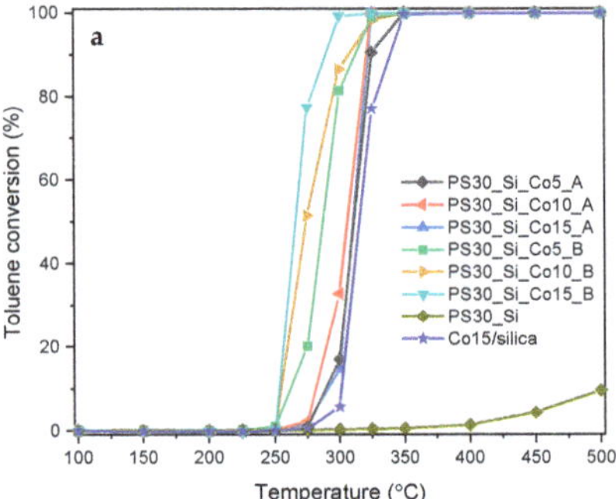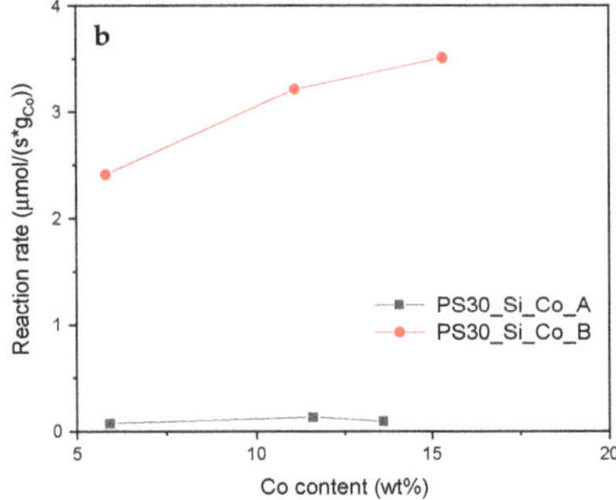

Figure 9. (**a**) Toluene conversion achieved at various temperatures over PS30_Si-based catalysts compared to the reference Co15/silica sample, and (**b**) rate of the toluene combustion at 275 °C calculated in relation to Co content.

Table 2. Temperatures at which 20, 50 and 90% toluene conversion was achieved in the toluene combustion over PS30_Si-based catalysts.

Sample	Temperature [°C]		
	T_{20}	T_{50}	T_{90}
PS30_Si_Co5_A	301	312	325
PS30_Si_Co10_A	290	306	322
PS30_Si_Co15_A	302	311	322
PS30_Si_Co5_B	274	287	312
PS30_Si_Co10_B	260	274	307
PS30_Si_Co15_B	256	266	290

It should therefore be emphasized that the procedure used to introduce the active phase is of crucial importance in profiling the catalytic activity of Co_3O_4@SiO_2 nanorattles. Due to the larger available nanospace for the arrangement of the active phase grains, much more favorable effects are obtained by the deposition of the Co_3O_4 precursor in the empty SiO_2 spheres. In this case, the absence of strong interactions between the spinel phase crystallites and the silica support is confirmed even for the content of 15 wt.% of Co. Furthermore, the oxide particles do not aggregate significantly and finally the highest reaction rate is reached.

In order to verify the real usefulness of the developed catalysts in the process of toluene combustion, the catalytic test was also carried out for the material prepared using commercial spherical silica gel, which was modified by introducing 15 wt.% of Co with an analogous procedure as for the PS30_Si_Co samples. As can be seen from Figure 9a, the reference Co15/silica sample shows a lower activity than all developed yolk–shell catalysts. Compared to the most active PS30_Si_Co15_B material, this difference is huge—the T_{50} value is reached at a temperature approx. 50 °C lower.

On the other hand, additional measurements for the most active PS30_Si_Co15_B catalyst were performed to show its stability and behavior at much higher toluene content (the toluene content in the feed was increased from 1000 ppm to 2500 ppm). After the catalytic run in the temperature range of 100–500 °C in the presence of 1000 ppm of toluene in the flowing air (cycle 1), the reactor was cooled down to 100 °C and the measurement was repeated (cycle 2). After cycle 2, the toluene concentration in the feed was increased to 2500 ppm and two subsequent polythermal runs (cycle 3 and 4) were carried out. The results presented in Figure 10 clearly confirm high stability of the developed material in the repeated runs. Only a slight decrease in the toluene conversion at 275 °C is found. Moreover, an increase in the toluene concentration results in a noticeable decrease in its conversion from 77.4% (cycle 1 at 1000 ppm) to 38.4% (cycle 3 at 2500 ppm) at this temperature. At 300 °C these differences are not so important, and starting from 325 °C, practically complete conversion is achieved regardless of the toluene concentration.

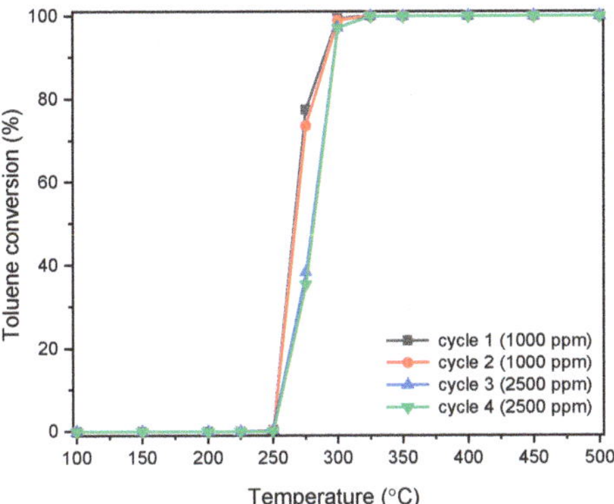

Figure 10. Catalytic activity in toluene combustion over PS30_Si_Co15_B in four successive reaction cycles (toluene concentration: cycles 1 and 2—1000 ppm, cycles 3 and 4—2500 ppm).

To conclude, the presented results clearly indicate the enormous potential resulting from the location of the oxide phase inside the SiO_2 shell, where properly protected, can actively, selectively and stably act in the total oxidation of VOCs.

3. Materials and Methods

3.1. Chemicals

Styrene (Sigma-Aldrich, Germany, >99.0%), acrylic acid (Acros Organics, France, 98%, extra pure), potassium persulfate (Chempur, Piekary Śląskie, Poland, pure p.a.), tetraethyl orthosilicate (TEOS, Sigma-Aldrich, China, ≥99.0%), ammonia solution (28–30%, J.T. Baker, Philipsburg, NJ, USA), isopropanol (Stanlab, Lublin, Poland, pure), cobalt(II) nitrate(V) hexahydrate (Honeywell, Seelze, Germany, 98%), toluene (Honeywell, Seelze, Germany, puriss p.a.) and nitrogen (Air Products, Świerzewo, Poland, grade 5.2).

3.2. Synthesis

Poly(styrene-*co*-acrylic acid) latex spheres were prepared by emulsion polymerization [47]. Briefly, 360 mL of distilled water was introduced into a 1000 mL three-neck round-bottom flask equipped with a mechanical stirrer (300 rpm), a reflux condenser and a N_2 dosing tube. Then, acrylic acid (1.6 g) and styrene (6.4 g) were added slowly by dropping it into the flask, and the mixture was deoxygenated by bubbling with nitrogen gas for 30 min. The temperature was raised to 70 °C and an aqueous solution of potassium persulfate (0.24 g in 40 g water) was poured to initialize the polymerization reaction at a constant mixing speed of 300 rpm. After 24 h the reaction mixture was allowed to cool down to room temperature.

In the next step, the surface of the obtained poly(styrene-*co*-acrylic acid) spheres (coded as PS30) was coated with a SiO_2 layer. For this purpose, 5 mL of the mixture containing the synthesized latex was placed along with 35 mL of distilled water and 200 mL of isopropanol in a 1000 mL vessel. After 15 min, 5 mL of ammonia solution and 4.5 mL of TEOS were added dropwise while vigorously stirring the mixture using a mechanical stirrer (250 rpm). The mixing was continued for another 3 h. Subsequently, the solid was centrifuged, rinsed twice with distilled water and once with isopropanol, and finally dried at 80 °C. The formed core–shell material is named as PS30_Si.

The synthesized composite was modified with Co using two different procedures (path A and B) shown schematically in Scheme 1. The series A was prepared by infiltrating the active phase precursor into the previously obtained PS30_Si material, drying and then calcinating. In the case of the series B, the copolymer core was first removed by calcination, then the active phase precursor was introduced. The resulting material was dried and calcined again.

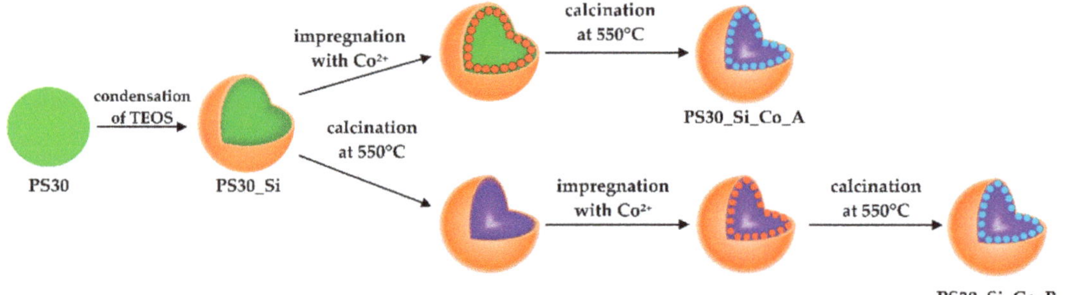

Scheme 1. Synthesis of Co_3O_4@SiO_2 composites using spherical poly(styrene-*co*-acrylic acid) template.

The preparations of both series were impregnated with aqueous solutions of $Co(NO_3)_2$ with concentrations that allowed Co loadings of 5, 10 and 15 wt.% in the final catalysts. Appropriate amounts of $Co(NO_3)_2 \cdot 6H_2O$ were dissolved in distilled water and added to 1 g of the support placed in crucibles. After thorough mixing, the samples were allowed to dry at 80 °C. Finally, the materials were calcined at 550 °C for 4 h (heating rate of 1 °C/min) in order to remove the polymer core and form the Co_3O_4 spinel phase.

3.3. Characterization

Surface morphology was investigated by SEM imaging using a Hitachi S-4700 field emission scanning electron microscope at an accelerating voltage of 20 kV. Samples were mounted on sticky carbon discs and coated with a gold layer. Secondary electron (SE) signal was used for observations.

Particle size distribution and average ζ-potential were determined by DLS measurements in a Zetasizer Nano ZS instrument (Malvern Instruments) equipped with a He-Ne laser source (λ = 633 nm). Prior to analysis, a suspension containing 0.1 wt.% of a material was prepared using deionized water and sonicated in an ultrasonic bath for 30 min.

Chemical composition was studied by XRF using an ARL Quant'x (Thermo Scientific) spectrometer, whereas polymer and inorganic contents in the PS30_Si composite were determined by thermal analysis. About 10 mg of the sample was tested in a TA Instruments SDT Q600 thermobalance in the range of 20–1000 °C at a linear heating rate of 20 °C/min in flowing air flow (100 mL/min).

The structure of samples was examined by XRD. The XRD patterns were collected on a Bruker D2 Phaser instrument using Cu Kα radiation (λ = 1.54184° Å) and a LYNXEYE detector within a 2θ range of 5–70° at a step of 0.02°.

Textural properties were calculated from low-temperature N_2 adsorption isotherms measured at -196 °C using a Micromeritics ASAP 2020 sorptometer. Samples were initially outgassed at 250 °C for 5 h under vacuum.

UV-Vis spectra were collected in a Thermo Scientific UV-VIS Evolution 200 double-beam spectrometer equipped with a xenon lamp and a diffuse reflectance accessory, enabling analyses of solids. The spectra were recorded in the wavelength range of 190–1100 nm, collecting 60 scans per min.

Surface analyses were performed by XPS in an ultra-high vacuum system manufactured by Prevac. The XPS spectra were collected using a monochromatized aluminum source AlKα (E = 1486.6 eV) and a hemispherical analyzer (VG SCIENTA R3000). The binding energy scale for the non-conductive samples was calibrated by referring to a position of Si 2p (E_b = 103.0 eV). The Shirley background and fitting with the mixed Gauss–Lorentz function (GL = 30) were used during interpretation of the spectra in the CasaXPS software.

H$_2$-TPR measurements were performed in a self-made equipment. An amount of 50 mg of a dried sample was placed in a quartz reactor and heated from room temperature to 900 °C at a linear temperature increase of 10 °C/min in a mixture of Ar/H$_2$ (95/5 vol.%), which was passed through the bed of the analyzed sample at a total flow rate of 40 mL/min. Hydrogen concentration in the gas mixture was analyzed by a thermal conductivity detector (TCD) and recalculated into consumption by the reduction reaction based on calibration with CuO as a reference.

3.4. Catalytic Activity

Catalytic combustion of toluene was studied as a model reaction in a quartz microreactor with an inner diameter of 8.0 mm. Before starting a measurement, a catalyst (0.1 g, particle size 160–315 μm) was degassed at 500 °C for 30 min in a stream of air (flow rate = 100 mL/min). After that time, the reactor was cooled to 100 °C and toluene dosing (at concentration of 1000 ppm) to flowing air was started. The organic substrate was introduced into the feed using a thermostated scrubber filled with the liquid compound kept at −7.5 °C ensuring its vapor pressure at the assumed level. The catalytic tests were carried out in the temperature range of 100–500 °C. The reactor was held at each temperature step for 80 min during which three analyses of the reaction products were performed. The reaction products were analyzed using a Bruker 450 gas chromatograph equipped with two capillary columns (Porapak S and Chromosorb WAW-DMCS), two flame ionization detectors, a thermal conductivity detector and a methanizer.

4. Conclusions

The main goal of this study was to develop yolk–shell materials containing Co$_3$O$_4$ nanoparticles, catalytically active in the total oxidation of VOCs. The introduction of a shell made of porous SiO$_2$ was to play a role in protecting the active phase against aggregation, which ensures a high dispersion of active sites and creates a nanospace for the reactants involved in the catalytic process. The bottom-up strategy was used for the construction of Co$_3$O$_4$@SiO$_2$ nanorattles, in which in the initial step a spherical poly(styrene-*co*-acrylic acid) template with a particle size in the range of 240–260 nm was synthesized by emulsion polymerization. On the surface of the copolymer scaffold, a SiO$_2$ nanolayer (with a thickness of approx. 20–30 nm) was deposited by hydrolysis and condensation of TEOS. The enrichment of the PS30@SiO$_2$ composite in Co was carried out by the wet impregnation method, with the active phase precursor being introduced either into SiO$_2$ spheres filled with the copolymer core (series A) or into hollow SiO$_2$ spheres after removal of the core by calcination (series B). The catalysts of both series showed high activity in total oxidation of toluene, exceeding that of the reference material containing 15 wt.% of Co on a commercial spherical silica gel. The catalysts of series B, however, were characterized by significantly lower temperatures of toluene conversion. Based on wide physicochemical characterization, including XRD, low-temperature adsorption of N$_2$, DLS, SEM, XRF, UV-Vis-DR, XPS and H$_2$-TPR, it was found that in the materials synthesized on the path B, the nanoparticles of the Co$_3$O$_4$ spinel phase were arranged loosely inside the SiO$_2$ shell. No strong interaction between the crystallites of the active phase and SiO$_2$ surface resulted in a weak stabilization effect observed for lattice oxygen in Co$_3$O$_4$. On the other hand, the high dispersion of the active phase, while facilitating the transport of the reagent molecules through the porous SiO$_2$ shell, created an ideal structure of a nanoreactor with high operating efficiency. It can therefore be concluded that the Co$_3$O$_4$@SiO$_2$ materials are an attractive alternative to conventional catalysts of VOCs combustion and can be considered as a platform for commercial applications in the future.

Author Contributions: Conceptualization, A.R. and P.K.; methodology, A.R., P.Ł. and P.K.; investigation, A.R., M.Ż., P.Ł., M.D. and M.M.; writing—original draft preparation, A.R., M.Ż. and P.K.; writing—review and editing, A.R. and P.K.; visualization, A.R. and M.Ż.; supervision, P.K. All authors have read and agreed to the published version of the manuscript.

Funding: This research was funded by National Science Centre, Poland, grant number 2018/02/X/ST5/03641. Some measurements were carried out with equipment purchased thanks to the financial support of the European Regional Development Fund in the framework of the Polish Innovation Economy Operational Program (contract no. POIG.02.01.00-12-023/08).

Data Availability Statement: The data presented in this study are available on request from the corresponding author.

Conflicts of Interest: The authors declare no conflict of interest. The funders had no role in the design of the study; in the collection, analyses, or interpretation of data; in the writing of the manuscript; or in the decision to publish the results.

References

1. Huang, H.; Xu, Y.; Feng, Q.; Leung, D.Y.C. Low temperature catalytic oxidation of volatile organic compounds: A review. *Catal. Sci. Technol.* **2015**, *5*, 2649–2669. [CrossRef]
2. Kuśtrowski, P.; Rokicińska, A.; Kondratowicz, T. Abatement of Volatile Organic Compounds Emission as a Target for Various Human Activities Including Energy Production. *Adv. Inorg. Chem.* **2018**, *72*, 385–419.
3. Kamal, M.S.; Razzak, S.A.; Hossain, M.M. Catalytic oxidation of volatile organic compounds (VOCs)—A review. *Atmos. Environ.* **2016**, *140*, 117–134. [CrossRef]
4. Zhang, Z.; Jiang, Z.; Shangguan, W. Low-temperature catalysis for VOCs removal in technology and application: A state-of-the-art review. *Catal. Today* **2016**, *264*, 270–278. [CrossRef]
5. Rokicińska, A.; Berniak, T.; Drozdek, M.; Kuśtrowski, P. In Search of Factors Determining Activity of Co_3O_4 Nanoparticles Dispersed in Partially Exfoliated Montmorillonite Structure. *Molecules* **2021**, *26*, 3288. [CrossRef]
6. Saqer, S.M.; Kondarides, D.I.; Verykios, X.E. Catalytic oxidation of toluene over binary mixtures of copper, manganese and cerium oxides supported on γ-Al_2O_3. *Appl. Catal. B* **2011**, *103*, 275–286. [CrossRef]
7. Carrillo, A.M.; Carriazo, J.G. Cu and Co oxides supported on halloysite for the total oxidation of toluene. *Appl. Catal. B* **2015**, *164*, 443–452. [CrossRef]
8. Gosecka, M.; Gosecki, M. Characterization methods of polymer core–shell particles. *Colloid Polym. Sci.* **2015**, *293*, 2719–2740. [CrossRef]
9. Ramli, R.A.; Laftah, W.A.; Hashim, S. Core–shell polymers: A review. *RSC Adv.* **2013**, *3*, 15543–15565. [CrossRef]
10. Lu, W.; Guo, X.; Luo, Y.; Li, Q.; Zhu, R.; Pang, H. Core-Shell Materials for Advanced Batteries. *Chem. Eng. J.* **2019**, *355*, 208–237. [CrossRef]
11. Li, Z.; Li, M.; Bian, Z.; Kathiraser, Y.; Kawi, S. Design of highly stable and selective core/yolk–shell nanocatalysts—A review. *Appl. Catal. B* **2016**, *188*, 324–341. [CrossRef]
12. El-Toni, A.M.; Habila, M.A.; Puzon Labis, J.; Al Othman, Z.A.; Alhoshan, M.; Elzatahry, A.A.; Zhang, F. Design, synthesis and applications of core–shell, hollow core, and nanorattle multifunctional nanostructures. *Nanoscale* **2016**, *8*, 2510–2531. [CrossRef] [PubMed]
13. Adijanto, L.; Bennett, D.A.; Chen, C.; Yu, A.S.; Cargnello, M.; Fornasiero, P.; Gorte, R.J.; Vohs, J.M. Exceptional Thermal Stability of $Pd@CeO_2$ Core–Shell Catalyst Nanostructures Grafted onto an Oxide Surface. *Nano Lett.* **2013**, *13*, 2252–2257. [CrossRef] [PubMed]
14. Lu, C.; Ge, C.; Wang, A.; Yin, H.; Ren, M.; Zhang, Y.; Yu, L.; Jiang, T. Synthesis of porous hollow silica spheres using functionalized polystyrene latex spheres as templates. *Korean J. Chem. Eng.* **2011**, *28*, 1458–1463. [CrossRef]
15. Ge, C.; Zhang, D.; Wang, A.; Yin, H.; Ren, M.; Liu, Y.; Jiang, T.; Yu, L. Synthesis of porous hollow silica spheres using polystyrene–methyl acrylic acid latex template at different temperatures. *J. Phys. Chem. Solids* **2009**, *70*, 1432–1437. [CrossRef]
16. Fujiwara, M.; Shiokawa, K.; Tanaka, Y.; Nakahara, Y. Preparation and Formation Mechanism of Silica Microcapsules (Hollow Sphere) by Water/Oil/Water Interfacial Reaction. *Chem. Mater.* **2004**, *16*, 5420–5426. [CrossRef]
17. Lu, Y.; McLellan, J.; Xia, Y. Synthesis and Crystallization of Hybrid Spherical Colloids Composed of Polystyrene Cores and Silica Shells. *Langmuir* **2004**, *20*, 3464–3470. [CrossRef]
18. Zou, H.; Wu, S.; Shen, J. Preparation of Silica-Coated Poly(styrene-co-4-vinylpyridine) Particles and Hollow Particles. *Langmuir* **2008**, *24*, 10453–10461. [CrossRef]
19. Du, B.; Cao, Z.; Li, Z.; Mei, A.; Zhang, X.; Nie, J.; Xu, J.; Fan, Z. One-Pot Preparation of Hollow Silica Spheres by Using Thermosensitive Poly(N-isopropylacrylamide) as a Reversible Template. *Langmuir* **2009**, *25*, 12367–12373. [CrossRef]
20. Guo, Y.; Davidson, R.A.; Peck, K.A.; Guo, T. Encapsulation of multiple large spherical silica nanoparticles in hollow spherical silica shells. *J. Colloid Interface Sci.* **2015**, *445*, 112–118. [CrossRef]
21. Du, P.; Liu, P. Novel Smart Yolk/Shell Polymer Microspheres as a Multiply Responsive Cargo Delivery System. *Langmuir* **2014**, *30*, 3060–3068. [CrossRef]
22. Blas, H.; Save, M.; Pasetto, P.; Boissiere, C.; Sanchez, C.; Charleux, B. Elaboration of Monodisperse Spherical Hollow Particles with Ordered Mesoporous Silica Shells via Dual Latex/Surfactant Templating: Radial Orientation of Mesopore Channels. *Langmuir* **2008**, *24*, 13132–13137. [CrossRef]

23. Nakashima, Y.; Takai, C.; Razavi-Khosroshahi, H.; Shirai, T.; Fuji, M. Effects of primary- and secondary-amines on the formation of hollow silica nanoparticles by using emulsion template method. *Colloids Surf. A Physicochem. Eng. Asp.* **2016**, *506*, 849–854. [CrossRef]

24. Telford, A.M.; Pham, B.T.T.; Neto, C.; Hawkett, B.S. Micron-sized polystyrene particles by surfactant-free emulsion polymerization in air: Synthesis and mechanism. *J. Polym. Sci. A Polym. Chem.* **2013**, *51*, 3997–4002. [CrossRef]

25. Prokopov, N.I.; Gritskova, I.A.; Kiryutina, O.P.; Khaddazh, M.; Tauer, K.; Kozempel, S. The Mechanism of Surfactant-Free Emulsion Polymerization of Styrene. *Polym. Sci. Ser. B* **2010**, *52*, 339–345. [CrossRef]

26. Agrawal, M.; Pich, A.; Gupta, S.; Zafeiropoulos, N.E.; Simon, P.; Stamm, M. Synthesis of Novel Tantalum Oxide Sub-micrometer Hollow Spheres with Tailored Shell Thickness. *Langmuir* **2008**, *24*, 1013–1018. [CrossRef] [PubMed]

27. Nakashima, Y.; Takai, C.; Razavi-Khosroshahi, H.; Fuji, M. Influence of the PAA concentration on PAA/NH$_3$ emulsion template method for synthesizing hollow silica nanoparticles. *Colloids Surf. A* **2018**, *546*, 301–306. [CrossRef]

28. Farooq, U.; Tweheyo, M.T.; Sjöblom, J.; Øye, G. Surface Characterization of Model, Outcrop and Reservoir Samples in Low Salinity Aqueous Solutions. *J. Dispers. Sci. Technol.* **2011**, *32*, 519–531. [CrossRef]

29. Bhattacharjee, S. DLS and zeta potential—What they are and what they are not? *J. Control. Release* **2016**, *235*, 337–351. [CrossRef] [PubMed]

30. Leroy, P.; Devau, N.; Revil, A.; Bizi, M. Influence of surface conductivity on the apparent zeta potential of amorphous silica nanoparticles. *J. Colloid Interface Sci.* **2013**, *410*, 81–93. [CrossRef]

31. Bao, Y.; Wang, T.; Kang, Q.; Shi, C.; Ma, J. Micelle-template synthesis of hollow silica spheres for improving water vapor permeability of waterborne polyurethane membrane. *Sci. Rep.* **2017**, *7*, 46638. [CrossRef] [PubMed]

32. Dutta, P.; Seehra, M.S.; Thota, S.; Kumar, J. A comparative study of the magnetic properties of bulk and nanocrystalline Co$_3$O$_4$. *J. Phys. Condens. Matter* **2008**, *20*, 015218. [CrossRef]

33. Popova, M.; Ristić, A.; Mavrodinova, V.; Maučec, D.; Mindizova, L.; Novak Tušar, N. Design of Cobalt Functionalized Silica with Interparticle Mesoporosity as a Promising Catalyst for VOCs Decomposition. *Catal. Lett.* **2014**, *144*, 1096–1100. [CrossRef]

34. Lin, L.Y.; Bai, H. Salt-induced formation of hollow and mesoporous CoO$_x$/SiO$_2$ spheres and their catalytic behavior in toluene oxidation. *RSC Adv.* **2016**, *6*, 24304–24313. [CrossRef]

35. Taghavimoghaddam, J.; Knowles, G.P.; Chaffee, A.L. Mesoporous Silica SBA-15 Supported Co$_3$O$_4$ Nanorods as Efficient Liquid Phase Oxidative Catalysts. *Top. Catal.* **2012**, *55*, 571–579. [CrossRef]

36. Rokicińska, A.; Drozdek, M.; Bogdan, E.; Węgrzynowicz, A.; Michorczyk, P.; Kuśtrowski, P. Combustion of toluene over cobalt-modified MFI zeolite dispersed on monolith produced using 3D printing technique. *Catal. Today* **2021**, *375*, 369–376. [CrossRef]

37. Klegova, A.; Inayat, A.; Indyka, P.; Gryboś, J.; Sojka, Z.; Pacultová, K.; Schwieger, W.; Volodarskaja, A.; Kuśtrowski, P.; Rokicińska, A.; et al. Cobalt mixed oxides deposited on the SiC open-cell foams for nitrous oxide decomposition. *Appl. Catal. B* **2019**, *255*, 117745. [CrossRef]

38. Rokicińska, A.; Natkański, P.; Dudek, B.; Drozdek, M.; Lityńska-Dobrzyńska, L.; Kuśtrowski, P. Co$_3$O$_4$-pillared montmorillonite catalysts synthesized by hydrogel-assisted route for total oxidation of toluene. *Appl. Catal. B* **2016**, *195*, 59–68. [CrossRef]

39. Ataloglou, T.; Fountzoula, C.; Bourikas, K.; Vakros, J.; Lycourghiotis, A.; Kordulis, C. Cobalt oxide/γ-alumina catalysts prepared by equilibrium deposition filtration: The influence of the initial cobalt concentration on the structure of the oxide phase and the activity for complete benzene oxidation. *Appl. Catal. A* **2005**, *288*, 1–9. [CrossRef]

40. Iablokov, V.; Barbosa, R.; Pollefeyt, G.; Van Driessche, I.; Chenakin, S.; Kruse, N. Catalytic CO Oxidation over Well-Defined Cobalt Oxide Nanoparticles: Size-Reactivity Correlation. *ACS Catal.* **2015**, *5*, 5714–5718. [CrossRef]

41. Li, G.; Zhang, C.; Wang, Z.; Huang, H.; Peng, H.; Li, X. Fabrication of mesoporous Co$_3$O$_4$ oxides by acid treatment and their catalytic performances for toluene oxidation. *Appl. Catal. A* **2018**, *550*, 67–76. [CrossRef]

42. Zhang, Q.; Mo, S.; Chen, B.; Zhang, W.; Huang, C.; Ye, D. Hierarchical Co$_3$O$_4$ nanostructures in-situ grown on 3D nickel foam towards toluene oxidation. *Mol. Catal.* **2018**, *454*, 12–20. [CrossRef]

43. Chen, S.; Xie, H.; Zhou, G. Tuning the pore structure of mesoporous Co$_3$O$_4$ materials for ethanol oxidation to acetaldehyde. *Ceram. Int.* **2019**, *45*, 24609–24617. [CrossRef]

44. Shamma, E.H.; Said, S.; Riad, M.; Mikhail, S.; Ismail, A.E.H. Modernization of the bulk Co$_3$O$_4$ to produce meso-porous Co$_3$O$_4$ nanoparticles with enhanced structural and morphological characteristics. *J. Porous Mater.* **2020**, *27*, 647–657. [CrossRef]

45. Mandal, S.; Rakibuddin, M.; Ananthakrishnan, R. Strategic Synthesis of SiO$_2$-Modified Porous Co$_3$O$_4$ Nano-Octahedra Through the Nanocoordination Polymer Route for Enhanced and Selective Sensing of H$_2$ Gas over NO$_x$. *ACS Omega* **2018**, *3*, 648–661. [CrossRef]

46. Rokicińska, A.; Drozdek, M.; Dudek, B.; Gil, B.; Michorczyk, P.; Brouri, D.; Dzwigaj, S.; Kuśtrowski, P. Cobalt-containing BEA zeolite for catalytic combustion of toluene. *Appl. Catal. B* **2017**, *212*, 59–67. [CrossRef]

47. Yan, R.; Zhang, Y.; Wang, X.; Xu, J.; Wang, D.; Zhang, W. Synthesis of porous poly(styrene-co-acrylic acid) microspheres through one-step soap-free emulsion polymerization: Whys and wherefores. *J. Colloid Interface Sci.* **2012**, *368*, 220–225. [CrossRef]

Article

Catalytic Ozonation of Toluene over Acidic Surface Transformed Natural Zeolite: A Dual-Site Reaction Mechanism and Kinetic Approach

Serguei Alejandro-Martín [1], Héctor Valdés [2,*] and Claudio A. Zaror [3]

1 Wood Engineering Department, Faculty of Engineering, Universidad del Bío-Bío, Concepción 4030000, Chile; salejandro@ubiobio.cl

2 Clean Technologies Laboratory, Faculty of Engineering, Universidad Católica de la Santísima Concepción, Concepción 4030000, Chile

3 Chemical Engineering Department, Faculty of Engineering, Universidad de Concepción, Concepción 4030000, Chile; czaror@udec.cl

* Correspondence: hvaldes@ucsc.cl

Abstract: Volatile organic compounds (VOCs) are responsible for damage to health due to their carcinogenic effects. Catalytic ozonation using zeolite appears as a valuable process to eliminate VOCs from industrial emissions at room temperature. For full-scale application of this new abatement technology, an intrinsic reaction rate equation is needed for an effective process design and scale-up. Results obtained here provide a mechanistic approach during the initial stage of catalytic ozonation of toluene using an acidic surface transformed natural zeolite. In particular, the contribution of Lewis and Brønsted acid sites on the surface reaction mechanism and overall kinetic rate are identified through experimental data. The least-squares non-linear regression method allows the rate-determining step to be established, following a Langmuir–Hinshelwood surface reaction approximation. Experimental evidence suggest that ozone is adsorbed and decomposed at Lewis acid sites, forming active atomic oxygen that leads to the oxidation of adsorbed toluene at Brønsted acid sites.

Keywords: catalytic ozonation; Lewis and Brønsted acid sites; natural zeolite; reaction mechanism; toluene

Citation: Alejandro-Martín, S.; Valdés, H.; Zaror, C.A. Catalytic Ozonation of Toluene over Acidic Surface Transformed Natural Zeolite: A Dual-Site Reaction Mechanism and Kinetic Approach. *Catalysts* **2021**, *11*, 958. https://doi.org/10.3390/catal11080958

Academic Editors: Fernando J. Beltrán Novillo and Narendra Kumar

Received: 2 July 2021
Accepted: 6 August 2021
Published: 10 August 2021

Publisher's Note: MDPI stays neutral with regard to jurisdictional claims in published maps and institutional affiliations.

1. Introduction

Volatile organic compounds (VOCs), such as benzene, toluene, ethylbenzene, and xylene (BTEX) are emitted from a wide variety of sources, such as internal combustion engines, power stations, refineries, the food processing industry, the forestry industry, wastewater treatment facilities, and the dry cleaning industry, among other industrial sectors [1,2]. VOCs are also released from indoor sources, including: construction materials, damaged furniture, aerosol sprays, deodorants, air fresheners, cleaning products, disinfectants and insecticides [3]. BTEX are considered priority pollutants due to their toxic, mutagenic, and carcinogenic effects [4], and their role in the formation of secondary atmospheric pollutants such as ozone and airborne particulate matter [5].

During the last decades, great efforts have been made to remove VOCs from both indoor and outdoor emissions, using recovery methods (e.g., absorption, adsorption, membrane separation, and condensation) or destructive methods (e.g., thermal, catalytic, or biological oxidation) [6]. However, the disposal of VOC-saturated solvents or exhausted adsorbents from recovery processes has impaired large-scale implementation. Moreover, conventional destruction methods such as thermal oxidation might generate undesirable or noxious by-products, such as dioxins and carbon monoxide, and must be applied below the lower explosive level (LEL) of target VOCs. In the case of biological oxidation, special attention should be paid to avoid microorganism deactivation [7]. Alternatively, catalytic oxidation is claimed to be the most effective and economically feasible technique for the oxidation of VOCs into less harmful compounds and, eventually, into carbon dioxide and water [8,9]. Catalytic oxidation could be applied to treat waste gas streams in a wide range of VOCs concentrations and flow rates at temperatures around 473 K. This technique can be considered as environmentally friendly due to its low energy demand and lower generation of harmful products, using appropriate catalysts [9–12].

Recently, a new catalytic process to oxidize VOCs using ozone has attracted increasing interest [13–18]. Ozone-based catalytic oxidation allows removing persistent VOCs from industrial emissions at room temperature, reducing operational cost [19–22]. Synthetic zeolites and metal-supported catalysts have been used as catalysts in the catalytic ozonation of VOCs [19,23–28]. Additionally, natural zeolites have started to be applied in combination with ozone in this abatement process [29,30]. Thus, some studies have indicated the positive contribution of modified natural zeolite on the catalytic treatment [31,32]. Even though the influence of surface acid sites of natural zeolites has been assessed in previous studies, there are still some doubts about the reaction mechanism that takes place among ozone and VOC interactions at the zeolite surface and there is also a lack of a kinetic expression to allow process scale-up.

For commercial-scale implementation of catalytic ozonation of VOCs using zeolite at room temperature, accurate kinetic rate expressions are needed to allow reactor design and optimization. In this study, the contribution of Lewis and Brønsted acid sites on the surface reaction mechanism is established and an overall kinetic model, representing the catalytic ozonation of toluene over acidic surface-transformed natural zeolite at room temperature, is proposed.

2. Results and Discussion

2.1. FTIR Evidence of the Heterogeneous Catalytic Ozonation of Toluene

Figure 1 displays FTIR spectra of uncovered acidic transformed natural zeolite sample (dashed line), upon contact with toluene (dotted line), and after catalytic ozonation experiments (solid line). An IR vibration band around 3584 cm^{-1} is observed in the uncovered and toluene-saturated samples. Such an IR band is related to acidic bridging hydroxyl groups (Si(OH)Al) on natural zeolite, as reported elsewhere [33,34].

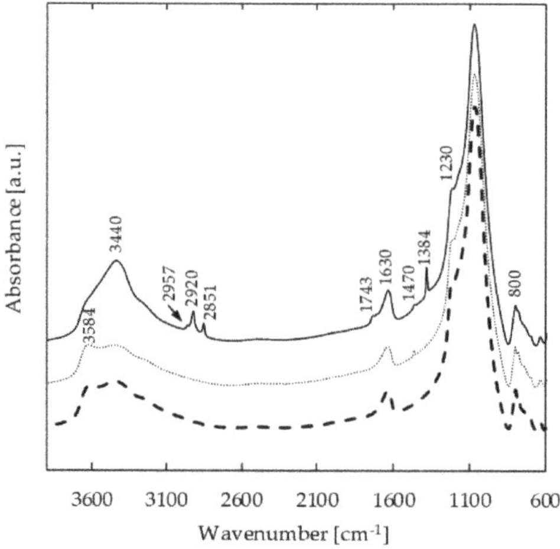

Figure 1. In situ FTIR spectra of acidic transformed natural zeolite (AZ) samples: uncovered AZ sample (dashed line); AZ sample saturated with toluene (dotted line); AZ sample after catalytic ozonation (solid line).

The IR spectrum after toluene adsorption shows an IR band at 1470 cm^{-1} that is linked to adsorbed toluene molecules. In particular, the IR bands around 1450–1540 cm^{-1} (considering a maximum at 1477) have been associated with the stretching vibrations of C=C bonds of the aromatic ring of adsorbed toluene on zeolites [35]. After catalytic ozonation of toluene, chemical surface intermediaries are identified in the collected spectrum (solid line). A typical absorption band at 1384 cm^{-1} is registered. This IR band is associated with active oxygen species formed at the zeolite surface [36–39]. Therefore, it can be related to ozone adsorption and decomposition at Lewis acid sites. It has been suggested that ozone is adsorbed and decomposed into atomic oxygen (O$^{\bullet}$) at Lewis acid sites, as represented by Equation (1) [37–44].

$$O_{3\,(g)} + s_L \rightleftharpoons O^{\bullet} - s_L + O_{2(g)} \tag{1}$$

The IR bands at 1057 and 1027 cm^{-1}, as well as the bands around 800 cm^{-1} and 1168 cm^{-1}, have been ascribed to molecular ozone adsorption, ozonide species (O_3^-), and superoxide (O_2^-), respectively [33]. However, no significant differences on these adsorption bands (before and after ozone exposure) can be observed here, probably due to a rapid generation of atomic oxygen (1384 cm^{-1}) at strong Lewis surface acid sites.

Additionally, an IR band at 1743 cm^{-1} is registered after catalytic ozonation (see solid line in Figure 1). IR bands around 1718–1752 cm^{-1} have been associated with adsorbed formic acid [45,46]. Additionally, some other distinctive peaks can be seen at 2851, 2920, and 2957 cm^{-1} after catalytic ozonation. IR bands around 2899 and 2935 cm^{-1} have been related to the (C–H) stretching vibrations of formate species [47]. Thus, the observed IR bands could be due to the shift of such IR bands associated with adsorbed toluene oxidation by-products. Complementary results obtained using HPLC analyses indicate the presence of formic acid, acetaldehyde, benzoic acid, and benzaldehyde as the main intermediate surface compounds after catalytic ozonation of toluene [48]. Furthermore, different studies of catalytic ozonation of toluene using different catalysts have shown that ozone is decomposed into active oxygen species, interacting with adsorbed toluene [24,49,50]. Therefore, it has been indicated that toluene is oxidized via methyl interaction. The lower dissociation energy of the methyl C–H bond (3.7 eV) could support that statement. This value is lower than the dissociation energies of other C–H bonds of toluene molecule (C–H bond of aromatic ring 4.3 eV; C–C bond of methyl 4.4 eV; C–C bond of aromatic ring 5.0–5.3 eV; C=C bond of the ring 5.5 eV), which justifies the schematic reaction route displayed in Figure 2. Thus, benzaldehyde, benzoic acid, and formic acid have been identified as primary reaction intermediates of catalytic ozonation of toluene in agreement with the results obtained in this work [15,48,51,52].

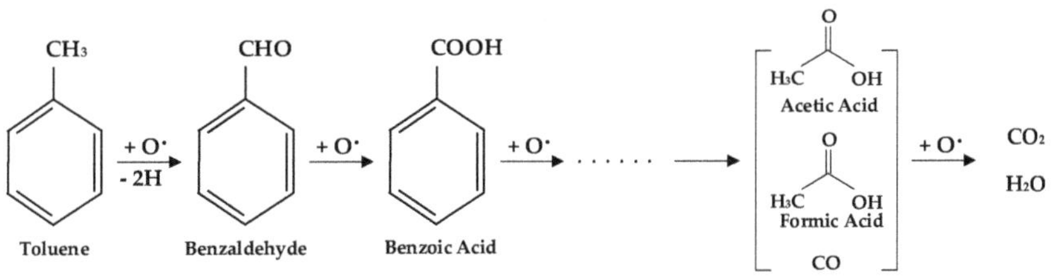

Figure 2. Proposed initial schematic route of the reaction of adsorbed toluene molecules with immediately adjacent adsorbed active oxygen species, forming toluene oxidation by-products. Reprinted with permission from references [51,53,54]. Copyright 2021 Elsevier.

IR evidence obtained here suggests that a surface reaction occurs between adsorbed toluene at Brønsted acid sites and active oxygen species generated after ozone adsorption and decomposition at Lewis acid sites. Such a surface reaction mechanism is in agreement with previous reports, that claim the adsorption of toluene molecules at Brønsted acid sites and the adsorption of ozone molecules at Lewis acid sites followed by its transformation into active oxygen species [32,48,55–57]. After that, it has been indicated that active oxygen species interact with adsorbed toluene molecules, leading to the formation of toluene oxidation by-products, as reported elsewhere [58,59]. Previously published works confirm the surface reaction mechanism proposed here, using not only natural zeolite mainly composed of clinoptilolite and mordenite and modified natural zeolite [32,48]; but also using synthetic zeolites such as faujasite, mordenite, and ZSM-5 [55–57] under the same experimental conditions.

2.2. Kinetic Approaches of Catalytic Ozonation of Toluene

In the absence of mass transfer limitations, the heterogeneous catalytic ozonation of toluene using acidic transformed natural zeolite can be represented by a general and simplified reaction, as follows [58,59]:

$$C_7H_8 + 6O_3 \xrightarrow{cat} 7CO_2 + 4H_2O \tag{2}$$

Firstly, a power law kinetic model is used here as an initial approximation for the estimation of the overall kinetic rate expression of catalytic ozonation of toluene, r_{Tol} ($\mu mol \cdot dm^{-3} s^{-1}$):

$$-r_{Tol} = k_{app} C_{C_7H_8}^{\alpha} \cdot C_{O_3}^{\beta} \tag{3}$$

where k_{app} is the apparent reaction rate constant $((dm^3 \cdot \mu mol)^{1+\alpha+\beta} \cdot s^{-1})$; $C_{C_7H_8}$ and C_{O_3} stand for toluene and ozone concentrations ($\mu mol \cdot dm^{-3}$), respectively; α and β are orders of reaction related to toluene and ozone, respectively. Table 1 lists such kinetic parameters obtained after applying the NLR technique, using the tools of the Microsoft Excel Solver add-in [60]. Results are compared with others reported in the literature for catalytic ozonation of toluene using different catalysts.

As can be seen, the order of reaction obtained for toluene is lower than the order of reaction for ozone, which is evidence of the higher ozone dependence during catalytic ozonation of toluene promoted by acidic transformed natural zeolite. Moreover, the obtained apparent activation energy (17.21 kJ·mol^{-1}) is lower than the reported activation energy in the toluene–ozone homogenous reaction (55.5 kJ·mol^{-1}) [61]. Such results clearly indicate that active surface sites of acidic transformed natural zeolite play a significant role during the heterogeneous reaction.

Table 1. Power law kinetic parameters for catalytic ozonation of toluene.

Catalysts	Reaction Temperature (K)	$k_{app} \times 10^3$ $(dm^3 \cdot \mu mol)^{1+\alpha+\beta} \cdot s^{-1})$	$E_{A\,app}$ $(kJ \cdot mol^{-1})$	α	β	R^2	Reference
AZ	293	5.07	17.21	0.66	0.82	0.97	This study
MnO$_2$/graphene	295	3.25	29.3	0.63	0.55	0.91	[62]
MnOx/γ-alumina	343	n.r	31	−1	2	n.r	[63]
MnOx/γ-alumina	298	n.r	33	0.18	0.56	0.98	[64]

n.r: non-reported.

Secondly, a kinetic expression that takes into account the contribution of acidic surface sites of transformed natural zeolite during the catalytic ozonation of toluene is postulated following the dual-site Langmuir–Hinshelwood (dsL-H) surface reaction approximation and is validated using the experimental data collected during this study. It is assumed that all chemical species react after their adsorption at acidic surface sites, and each active site is occupied by only one reactive specie [65,66]. Thus, previous experimental results can be taken as shreds of evidence of the existence of two main active acidic surface sites where selective adsorption of toluene and ozone occur. Toluene has proved its affinity toward Brønsted acid sites and ozone for Lewis acid sites [32,67,68]. Hence, the surface reaction mechanism could be described by three elementary steps, including a dual-site surface reaction as the rate-determining step (RDS), as follows:

1. Ozone adsorption and decomposition at Lewis acid sites (S_L) could be described by a simplified adsorption-desorption rate equation, r_1(μmol·g^{-1}·s^{-1}), as follows:

$$O_{3\,(g)} + s_L \underset{k_{-1}}{\overset{k_1}{\rightleftharpoons}} O^{\bullet} - s_L + O_{2(g)} \quad r_1 = k_1 C_{O_3}[s_L] - k_{-1}[O^{\bullet} - s_L]C_{O_2} \tag{4}$$

where $[s_L]$ and $[O^{\bullet} - s_L]$ represent the total vacant Lewis acid sites of acidic transformed natural zeolite or the total surface concentration of Lewis acid sites (μmol·g^{-1}), and active atomic oxygen specie adsorbed at Lewis acid sites (μmol·g^{-1}), respectively. k_1 and k_{-1} represent forward and reverse rate constants of active atomic oxygen surface specie formation (dm^3·s^{-1}·μmol^{-1}), C_{O_3} is the concentration of ozone (μmol·dm^{-3}) and C_{O_2} is the concentration of oxygen (μmol·dm^{-3}).

2. Toluene adsorption at Brønsted acid sites (S_B) could be represented by a simplified adsorption-desorption rate equation, r_2(μmol·g^{-1}·s^{-1}), as follows:

$$C_7H_{8(g)} + s_B \underset{k_{-2}}{\overset{k_2}{\rightleftharpoons}} C_7H_8 - s_B \quad r_2 = k_2 C_{C_7H_8}[s_B] - k_{-2}[C_7H_8 - s_B] \tag{5}$$

where $[s_B]$ and $[C_7H_8 - s_B]$ represent the total vacant Brønsted acid sites of acidic transformed natural zeolite or the total surface concentration of Brønsted acid sites (μmol·g^{-1}), and toluene-adsorbed specie at Brønsted acid sites (μmol·g^{-1}), respectively. k_2 represents forward rate constant of toluene surface complex formation (dm^3·s^{-1}·μmol^{-1}), k_{-2} represents the reverse rate constant of toluene surface complex formation (s^{-1}), $C_{C_7H_8}$ is the concentration of toluene (μmol·dm^{-3}).

3. Dual-site surface reaction (rate-determining step) r_3(μmol·g^{-1}·s^{-1}), between adsorbed toluene molecules at Brønsted acid sites and active oxygen species adsorbed at Lewis acid sites, leading to the formation of toluene oxidation by-products, as follows:

$$C_7H_8 - s_B + O^{\bullet} - s_L \overset{k_3}{\rightarrow} C_7H_6O - s_B + s_L \quad r_3 = k_3[O^{\bullet} - s_L][C_7H_8 - s_B] \tag{6}$$

where $[C_7H_8 - s_B]$ and $[O^{\bullet} - s_L]$ stand for the amount of toluene adsorbed at Brønsted acid sites (μmol·g^{-1}), and the amount of active oxygen species adsorbed at Lewis acid sites (μmol·g^{-1}), respectively. k_3 is the disappearance rate constant of the adsorbed toluene surface complex (g·s^{-1}·μmol^{-1}).

The unknown surface concentrations of vacant sites of Lewis $[S_L]$ and Brønsted $[S_B]$ acidic sites can be calculated from site balances, as follows:

$$C_{L_t} = [s_L] + [O^{\bullet} - s_L] \tag{7}$$

$$C_{B_t} = [s_B] + [C_7H_8 - s_B] \tag{8}$$

where C_{L_t} and C_{B_t} are the total site concentration of Lewis acid sites (μmol·g^{-1}) and Brønsted acid sites (μmol·g^{-1}), respectively; $[O^{\bullet} - s_L]$ and $[C_7H_8 - s_B]$ represent adsorbed active oxygen specie (μmol·g^{-1}) and adsorbed toluene molecule (μmol·g^{-1}) at Lewis and Brønsted acid sites, respectively.

Here, use is made of the steady-state approximation for the rate of ozone adsorption and its decomposition into active atomic oxygen and for the rate of toluene adsorption. This assumes that $r_1 \simeq 0$ and $r_2 \simeq 0$ or the right-hand sides of Equations (4) and (5) are in a pseudo-equilibrium or stationary state.

Thus, the unknown concentration of active atomic oxygen specie adsorbed at Lewis acid sites can be estimated from Equation (4), assuming $r_1 \simeq 0$, as follows:

$$[O^{\bullet} - s_L] = \frac{k_1 C_{O_3}[s_L]}{k_{-1} C_{O_2}} = K_1 \frac{C_{O_3}[s_L]}{C_{O_2}} \tag{9}$$

Since the inlet concentration of oxygen in all experiments was kept in excess, the oxygen concentration can be considered to remain constant. Thus, a new constant can be set as indicated by Equation (10):

$$K_1' = \frac{K_1}{C_{O_2}} \tag{10}$$

Then, the unknown concentration of adsorbed active oxygen species can be finally obtained, as follows:

$$[O^\bullet - s_L] = K_1' C_{O_3}[s_L] \tag{11}$$

Substituting Equation (11) into Equation (7) yields Equation (13):

$$C_{L_t} = [s_L] + K_1' C_{O_3}[s_L] = [s_L]\left(1 + K_1' C_{O_3}\right) \tag{12}$$

$$[s_L] = \frac{C_{L_t}}{\left(1 + K_1' C_{O_3}\right)} \tag{13}$$

In the case of the unknown concentration of adsorbed toluene molecules at Brønsted acid sites, it can be estimated from Equation (5) at a pseudo-equilibrium condition ($r_2 \simeq 0$), leading to:

$$[C_7H_8 - s_B] = \frac{k_2}{k_{-2}} C_{C_7H_8}[s_B] = K_2 C_{C_7H_8}[s_B] \tag{14}$$

Then, Equation (14) can be combined with Equation (8) to obtain Equation (16):

$$C_{B_t} = [s_B] + K_2 C_{C_7H_8}[s_B] = [s_B]\left(1 + K_2 C_{C_7H_8}\right) \tag{15}$$

$$[s_B] = \frac{C_{B_t}}{\left(1 + K_2 C_{C_7H_8}\right)} \tag{16}$$

After substituting Equations (11) and (14) in Equation (6) and then incorporating Equations (13) and (16), this yields Equation (18):

$$- r_3 = k_3[C_7H_8 - s_B][O^\bullet - s_L] = k_3 K_2 C_{C_7H_8}[s_B]K_1' C_{O_3}[s_L] \tag{17}$$

$$- r_3 = K_1' K_2 k_3 C_{L_t} C_{B_t} \frac{C_{O_3} C_{C_7H_8}}{\left(1 + K_1' C_{O_3}\right)\left(1 + K_2 C_{C_7H_8}\right)} \tag{18}$$

In Equation (18) the values of K_2 and C_{B_t} can be obtained from adsorption experiments since K_2 is the Langmuir adsorption equilibrium constant of toluene surface complex formation at Brønsted acid sites ($dm^3 \cdot \mu mol^{-1}$) and C_{B_t} is the maximum toluene uptake, corresponding to the saturation of Brønsted acid sites or monolayer coverage ($\mu mol \cdot g^{-1}$). Such equilibrium parameters come by substituting Equation (8) into Equation (14), resulting in Equation (19), as follows:

$$[C_7H_8 - s_B] = \frac{K_2 C_{C_7H_8} C_{B_t}}{1 + K_2 C_{C_7H_8}} \tag{19}$$

where $C_{C_7H_8}$ is the concentration of toluene at the equilibrium ($\mu mol \cdot dm^{-3}$), $[C_7H_8 - s_B]$ is the amount of toluene adsorbed at Brønsted acid sites ($\mu mol \cdot g^{-1}$) at equilibrium, C_{B_t} is the maximum toluene uptake, corresponding to the saturation of Brønsted acid sites or monolayer coverage ($\mu mol \cdot g^{-1}$). Thus, equilibrium parameters are gathered experimentally from a set of breakthrough curves of q-concentration data obtained using Equation (26).

Figure 3 depicts the adsorption equilibrium data of toluene at 293 K over acidic transformed natural zeolite, computed from breakthrough curves using Equation (26). As can be seen, the Langmuir adsorption isotherm adjusts very well to the experimental data ($R^2 = 0.9959$). The maximum toluene uptake by Brønsted acid sites, C_{B_t} (0.23429 $\mu mol \cdot g^{-1}$), and the adsorption equilibrium constant, K_2 (3.81 $dm^3 \cdot \mu mol^{-1}$) were obtained from the intercept and the slope of the linearized form of Langmuir isotherm expression, respectively.

Such equilibrium parameters agree with values found in previous works using natural and modified natural zeolite mainly composed by clinoptilolite and mordenite [39].

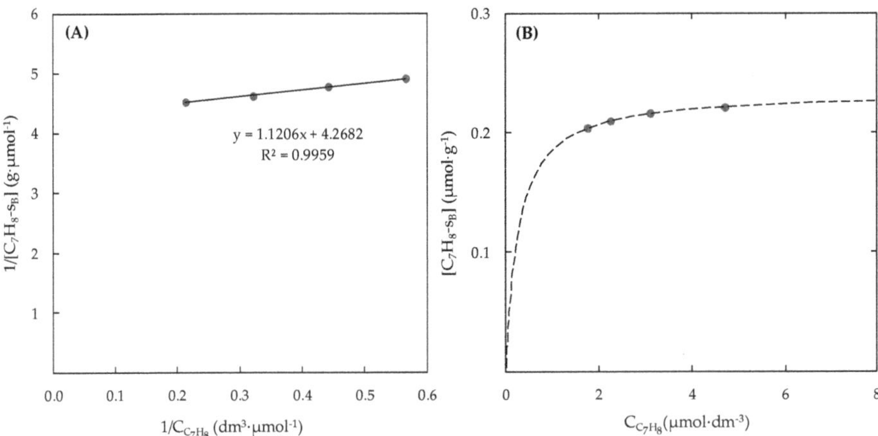

Figure 3. Adsorption equilibrium data of toluene at 293 K over acidic transformed natural zeolite: (**A**) linearized form of Langmuir adsorption model from Equation (22). (**B**) Langmuir adsorption model using Equation (22). (●) stands for experimental q values calculated from Equation (3); (−) represents linearized form of Langmuir adsorption model; (- -) stands for Langmuir adsorption model.

If an apparent constant k'_{app} is defined as:

$$k'_{app} = K'_1 K_2 k_3 C_{L_t} C_{B_t} \tag{20}$$

Then, the overall kinetic expression can be rewritten, as follows:

$$-r_3 = k'_{app} \frac{C_{O_3} C_{C_7 H_8}}{\left(1 + K'_1 C_{O_3}\right)\left(1 + K_2 C_{C_7 H_8}\right)} \tag{21}$$

Hence, substituting the obtained experimental equilibrium constant of toluene adsorption into Equation (21), gives:

$$-r_3 = k'_{app} \frac{C_{O_3} C_{C_7 H_8}}{\left(1 + K'_1 C_{O_3}\right)\left(1 + 3.81 C_{C_7 H_8}\right)} \tag{22}$$

Table 2 summarizes experimental and estimated reaction rate values of catalytic ozonation of toluene using acidic transformed natural zeolite. During a set of experimental runs, the inlet concentration ratio of toluene/ozone was varied from 5.3 to 37.8. The NLR technique was conducted to obtain model parameters of Equation (22), using the tools of the Microsoft Excel Solver add-in [60].

Finally, the proposed overall kinetic rate expression of catalytic ozonation of toluene at room temperature, using acidic transformed natural zeolite, is represented by Equation (23):

$$-r_3 = 0.233 \frac{C_{O_3} C_{C_7 H_8}}{\left(1 + 0.063 C_{O_3}\right)\left(1 + 3.81 C_{C_7 H_8}\right)} \tag{23}$$

where the constants k'_{app}, K'_1, and K_2 take the values of 0.233 $(dm^3)^2 \cdot \mu mol^{-1} \cdot g^{-1} \cdot s^{-1}$, 0.063 $dm^3 \cdot \mu mol^{-1}$, and 3.81 $dm^3 \cdot \mu mol^{-1}$, respectively.

Figure 4 shows a visual comparison of the obtained experimental and estimated reaction rate values of catalytic ozonation of toluene. An excellent coefficient of determination ($R^2 = 0.97$) is obtained. A value of $R^2 = 0.91$ was reported previously by Hu et al. [62] using the dsL-H reaction approach in the catalytic ozonation of toluene over MnO_2/grapheme, suggesting that two active sites are involved in the reaction mechanism.

Table 2. Experimental and calculated reaction rate values.

Runs	Inlet Concentration ($\mu mol \cdot dm^{-3}$)		Conversion of Toluene (%)	$-r_3$ ($\mu mol \cdot g^{-1} \cdot s^{-1}$)		
	Toluene	Ozone		Experimental	Estimated by the Power Law Model	Estimated by the *dsL-H* Model
1	42.36	3.57	11.0	0.207	0.1769	0.1810
2	42.36	6.48	15.7	0.295	0.2876	0.2871
3	42.36	6.48	14.4	0.272	0.2876	0.2871
4	42.36	1.92	5.6	0.106	0.1054	0.1060
5	42.36	1.12	3.0	0.057	0.0676	0.0646
6	42.36	1.12	3.2	0.061	0.0676	0.0646
7	42.86	1.92	4.3	0.081	0.1063	0.1068
8	33.99	3.57	10.6	0.160	0.1519	0.1562
9	33.85	6.43	16.8	0.253	0.2461	0.2456
10	33.96	1.92	5.4	0.081	0.0910	0.0914
11	50.99	3.57	9.3	0.212	0.1990	0.2032
12	42.71	3.57	10.4	0.198	0.1768	0.1820
13	42.36	1.79	4.3	0.081	0.0995	0.0995
14	42.56	7.59	16.4	0.311	0.3287	0.3218

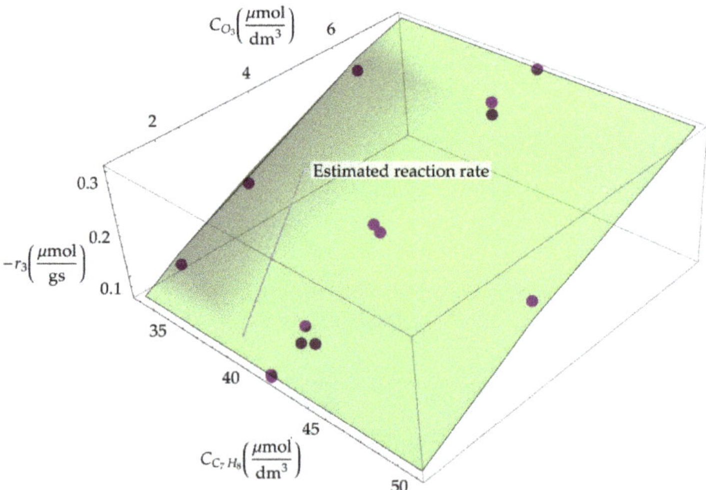

Figure 4. Fit between the proposed dsL-H kinetic reaction model and the experimental data reported in Table 2 for the catalytic ozonation of toluene using acidic surface-transformed natural zeolite, where the surface reaction between adsorbed species on adjacent active sites is the rate-determining step.

On the one hand, results obtained in this study provide evidence that the dsL-H reaction model shows a higher dependence of ozone toward Lewis acid sites during the catalytic ozonation of toluene. Such a result is in agreement with that obtained with the applied power law kinetic model. Our previous work [32], reported the influence of Lewis acid sites on the toluene ozonation reaction rate, using natural and modified zeolites. Experimental results evidenced a linear dependency (slope = 0.05) of the registered reaction rate and Lewis acid sites. Consequently, a higher surface concentration of atomic oxygen enhances catalytic ozonation of toluene, in concordance with experimental findings reported before [36,37,39,42–44]. On the other hand, the influence of Brønsted strong acid sites density over toluene ozonation reaction rate also showed a linear dependency, resulting in a lower contribution (slope = 0.01). Former studies reported the affinity of toluene for Brønsted acid sites [69,70]. However, this interaction might lower the reaction rate due to strong toluene adsorption at Brønsted acid sites. Therefore, surface acidic sites of the modified natural zeolite framework reveal a remarkable contribution to toluene ozonation at room temperature.

3. Materials and Methods

3.1. Materials

Chilean natural zeolite mainly composed of clinoptilolite (53%), mordenite (40%) and quartz (7%) was acquired from Minera FORMAS™, Santiago, RM, Chile. As-received zeolite sample was ground and sieved. Particle sizes in the range 0.3–0.425 mm were used in the whole study and they were washed with de-ionized water, oven dried at 398 K and stored in a desiccator. This zeolite exhibits an apparent density, 2.3 g·cm^{-3}; BET surface area, 205 m^2 g^{-1}; total pore volume, 0.11 cm^3 g^{-1}; cation exchange capacity, 2.05 meq·g^{-1}; with a Si/Al of 5.34; and it is mainly made up of SiO_2 (75.25%) and Al_2O_3 (14.1%) followed by CaO (4.57%), Fe_2O_3 (2.31%), Na_2O (1.89%), K_2O (0.74%), MgO (0.66%), TiO_2 (0.42%), and MnO (0.05%); with a concentration of Brønsted and Lewis acid sites of 0.4 μmol·g^{-1} and 31.1 μmol·g^{-1}, respectively [32]. Natural zeolite was used as a parent material to produce a zeolite sample with a higher content of acidic sites.

Ozone was generated in situ from instrumental dry air provided by AGA using an AZCOZON A-4 ozone generator (Vancouver, BC, Canada). A contaminated stream made of toluene vapors was used here as a target VOC. The stream enriched with toluene vapors was produced by continuously bubbling dry argon into a pure liquid toluene (confined in a flask at a controlled temperature of 261 K). Toluene was supplied by Merck (Darmstadt, Hessen, Germany) with a purity of 99.8%.

3.2. Transformation of As-Received Natural Zeolite into a Modified Zeolite Rich in Acidic Sites

Natural zeolite was modified by ion-exchange treatment using an ammonium sulphate solution (0.1 mol·dm^{-3}) with a volume/solid ratio of 10/1 at 363 K during 3 h. Then, it was rinsed with ultra-pure water for 4 h, replacing water after 2 and 3 h of contact time. After that, a second ion-exchange using the ammonium sulphate solution was applied using the same procedure and steps as described above. This modification method decreases the content of compensating cations, being ion-exchanged by ammonium cations. Then, ammonium-exchanged zeolite sample was oven-dried at 378 K for 24 h and stored in a desiccator until further use. Finally, a modified zeolite sample with a higher content of acid sites was produced just before the catalytic ozonation experiments by thermal outgassing at 823 K and it is identified here as AZ.

Nitrogen adsorption-desorption isotherms at 77 K, X-ray powder diffraction (XRD), X-ray fluorescence (XRF), scanning electron microscopy (SEM), energy-dispersive X-ray spectroscopy (EDS), and transmission electron microscopy (TEM) were conducted to characterize the transformed natural zeolite sample. Surface area (S_{BET}) was obtained from nitrogen adsorption-desorption isotherms measured in a Micromeritics Gemini 3175 sorptometer (Norcross, GA, USA). The crystalline structure was registered by XRD analysis in a Bruker AXS Model D4 ENDEAVOR diffractometer (Billerica, MA, USA); the presence of main compensating cations was determined by X-ray fluorescence, using a RIGAKU Model 3072 spectrometer (Akishima, Tokyo Metropolis, Japan). SEM studies were carried out in a JEOL (Akishima, Tokyo Metropolis, Japan) JSM-6380 microscope, operating at 20 kV. Samples were ground to 150 μm and a small amount was deposited in the apparatus. Chemical composition was measured by EDS coupled to the SEM microscope. TEM images were obtained with a JEOL (Akishima, Tokyo Metropolis, Japan), JEM 1200 EX-II device equipped with a Gatan 782 camera for Electron Microscope Erlangshen ES 500. Samples with a particle size of 150 μm were deposited over a mesh previously covered by carbon. Analyses were carried out using an accelerating voltage of 120 kV. Additionally, the nature and strength of acidic sites were investigated by in situ Fourier transform infrared spectroscopy

(FTIR), using pyridine as a probe molecule in adsorption and desorption assays followed by FTIR analysis (Py-FTIR), according to a procedure reported previously [32]. More details about characterization procedures can be found in other publications [67,71–74].

The XRD diffraction pattern of transformed natural zeolite showed no significant changes with respect to as-received natural zeolite, keeping the same crystalline structure after the applied ion-exchange treatments with ammonium sulphate (see Figure 5). Characteristic peaks of clinoptilolite (C), mordenite (M) and quartz (Q) are noticed with equal intensity in both difractograms and were identified according to JCPDS 39–183, JCPDS 29–1257 and JCPDS 461045, respectively. However, XRF analysis revealed a reduction in the amount of compensating cations without any effect on the Si/Al ratio (5.34). Thus, the modified zeolite sample is mainly composed by SiO_2 (79.26%) and Al_2O_3 (14.85%) followed by CaO (1.82%), Fe_2O_3 (2.53%), Na_2O (0.26%), K_2O (0.39%), MgO (0.37%), TiO_2 (0.47%), and MnO (0.05%). Table 3 lists the main physical-chemical characteristics of the acidic transformed natural zeolite sample. As expected, the surface area and the content of acidic sites in the form of both Brønsted and Lewis acid sites were increased with the applied modification treatment.

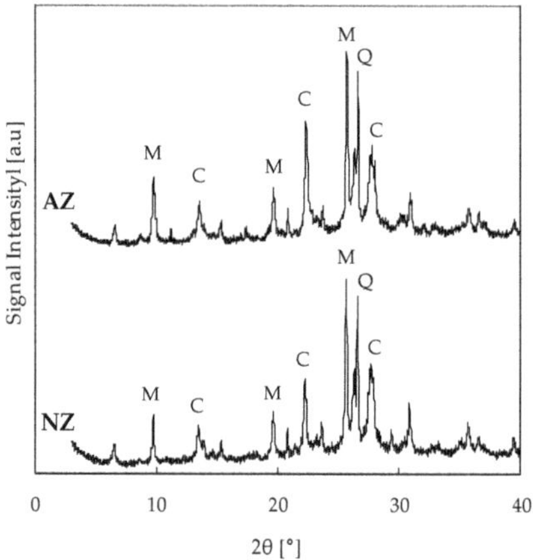

Figure 5. X-ray powder diffraction patterns of as-received natural zeolite (NZ) and acidic transformed natural zeolite (AZ) samples. Clinoptilolite (C), mordenite (M) and quartz (Q).

Table 3. Textural and acidic content of acidic transformed natural zeolite.

S_{BET} [a] $(m^2 \cdot g^{-1})$	Brønsted [b] Acid Sites $(\mu mol \cdot g^{-1})$	Lewis [b] Acid Sites $(\mu mol \cdot g^{-1})$	Total Acidity $(\mu mol \cdot g^{-1})$
261	179.8	282.8	462.6

[a] Obtained from nitrogen adsorption-desorption isotherm. [b] Determined by in situ Py-FTIR analyses, quantifying the remaining adsorbed pyridine after heating the saturated-samples up to 823 K.

SEM-EDS analyses (see Figure 6) corroborated the results obtained by XRD and XRF. A reduction in the content of compensating cations could be noticed in the as-received zeolite sample after the applied ion-exchange modification procedures.

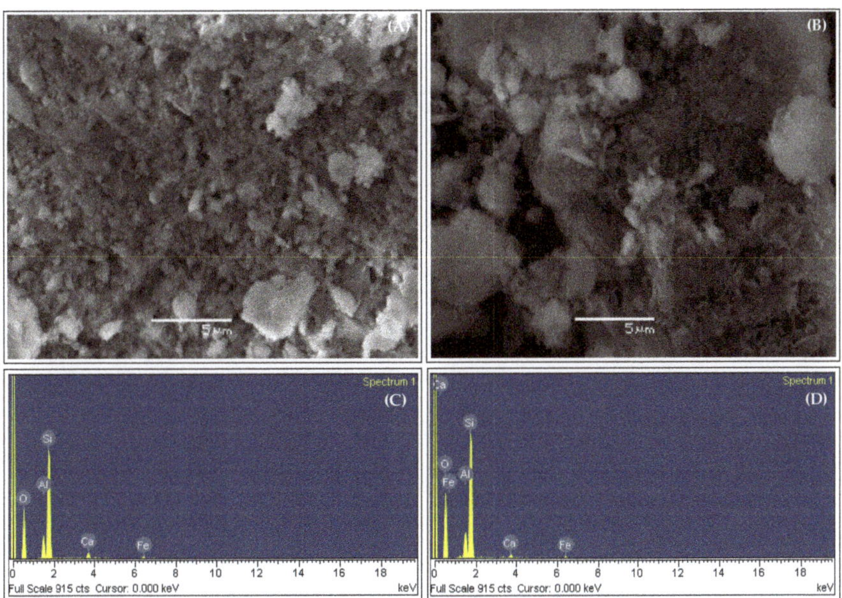

Figure 6. SEM-EDS results. (**A,C**) natural zeolite; (**B,D**) acidic transformed natural zeolite.

EDS results shown in Figure 6 indicate a 49.1% and a 40.1% of decrease in the amount of Ca and Fe cations, respectively. Also, the Si/Al ratio obtained with this technique remained almost constant, confirming that the zeolite structure was not significantly changed. Based on previous experimental results [32,67,68,71], a lower content of compensating cations has been related to a better ozone and toluene diffusion and accessibility to active surface sites of the zeolite framework, suggesting a better performance of acidic transformed natural zeolite during catalytic ozonation of VOCs.

TEM images of natural zeolite and acidic transformed natural zeolite samples are presented in Figure 7A,B, respectively. As can be seen, no substantial changes in the structure and morphology of zeolite samples are observed.

Figure 7. TEM images of (**A**) natural zeolite; (**B**) acidic transformed natural zeolite.

3.3. Experimental Description of Catalytic Ozonation Studies

Catalytic ozonation experiments were conducted at 293 K, using a U-type quartz packed-bed reactor (see Figure 8). Experimental conditions were set after preliminary experiments, following the Madon–Boudart criterion to assure the absence of mass transfer limitations [66]. Briefly, a mass of 150 mg of zeolite sample mixed with quartz particles of the same size with a mass ratio of 1:4 ($d_p \approx 0.36$ mm) was loaded and thermally outgassed (before reaction) at a heating rate of 1 °C·min^{-1} under argon flow (100 cm^3·min^{-1}) until reaching a temperature of 823 K. Then, isothermal conditions were maintained for two hours, before being cooled down to room temperature. After that, at a temperature of 293 K, a total gas flow of 100 cm^3·min^{-1} containing toluene and ozone was supplied over the reactor. Toluene and ozone concentrations were varied during the kinetic study, within ranges limited by physical chemical equilibria after dilutions of saturated streams and inert fresh streams. The concentrations of toluene and ozone were monitored on-line using a Perkin Elmer Clarus 500 chromatograph (Perkin Elmer, Waltham, MA, USA) and a BMT 964 ozone analyzer (BMT Messtechnik GmbH, Stahnsdorf, Germany), respectively. More details about experimental conditions and procedures can be found elsewhere [32]. All trials were performed in the absence of mass transfer limitations.

The reactor was operated in a differential mode. Thus, from a series of experimental runs, a set of rate-concentration data was obtained using the plug flow performance equation [65], as follows:

$$\frac{W}{F_{Tol_{in}}} = \frac{X_{Tol}}{(-r_{Tol})_{exp}} \tag{24}$$

where $F_{Tol_{in}}$ represents the feed rate of toluene (µmol·s^{-1}), $(-r_{Tol})_{exp}$ is the rate of catalytic oxidation reaction of toluene by ozone (µmol·g^{-1}·s^{-1}), W is the mass of zeolite sample (g), and X_{Tol} is the fraction of toluene converted.

In order to obtain a kinetic reaction model that represents the catalytic ozonation of toluene promoted by transformed natural zeolite at room temperature, firstly, a power law kinetic model was assessed as an initial approximation. Secondly, a set of several elementary steps were proposed with individual rates and an overall kinetic rate expression was postulated, considering the rate-determining step (RDS) by taking into account the experimental findings obtained in this study and those reported previously [32]. In both cases, the proposed overall rate expressions were validated with the experimental data using the least-squares non-linear regression method (NLR). Parameters of the proposed kinetic expressions were gathered by minimizing the sum of the squares (σ^2) of the difference of obtained reaction rates for each experimental run, using the solver tool of Microsoft Excel [60] implemented through Equation (25).

$$s^2 = \sum_{i=1}^{14} \left[\left((-r_{Tol})_{exp} - (-r_{Tol})_{mod} \right)_i^2 \right] \tag{25}$$

where $(-r_{Tol})_{exp}$ and $(-r_{Tol})_{mod}$ stand for the rate of catalytic ozonation of toluene (µmol·g^{-1}·s^{-1}), obtained experimentally and using the proposed kinetic model, respectively.

Additionally, Langmuir adsorption equilibrium parameters were obtained from dynamic adsorption experiments conducted at 293 K, using the same experimental set-up without the inlet of ozone. Prior to adsorption experiments, zeolite samples were thermally outgassed as described before. The adsorbed amounts of toluene over acidic transformed natural zeolite at equilibrium (µmol·g^{-1}) for different inlet concentrations of toluene, were calculated from breakthrough curves using Equation (26), as reported elsewhere [68].

$$q = \frac{1}{W} \int_0^{t_s} \left(F_{Tol_{in}} - F_{Tol_t} \right) dt \tag{26}$$

where q is the adsorbed amount of toluene over acidic transformed natural zeolite at equilibrium ($\mu mol \cdot g^{-1}$). W is the zeolite mass (g), and t_s is the time needed to reach the zeolite saturation (s). $F_{Tol_{in}}$ and F_{Tol_t} are the feed rate of toluene ($\mu mol \cdot s^{-1}$), at the adsorber inlet and outlet streams as a function of time, respectively.

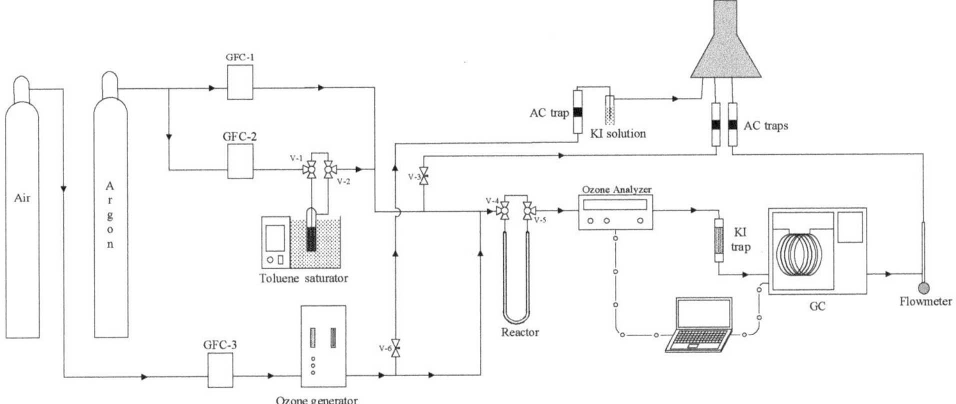

Figure 8. Experimental set-up for the catalytic ozonation of toluene using acidic surface transformed natural zeolite as catalyst at room temperature. Gas flow controllers (GFC); activated carbon (AC) traps; potassium iodine (KI) trap. Reprinted from Reference [32].

3.4. In Situ FTIR Study of Surface Interactions between Zeolite Sample and Toluene Oxidation By-Products

Oxidation by-products that remained adsorbed after catalytic ozonation of toluene were identified by FTIR spectroscopy in a Bruker Mod Tensor 27 spectrometer (Billerica, MA, USA). IR spectra were collected at an average of 100 scans with a resolution of 4 cm^{-1} using a self-pressed disc. Additionally, IR spectra before and after toluene adsorption were also recorded.

4. Conclusions

Results obtained in this study provide new insights on the surface reaction mechanism during the catalytic ozonation of toluene using acidic transformed natural zeolite. Experimental evidence confirmed that surface interactions take place between adsorbed toluene at Brønsted acid sites with active atomic oxygen, coming from ozone adsorption and decomposition at Lewis acid sites, leading to toluene oxidation. Experimental data proved that catalytic ozonation of toluene using acidic transformed natural zeolite is very well represented by the dual-site Langmuir–Hinshelwood (dsL-H) kinetic approximation where the rate-determining step results to be the surface reaction between adsorbed species on adjacent active sites. This kinetic expression can be used in the design and optimization of this new process technology. Acidic transformed natural zeolite could be applied as an effective and low-cost catalyst for the catalytic oxidation of VOCs using ozone at room temperature in a novel air pollution control technology. Further research using metal-supported modified natural zeolites in toluene oxidation is under way to complement information provided here in order to facilitate large scale implementation.

Author Contributions: Conceptualization, S.A.-M., H.V. and C.A.Z.; methodology S.A.-M. and H.V.; investigation, S.A.-M., H.V.; resources, H.V.; data curation, S.A.-M.; writing—original draft preparation, S.A.-M., H.V.; writing—review and editing, S.A.-M., H.V. and C.A.Z.; visualization, S.A.-M.; supervision, H.V. and C.A.Z.; project administration, H.V.; funding acquisition, H.V. All authors have read and agreed to the published version of the manuscript.

Funding: This research was funded by CONICYT, FONDECYT/Regular, grant number 1130560 and by the Universidad Católica de la Santísima Concepción, grant number DI-FME 02/2021.

Data Availability Statement: Data presented in this study are available in the tables displayed inside this article.

Acknowledgments: The authors would like to thank Víctor Solar from the Clean Technology Laboratory, Universidad Católica de la Santísima Concepción for his valued collaboration.

Conflicts of Interest: The authors declare no conflict of interest. The founding sponsors of this article had no role in the design of the study; in the collection, analyses, or interpretation of data; in the writing of the manuscript, and in the decision to publish the results.

References

1. Parmar, G.R.; Rao, N.N. Emerging Control Technologies for Volatile Organic Compounds. *Crit. Rev. Environ. Sci. Technol.* **2008**, *39*, 41–78. [CrossRef]
2. Baek, S.-W.; Kim, J.-R.; Ihm, S.-K. Design of dual functional adsorbent/catalyst system for the control of VOC's by using metal-loaded hydrophobic Y-zeolites. *Catal. Today* **2004**, *93–95*, 575–581. [CrossRef]
3. Fan, Z.; Lioy, P.; Weschler, C.; Fiedler, N.; Kipen, H.; Zhang, J. Ozone-Initiated Reactions with Mixtures of Volatile Organic Compounds under Simulated Indoor Conditions. *Environ. Sci. Technol.* **2003**, *37*, 1811–1821. [CrossRef]
4. Megías-Sayago, C.; Lara-Ibeas, I.; Wang, Q.; Le Calvé, S.; Louis, B. Volatile organic compounds (VOCs) removal capacity of ZSM-5 zeolite adsorbents for near real-time BTEX detection. *J. Environ. Chem. Eng.* **2020**, *8*, 103724. [CrossRef]
5. Kim, K.-H.; Kabir, E.; Kabir, S. A review on the human health impact of airborne particulate matter. *Environ. Int.* **2015**, *74*, 136–143. [CrossRef] [PubMed]
6. Zhang, X.; Gao, B.; Zheng, Y.; Hu, X.; Creamer, A.E.; Annable, M.D.; Li, Y. Biochar for volatile organic compound (VOC) removal: Sorption performance and governing mechanisms. *Bioresour. Technol.* **2017**, *245*, 606–614. [CrossRef]
7. Adnew, G.A.; Meusinger, C.; Bork, N.; Gallus, M.; Kyte, M.; Rodins, V.; Rosenørn, T.; Johnson, M.S. Gas-phase advanced oxidation as an integrated air pollution control technique. *AIMS Environ. Sci.* **2016**, *3*, 141–158. [CrossRef]
8. Ojala, S.; Pitkäaho, S.; Laitinen, T.; Koivikko, N.N.; Brahmi, R.; Gaálová, J.; Kucherov, A.; Matejova, L.; Päivärinta, S.; Hirschmann, C.; et al. Catalysis in VOC Abatement. *Top. Catal.* **2011**, *54*, 1224–1256. [CrossRef]
9. Tidahy, H.L.; Siffert, S.; Lamonier, J.F.; Cousin, R.; Zhilinskaya, E.A.; Aboukaïs, A.; Su, B.L.; Canet, X.; De Weireld, G.; Frère, M.; et al. Influence of the exchanged cation in Pd/BEA and Pd/FAU zeolites for catalytic oxidation of VOCs. *Appl. Catal. B Environ.* **2007**, *70*, 377–383. [CrossRef]
10. Joung, H.J.; Kim, J.H.; Oh, J.S.; You, D.W.; Park, H.O.; Jung, K.W. Catalytic oxidation of VOCs over CNT-supported platinum nanoparticles. *Appl. Surf. Sci.* **2014**, *290*. [CrossRef]
11. Antunes, A.P.; Ribeiro, M.F.; Silva, J.M.; Ribeiro, F.R.; Magnoux, P.; Guisnet, M. Catalytic oxidation of toluene over CuNaHY zeolites: Coke formation and removal. *Appl. Catal. B Environ.* **2001**, *33*, 149–164. [CrossRef]
12. Huang, H.; Xu, Y.; Feng, Q.; Leung, D.Y.C. Low temperature catalytic oxidation of volatile organic compounds: A review. *Catal. Sci. Technol.* **2015**, *5*, 2649–2669. [CrossRef]
13. Berenjian, A.; Khodiev, A. How ozone can affect volatile organic compounds. *Aust. J. Basic Appl. Sci.* **2009**, *3*, 385–388.
14. Rezaei, E.; Soltan, J.; Chen, N. Catalytic oxidation of toluene by ozone over alumina supported manganese oxides: Effect of catalyst loading. *Appl. Catal. B Environ.* **2013**, *136–137*, 239–247. [CrossRef]
15. Long, L.P.; Zhao, J.G.; Yang, L.X.; Fu, M.L.; Wu, J.L.; Huang, B.C.; Ye, D.Q. Room Temperature Catalytic Ozonation of Toluene over MnO_2/Al_2O_3. *Chin. J. Catal.* **2011**, *32*, 904–916. [CrossRef]
16. Rezaei, E.; Soltan, J. Low temperature oxidation of toluene by ozone over MnOx/alumina and MnOx/MCM-41 catalysts. *Chem. Eng. J.* **2012**, *198–199*, 482–490. [CrossRef]
17. Nawrocki, J.; Kasprzyk-Hordern, B. The efficiency and mechanisms of catalytic ozonation. *Appl. Catal. B Environ.* **2010**, *99*, 27–42. [CrossRef]
18. Rezaei, F.; Moussavi, G.; Bakhtiari, A.R.; Yamini, Y. Toluene removal from waste air stream by the catalytic ozonation process with MgO/GAC composite as catalyst. *J. Hazard. Mater.* **2016**, *306*, 348–358. [CrossRef]
19. Liu, Y.; Li, X.; Liu, J.; Shi, C.; Zhu, A. Ozone catalytic oxidation of benzene over AgMn/HZSM-5 catalysts at room temperature: Effects of Mn loading and water content. *Cuihua Xuebao Chin. J. Catal.* **2014**, *35*, 1465–1474. [CrossRef]
20. Huang, H.; Ye, X.; Huang, W.; Chen, J.; Xu, Y.; Wu, M.; Shao, Q.; Peng, Z.; Ou, G.; Shi, J.; et al. Ozone-catalytic oxidation of gaseous benzene over MnO2/ZSM-5 at ambient temperature: Catalytic deactivation and its suppression. *Chem. Eng. J.* **2015**, *264*, 24–31. [CrossRef]
21. Einaga, H.; Maeda, N.; Yamamoto, S.; Teraoka, Y. Catalytic properties of copper-manganese mixed oxides supported on SiO_2 for benzene oxidation with ozone. *Catal. Today* **2015**, *245*, 22–27. [CrossRef]
22. Gopi, T.; Swetha, G.; Shekar, S.C.; Krishna, R.; Ramakrishna, C.; Saini, B.; Rao, P.V.L.V.L. Ozone catalytic oxidation of toluene over 13X zeolite supported metal oxides and the effect of moisture on the catalytic process. *Arab. J. Chem.* **2016**, *12*, 4502–4513. [CrossRef]

23. Kim, H.-H.; Sugasawa, M.; Hirata, H.; Teramoto, Y.; Kosuge, K.; Negishi, N.; Ogata, A. Ozone-assisted catalysis of toluene with layered ZSM-5 and Ag/ZSM-5 zeolites. *Plasma Chem. Plasma Process.* **2013**, *33*, 1083–1098. [CrossRef]
24. Liu, B.; Ji, J.; Zhang, B.; Huang, W.; Gan, Y.; Leung, D.Y.C.; Huang, H. Catalytic ozonation of VOCs at low temperature: A comprehensive review. *J. Hazard. Mater.* **2021**, 126847. [CrossRef]
25. Wang, H.C.; Liang, H.S.; Chang, M.B. Ozone-enhanced catalytic oxidation of monochlorobenzene over iron oxide catalysts. *Chemosphere* **2011**, *82*, 1090–1095. [CrossRef] [PubMed]
26. Yang, C.; Miao, G.; Pi, Y.; Xia, Q.; Wu, J.; Li, Z.; Xiao, J. Abatement of various types of VOCs by adsorption/catalytic oxidation: A review. *Chem. Eng. J.* **2019**, *370*, 1128–1153. [CrossRef]
27. Guo, Y.; Wen, M.; Li, G.; An, T. Recent advances in VOC elimination by catalytic oxidation technology onto various nanoparticles catalysts: A critical review. *Appl. Catal. B Environ.* **2021**, *281*, 119447. [CrossRef]
28. Brodu, N.; Zaitan, H.; Manero, M.-H.; Pic, J.-S. Removal of volatile organic compounds by heterogeneous ozonation on microporous synthetic alumina silicate. *Water Sci. Technol.* **2012**, *66*, 2020–2026. [CrossRef]
29. Soylu, G.S.P.; Özçelik, Z.; Boz, İ.I. Total oxidation of toluene over metal oxides supported on a natural clinoptilolite-type zeolite. *Chem. Eng. J.* **2010**, *162*, 380–387. [CrossRef]
30. Nikolov, P.; Genov, K.; Konova, P.; Milenova, K.; Batakliev, T.; Georgiev, V.; Kumar, N.; Sarker, D.K.; Pishev, D.; Rakovsky, S. Ozone decomposition on Ag/SiO2 and Ag/clinoptilolite catalysts at ambient temperature. *J. Hazard. Mater.* **2010**, *184*, 16–19. [CrossRef]
31. Valdés, H.; Tardón, R.F.; Zaror, C.A. Role of surface hydroxyl groups of acid-treated natural zeolite on the heterogeneous catalytic ozonation of methylene blue contaminated waters. *Chem. Eng. J.* **2012**, *211–212*, 388–395. [CrossRef]
32. Alejandro-Martín, S.; Valdés, H.; Manero, M.-H.; Zaror, C. Catalytic Ozonation of Toluene Using Chilean Natural Zeolite: The Key Role of Brønsted and Lewis Acid Sites. *Catalysts* **2018**, *8*, 211. [CrossRef]
33. Valdés, H.; Ulloa, F.J.; Solar, V.A.; Cepeda, M.S.; Azzolina-Jury, F.; Thibault-Starzyk, F. New insight of the influence of acidic surface sites of zeolite on the ability to remove gaseous ozone using operando DRIFTS studies. *Microporous Mesoporous Mater.* **2020**, *294*, 109912. [CrossRef]
34. Wu, J.; Su, T.; Jiang, Y.; Xie, X.; Qin, Z.; Ji, H. In situ DRIFTS study of O_3 adsorption on CaO, γ-Al_2O_3, CuO, α-Fe_2O_3 and ZnO at room temperature for the catalytic ozonation of cinnamaldehyde. *Appl. Surf. Sci.* **2017**, *412*, 290–305. [CrossRef]
35. Roque Malherbe, R.; Wendelbo, R. Study of Fourier transform infrared-temperature-programmed desorption of benzene, toluene and ethylbenzene from H-ZSM-5 and H-Beta zeolites. *Thermochim. Acta* **2003**, *400*, 165–173. [CrossRef]
36. Einaga, H.; Futamura, S. Catalytic oxidation of benzene with ozone over alumina-supported manganese oxides. *J. Catal.* **2004**, *227*, 304–312. [CrossRef]
37. Einaga, H.; Ogata, A. Benzene oxidation with ozone over supported manganese oxide catalysts: Effect of catalyst support and reaction conditions. *J. Hazard. Mater.* **2009**, *164*, 1236–1241. [CrossRef] [PubMed]
38. Dhandapani, B.; Oyama, S.T. Gas phase ozone decomposition catalysts. *Appl. Catal. B Environ.* **1997**, *11*, 129–166. [CrossRef]
39. Li, W.; Oyama, S.T. Mechanism of Ozone Decomposition on a Manganese Oxide Catalyst. 2. Steady-State and Transient Kinetic Studies. *J. Am. Chem. Soc.* **1998**, *120*, 9047–9052. [CrossRef]
40. Einaga, H.; Futamura, S. Comparative study on the catalytic activities of alumina-supported metal oxides for oxidation of benzene and cyclohexane with ozone. *React. Kinet. Catal. Lett.* **2004**, *81*, 121–128. [CrossRef]
41. Oyama, S.T. Chemical and Catalytic Properties of Ozone. *Catal. Rev. Sci. Eng.* **2000**, *42*, 279–322. [CrossRef]
42. Kasprzyk-Hordern, B.; Ziólek, M.; Nawrocki, J. Catalytic ozonation and methods of enhancing molecular ozone reactions in water treatment. *Appl. Catal. B Environ.* **2003**, *46*, 639–669. [CrossRef]
43. Bulanin, K.M.; Lavalley, J.C.; Tsyganenko, A.A. IR spectra of adsorbed ozone. *Colloids Surf. A Physicochem. Eng. Asp.* **1995**, *101*, 153–158. [CrossRef]
44. Bulanin, K.M.; Lavalley, J.C.; Tsyganenko, A.A. Infrared Study of Ozone Adsorption on CaO. *J. Phys. Chem. B* **1997**, *101*, 2917–2922. [CrossRef]
45. Hu, M.; Hui, K.S.; Hui, K.N. Role of graphene in MnO_2/graphene composite for catalytic ozonation of gaseous toluene. *Chem. Eng. J.* **2014**, *254*, 237–244. [CrossRef]
46. Jeon, J.Y.; Kim, H.Y.; Woo, S.I. Mechanistic study on the SCR of NO by C_3H_6 over Pt/V/MCM-41. *Appl. Catal. B Environ.* **2003**, *44*, 301–310. [CrossRef]
47. Chen, X.-M.; Yang, X.-F.; Zhu, A.-M.; Fan, H.-Y.; Wang, X.-K.; Xin, Q.; Zhou, X.-R.; Shi, C. In situ DRIFTS study on the partial oxidation of ethylene over Co-ZSM-5 catalyst. *Catal. Commun.* **2009**, *10*, 428–432. [CrossRef]
48. Alejandro, S.; Valdés, H.; Manéro, M.-H.; Zaror, C.A. Oxidative regeneration of toluene-saturated natural zeolite by gaseous ozone: The influence of zeolite chemical surface characteristics. *J. Hazard. Mater.* **2014**, *274*, 212–220. [CrossRef]
49. Rezaei, E.; Soltan, J.; Chen, N.; Lin, J. Effect of noble metals on activity of MnOx/γ-alumina catalyst in catalytic ozonation of toluene. *Chem. Eng. J.* **2013**, *214*, 219–228. [CrossRef]
50. Kim, J.; Kwon, E.E.; Lee, J.E.; Jang, S.-H.; Jeon, J.-K.; Song, J.; Park, Y.-K. Effect of zeolite acidity and structure on ozone oxidation of toluene using Ru-Mn loaded zeolites at ambient temperature. *J. Hazard. Mater.* **2021**, *403*, 123934. [CrossRef]
51. Van Durme, J.; Dewulf, J.; Sysmans, W.; Leys, C.; Van Langenhove, H. Abatement and degradation pathways of toluene in indoor air by positive corona discharge. *Chemosphere* **2007**, *68*, 1821–1829. [CrossRef] [PubMed]

52. Li, J.; Na, H.; Zeng, X.; Zhu, T.; Liu, Z. In situ DRIFTS investigation for the oxidation of toluene by ozone over Mn/HZSM-5, Ag/HZSM-5 and Mn-Ag/HZSM-5 catalysts. *Appl. Surf. Sci.* **2014**, *311*, 690–696. [CrossRef]

53. Huang, H.; Li, W. Destruction of toluene by ozone-enhanced photocatalysis: Performance and mechanism. *Appl. Catal. B Environ.* **2011**, *102*, 449–453. [CrossRef]

54. Huang, H.; Ye, D.; Leung, D.Y.C.; Feng, F.; Guan, X. Byproducts and pathways of toluene destruction via plasma-catalysis. *J. Mol. Catal. A Chem.* **2011**, *336*, 87–93. [CrossRef]

55. Brodu, N.; Manero, M.-H.; Andriantsiferana, C.; Pic, J.-S.; Valdés, H. Role of Lewis acid sites of ZSM-5 zeolite on gaseous ozone abatement. *Chem. Eng. J.* **2013**, *231*, 281–286. [CrossRef]

56. Brodu, N.; Manero, M.-H.; Andriantsiferana, C.; Pic, J.-S.; Valdés, H. Gaseous ozone decomposition over high silica zeolitic frameworks. *Can. J. Chem. Eng.* **2018**, *96*, 1911–1918. [CrossRef]

57. Brodu, N.; Sochard, S.; Andriantsiferana, C.; Pic, J.-S.; Manero, M.-H. Fixed-bed adsorption of toluene on high silica zeolites: Experiments and mathematical modelling using LDF approximation and a multisite model. *Environ. Technol.* **2015**, *36*, 1807–1818. [CrossRef]

58. Chao, C.Y.H.; Kwong, C.W.; Hui, K.S. Potential use of a combined ozone and zeolite system for gaseous toluene elimination. *J. Hazard. Mater.* **2007**, *143*, 118–127. [CrossRef]

59. Einaga, H.; Futamura, S. Catalytic oxidation of benzene with ozone over Mn ion-exchanged zeolites. *Catal. Commun.* **2007**, *8*, 557–560. [CrossRef]

60. Harris, D.C. Nonlinear least-squares curve fitting with Microsoft Excel Solver. *J. Chem. Educ.* **1998**, *75*, 119. [CrossRef]

61. Toby, S.; Van de Burgt, L.J.; Toby, F.S. Kinetics and chemiluminescence of ozone-aromatic reactions in the gas phase. *J. Phys. Chem.* **1985**, *89*, 1982–1986. [CrossRef]

62. Hu, M.; Yao, Z.; Hui, K.N.; Hui, K.S. Novel mechanistic view of catalytic ozonation of gaseous toluene by dual-site kinetic modelling. *Chem. Eng. J.* **2017**, *308*, 710–718. [CrossRef]

63. Rezaei, E.; Soltan, J. EXAFS and kinetic study of MnOx/γ-alumina in gas phase catalytic oxidation of toluene by ozone. *Appl. Catal. B Environ.* **2014**, *148–149*, 70–79. [CrossRef]

64. Aghbolaghy, M.; Soltan, J.; Chen, N. Low Temperature Catalytic Oxidation of Binary Mixture of Toluene and Acetone in the Presence of Ozone. *Catal. Lett.* **2018**, *148*, 3431–3444. [CrossRef]

65. Scott, H.F. *Elements of Chemical Reaction Engineering*, 5th ed.; Prentice Hall: Boston, MA, USA, 2016; Volume 54, ISBN 0133887820.

66. Vannice, M.A. *Kinetics of Catalytic Reactions*; Springer: Boston, MA, USA, 2005; ISBN 978-0-387-24649-9.

67. Alejandro, S.; Valdés, H.; Zaror, C.A. Natural Zeolite Reactivity towards Ozone: The Role of Acid Surface Sites. *J. Adv. Oxid. Technol.* **2011**, *14*, 182–189. [CrossRef]

68. Alejandro, S.; Valdés, H.; Manero, M.-H.; Zaror, C.A. BTX abatement using Chilean natural zeolite: The role of Bronsted acid sites. *Water Sci. Technol.* **2012**, *66*, 1759–1769. [CrossRef]

69. Roque-Malherbe, R.; Ivanov, V. Codiffusion and counterdiffusion of para-xylene and ortho-xylene in a zeolite with 10 MR/12 MR interconnected channels. An example of molecular traffic control. *J. Mol. Catal. A Chem.* **2009**, *313*, 7–13. [CrossRef]

70. Roque-Malherbe, R. Complementary approach to the volume filling theory of adsorption in zeolites. *Microporous Mesoporous Mater.* **2000**, *41*, 227–240. [CrossRef]

71. Valdés, H.; Alejandro, S.; Zaror, C.A. Natural zeolite reactivity towards ozone: The role of compensating cations. *J. Hazard. Mater.* **2012**, *227–228*, 34–40. [CrossRef]

72. Valdés, H.; Farfán, V.J.; Manoli, J.A.; Zaror, C.A. Catalytic ozone aqueous decomposition promoted by natural zeolite and volcanic sand. *J. Hazard. Mater.* **2009**, *165*, 915–922. [CrossRef] [PubMed]

73. Guisnet, M.; Ayrault, P.; Coutanceau, C.; Fernanda Alvarez, M.; Datka, J. Acid properties of dealuminated beta zeolites studied by IR spectroscopy. *J. Chem. Soc. Faraday Trans.* **1997**, *93*, 1661–1665. [CrossRef]

74. Abreu, N.J.; Valdés, H.; Zaror, C.A.; Azzolina-Jury, F.; Meléndrez, M.F. Ethylene adsorption onto natural and transition metal modified Chilean zeolite: An operando DRIFTS approach. *Microporous Mesoporous Mater.* **2019**, *274*, 138–148. [CrossRef]

 *catalysts*

Article

Photocatalytic Study of Cyanide Oxidation Using Titanium Dioxide (TiO$_2$)-Activated Carbon Composites in a Continuous Flow Photo-Reactor

Stalin Coronel, Diana Endara *, Ana Belén Lozada, Lucía E. Manangón-Perugachi and Ernesto de la Torre

Department of Extractive Metallurgy, Escuela Politécnica Nacional, Ladrón de Guevara E11-253, Quito 170517, Ecuador; stalin.coronel@epn.edu.ec (S.C.); ana.lozada@epn.edu.ec (A.B.L.); lucia.manangon@epn.edu.ec (L.E.M.-P.); ernesto.delatorre@epn.edu.ec (E.d.l.T.)
* Correspondence: diana.endara@epn.edu.ec; Tel.: +593-(9)9854-9231

Abstract: The photocatalytic oxidation of cyanide by titanium dioxide (TiO$_2$) supported on activated carbon (AC) was evaluated in a continuous flow UV photo-reactor. The continuous photo-reactor was made of glass and covered with a wood box to isolate the fluid of external conditions. The TiO$_2$-AC synthesized by the impregnation of TiO$_2$ on granular AC composites was characterized by inductively coupled plasma optical emission spectrometry (ICP-OES), Scanning Electron Microscopy (SEM), and nitrogen adsorption-desorption isotherms. Photocatalytic and adsorption tests were conducted separately and simultaneously. The results showed that 97% of CN$^-$ was degraded within 24 h due to combined photocatalytic oxidation and adsorption. To estimate the contribution of only adsorption, two-stage tests were performed. First, 74% cyanide ion degradation was reached in 24 h under dark conditions. This result was attributed to CN$^-$ adsorption and oxidation due to the generation of H$_2$O$_2$ on the surface of AC. Then, 99% degradation of cyanide ion was obtained through photocatalysis during 24 h. These results showed that photocatalysis and the continuous photo-reactor's design enhanced the photocatalytic cyanide oxidation performance compared to an agitated batch system. Therefore, the use of TiO$_2$-AC composites in a continuous flow photo-reactor is a promising process for the photocatalytic degradation of cyanide in aqueous solutions.

Keywords: cyanide; activated carbon; titanium dioxide; composites; continuous flow; photocatalysis; adsorption

Citation: Coronel, S.; Endara, D.; Lozada, A.B.; Manangón-Perugachi, L.E.; de la Torre, E. Photocatalytic Study of Cyanide Oxidation Using Titanium Dioxide (TiO$_2$)-Activated Carbon Composites in a Continuous Flow Photo-Reactor. *Catalysts* **2021**, *11*, 924. https://doi.org/10.3390/catal11080924

Academic Editors: José Ignacio Lombraña, Héctor Valdés, Cristian Ferreiro and Giuseppina Iervolino

Received: 16 May 2021
Accepted: 27 July 2021
Published: 30 July 2021

Publisher's Note: MDPI stays neutral with regard to jurisdictional claims in published maps and institutional affiliations.

1. Introduction

Cyanide is a highly toxic pollutant, which, even at low concentration, may cause human health and environmental problems [1]. Cyanide is rapidly and extensively absorbed by the human body through the oral inhalation and dermal routes. It prevents the transport of oxygen, which affects the cellular respiration process, leading to suffocation in the worst case and eventually to death [2].

Cyanide is present in industrial wastewaters such as coal gasification, electroplating, plastics, pharmaceuticals, and the mining industry. These wastewaters are discharged in the water bodies causing serious threats to the environment [1,3,4].

In Ecuador, the artisanal and small-scale gold mining activities make significant contributions to mineral production [5,6]. Large-scale gold mining projects as "Fruta del Norte" and "Cascabel" have been developed in the last years. Both small- and large-scale gold mining industries use cyanidation to recover gold from ores.

In Ecuador, cyanidation and carbon in pulp (CIP) processes use an aqueous 500 mg/L NaCN solution. The Ecuadorian legislation TULSMA establishes a discharge limit of 1 mg/L total cyanide into sewers and 0.1 mg/L into surface fresh waters [7]. Cyanides exist in the form of free ions (CN$^-$) and also can form complexes, which makes its treatment more difficult. For example, a study with gold mining wastewater that contained copper and zinc was treated with zeolite, achieving a 93.97% degradation efficiency of total cyanides [8]. Due to the hazard of cyanide, its treatment is very important. Several physical, biological, and chemical treatment processes have been applied to remove cyanide. The most commonly used processes include chemical oxidation, alkaline chlorination, hydrogen peroxide oxidation, INCO process (purification with SO_2 and air), oxidation with Caro acid, and ozonation. Chlorination is the most used process, and it is very efficient in the elimination of cyanide; however, it has certain disadvantages, such as the generation of toxic intermediate compounds which must be treated, making the process expensive [9,10].

Advanced oxidation processes (AOPs) have been studied extensively in the removal of different contaminants. Hydroxyl radicals are considered the most reactive oxygenated species within AOPs due to their high oxidation potential and their non-selective nature. Photocatalysis is an effective technique for treating toxic substances, including cyanide. [11,12].

Several semiconductor materials have been tested as photocatalysts for the removal of aqueous pollutants. However, difficulties related to the stability of the photocatalyst under irradiation in water have been evidenced. It is accepted that titanium dioxide TiO_2 in anatase phase is the most reliable photocatalyst for aqueous pollutant removal [13].

In heterogeneous photocatalysis based on semiconductors, the photocatalyst TiO_2 is excited by absorbing incident UV radiation. Momentarily, the electron of the valence band is transferred to the conduction band, and the electron/hole pair (e^-/h^+) is formed. Then, the electron/hole pair reacts with water and dissolved molecular oxygen to generate hydroxyl and superoxide radicals, which are responsible for the photocatalytic oxidation of free cyanide [14].

Direct and indirect mechanisms of photocatalytic oxidation of free cyanide with TiO_2 have been proposed. In the direct mechanism, free cyanide present in the solution is oxidized directly through the transfer of electrons from the holes of the valence band. The indirect mechanism occurs through the adsorbed $OH^\bullet$ radicals, implying that the contaminants are first adsorbed on the photocatalyst surface and then react with the excited superficial e^-/h^+ pairs or the $OH^\bullet$ radicals [15].

The photocatalytic oxidation of cyanide starts with the formation of cyanide radical, which dimerizes to form cyanogen. Then, the cyanogen undergoes transformation in an alkaline medium to give cyanide and cyanate. Finally, cyanate oxidizes to form nitrite (NO_2^-), nitrate (NO_3^-), carbonate (CO_3^{2-}), carbon dioxide (CO_2), and nitrogen (N_2). These reactions are shown in the Equations (1)–(5) [16].

$$CN^- \xrightarrow{h^+/OH^\bullet} CN^\bullet \tag{1}$$

$$2CN^\bullet \rightarrow (CN)_2 \tag{2}$$

$$(CN)_2 + 2OH^- \rightarrow CN^- + CNO^- + H_2O \tag{3}$$

$$CNO^- + 8OH^- + 8h^+ \rightarrow NO_3^- + CO_2 + 4H_2O \tag{4}$$

$$CNO^- + 2H_2O + 3h^+ \rightarrow CO_3^{2-} + \frac{1}{2}N_2 + 4H^+ \tag{5}$$

TiO_2 has been widely used as photocatalyst due to its non-toxicity, low cost, chemical stability, and its favorable chemical and physical properties. It can be reused several times without reducing its catalytic activity. TiO_2 has been tested in the photocatalytic degradation of wastewater under ultraviolet (UV) light. However, low efficiencies in photodegradation have been achieved due to its rapid unfavorable charge carrier recombination reaction in TiO_2 and the high band gap energy of 3.2 eV, which limits its application from using solar energy [4,15]. In addition, filtration and separation processes are required at the end of the treatment, which increases the costs due to the granulometric size of TiO_2 (74 microns) [17].

Some strategies have been proposed in order to overcome these drawbacks. These include the use of supports such as silica [18], zeolites [19], and activated carbon. For example, the use of activated carbon (AC) as support achieves minimal losses of TiO_2. Studies show that AC improves the efficiency of the photocatalytic process thanks to its high adsorption capacity given by its porous structure. AC is a good support, due to its granulometry, hardness, and high surface area [17].

AC develops a synergistic adsorption-degradation effect according to its surface chemistry [17]. AC can oxidize cyanide through the adsorption of molecular oxygen on the AC surface. Adsorbed oxygen reacts with functional groups characteristic of the AC surface to form hydrogen peroxide (H_2O_2) that finally reacts with cyanide ion (CN^-) to form cyanate, according to Equation (6) [20]. The cyanide oxidation process with AC reached oxidation percentages between 60 and 70% after 8 h [21]. Although AC could adsorb cyanide, preliminary tests have shown that the adsorption percentage of cyanide is less than 5% [22].

$$CN^-_{(aq)} + H_2O_{2(aq)} \rightleftarrows CNO^-_{(aq)} + H_2O_{(aq)} \tag{6}$$

The optimization of the photocatalytic material and the study of the influence of factors such as pH, hydroxyl radicals concentration, or the organic compounds concentration on the cyanide photodegradation have been analyzed to improve the photocatalytic activity [10,14]. Nevertheless, the design of the photocatalytic reactor has been less studied, and there is more information available about the optimization of photocatalytic catalyst. For this reason, new alternatives referring to the photo-reactors configuration are necessary to improve the photocatalytic processes [23].

Batch photo-reactors require long times to achieve significant cyanide removal percentages (>90%); for this reason, the design of photocatalytic reactors in a different configuration than batch is essential. In a batch reactor, the UV light has contact mainly at the surface level, without considerable entry and dissipation within the fluid. The thickness of liquid formed from the base of the reactors decreases the activation of the photocatalyst and consequently the degradation of cyanide [17,24,25].

Another disadvantage of a batch reactor is the difficult separation of catalyst after the degradation process. This could be tackled by the implementation of a continuous reactor in the treatment of pollutants [26]. In a previous study, a multi-phase continuous flow reactor was tested in the photocatalytic oxidation of cyanide using TiO_2 as a photocatalyst. Then, the reactor was scaled up to degrade cyanide on an industrial level [27].

This investigation is oriented to the use of composites of titanium dioxide impregnated on active carbon (TiO_2-AC) as photocatalyst for the degradation of cyanide in a continuous flow photo-reactor. The use of the TiO_2-AC composite and the design of the continuous photo-reactor could enhance the photocatalytic oxidation performance. In addition, this strategic photo-reactor configuration can replace the conventional agitated batch system and reaches high cyanide degradation percentages.

2. Results

2.1. Physical and Chemical Characterization of GCR-20 Activated Carbon and TiO_2-AC Composite

The granular composite used as photocatalyst was obtained by the wet impregnation of TiO_2 over AC. A porous network was observed in SEM micrographs of AC and TiO_2-AC composite (Figure S1, Supplementary Materials). The content of TiO_2 on the AC was analyzed by ICP-OES, and the impregnation was 0.27% w/w.

The textural properties determined by N_2 physisorption and BET (Brunauer-Emmett_Teller) modeling listed in Table 1 show that AC support and TiO_2-AC catalyst had more than 900 m^2/g of specific surface area. It indicates that there exists available AC porosity for the adsorption of cyanide ion. ASTM (American Society for Testing and Materials) analysis results for d_{80} particle size, humidity, volatile, ashes, and fixed carbon listed in Table 2 show that the support material is resistant to thermal and mechanical environments, which are conditions that make the AC a good support and synergic effect material to oxidize cyanide, since it enables work with a clear fluid process. Meanwhile, the granular catalyst (3.10 mm) was immobilized avoiding following recovery operations [28].

Table 1. Textural properties determined by N_2 physisorption and BET modeling of AC and TiO_2-AC catalyst.

Material	S_{BET} (m^2/g)	Pore Volume (cm^3/g)	Φ (Å)
AC	1336	0.618	58.92
TiO_2-AC	902	0.504	33.97

Table 2. Physical and chemical properties of AC.

Parameter	Value
Particle size d_{80} (mm)	3.10
Humidity (%)	6.82
Volatile (%)	5.79
Ashes (%)	7.85
Fixed Carbon (%)	79.55

2.2. Photo-Reactor Construction

The investigation was performed in a continuous flow glass reactor with irradiation of UV lamps. TiO_2-AC catalysts were added into the reactor using nylon material nets to support the composite in several beds. We selected a reactor design with the maximum proximity between UV lamps and a stable and consistent flow into the reactor. The configuration of the used material is summarized in Scheme 1.

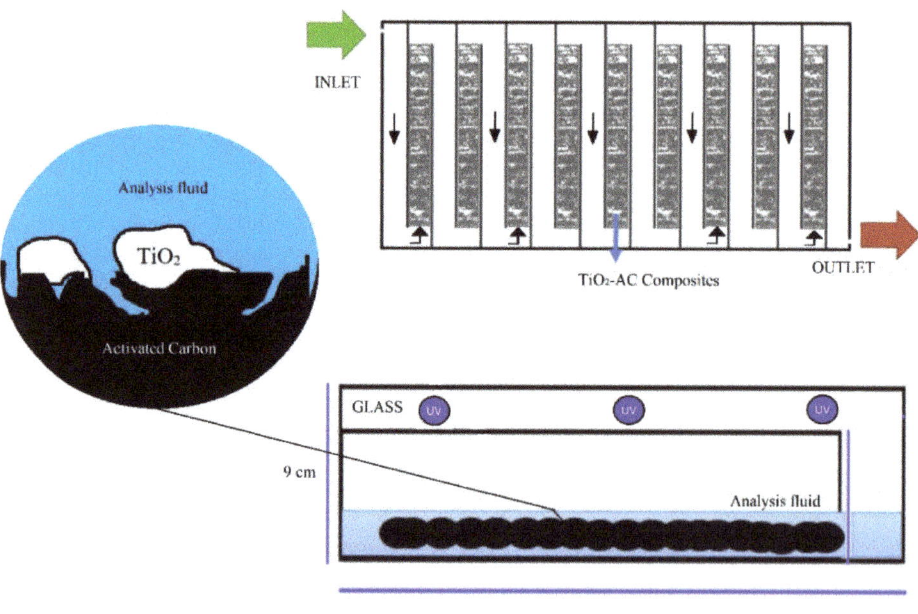

Scheme 1. Distribution of immobilized beds of AC or TiO_2-AC during the adsorption process and photocatalytic degradation of the cyanide ion in the continuous flow photo-reactor.

Then, the main configuration conditions were a continuous flow of 6.60 mL/s of cyanide solution generated by a 120 rpm peristaltic geopump, 8 mm height layer liquid, nine immobilized beds of AC or TiO_2-AC, and a total recirculation volume of 5 L.

Since the adsorption and photocatalytic degradation of cyanide can occur simultaneously, two stage-tests were performed in order to study adsorption and photocatalytic degradation individually. In addition, a simultaneous process was carried out.

2.3. Cyanide Adsorption Tests

Through tests of adsorption of CN^- under dark conditions (no UV irradiation) on the AC and TiO_2-AC composites, a required adsorption equilibrium time of 24 h was determined.

We found that the cyanide adsorption study had a better fit to the linearized mathematical model of the Langmuir isotherm for both AC and TiO_2-AC (results summarized in Table 3). The results indicated that the TiO_2-AC maximum cyanide adsorption capacity was lower (1/3) than AC. This finding agreed with the textural properties, since the specific surface area of AC decreased once TiO_2 was impregnated. Nonetheless, the adsorption and the energy associated to the process had similar behavior, as shown in Figure 1.

Table 3. Parameters calculated for Langmuir isotherm model.

Parameter	AC	TiO_2-AC
q_{max} (mg·g^{-1})	155.17	52.33
b (L·mg^{-1})	0.013	0.015
R^2	0.99	0.95

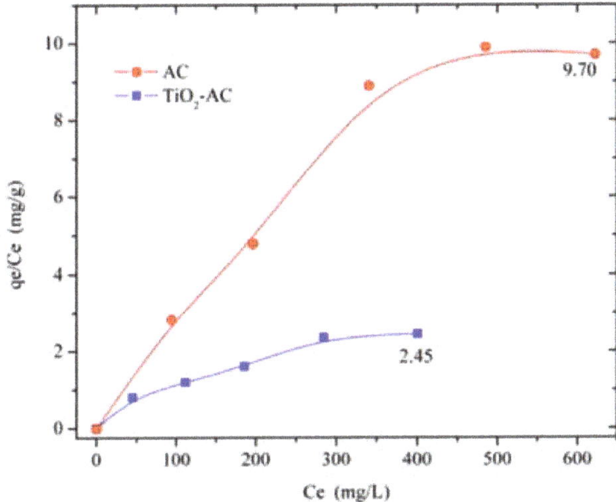

Figure 1. Langmuir adsorption isotherms of cyanide (T = 20 °C; 1 g TiO_2-AC/L).

Individually, the adsorption of cyanide ion tests were performed for three concentrations 30 g/L, 45 g/L, and 60 g/L of TiO_2-AC in 500 mg/L synthetic NaCN solutions, under dark conditions in the continuous flow reactor during 24 h at a pH of 10.5 and 20 °C. Another assay was carried out with 60 g/L of AC at the same conditions. The results listed in Table 4 demonstrate that AC (60 g/L) reached 78% of cyanide degradation due to the adsorption process. It is assumed that adsorption and oxidation processes simultaneously occurred due to oxidant species formed during the process with the dissolved oxygen and functional groups of the AC surface.

Table 4. Kinetic modeling of cyanide ion adsorption (pseudo-second order).

Parameter	AC 60 g/L	TiO$_2$-AC 30 g/L	TiO$_2$-AC 45 g/L	TiO$_2$-AC 60 g/L
K$_{app}$ (g·g^{-1}min^{-1})	5.42×10^{-6}	1.66×10^{-6}	3.98×10^{-6}	4.43×10^{-6}
R^2	0.98	0.83	0.98	0.98
CN$^-$ Degradation due to adsorption (%)	78.06	57.41	71.34	74.95

On the other hand, TiO$_2$-AC catalyst in 60 g/L concentration reached 74% of CN$^-$ degradation. The cyanide adsorption in the performed tests show similar trends (Figure 2a) and kinetics show that adsorption fits as a pseudo-second-order model through higher correlation coefficients calculation (Table 4). These results indicate that a chemisorption is possible, as suggested by Eskandari [29].

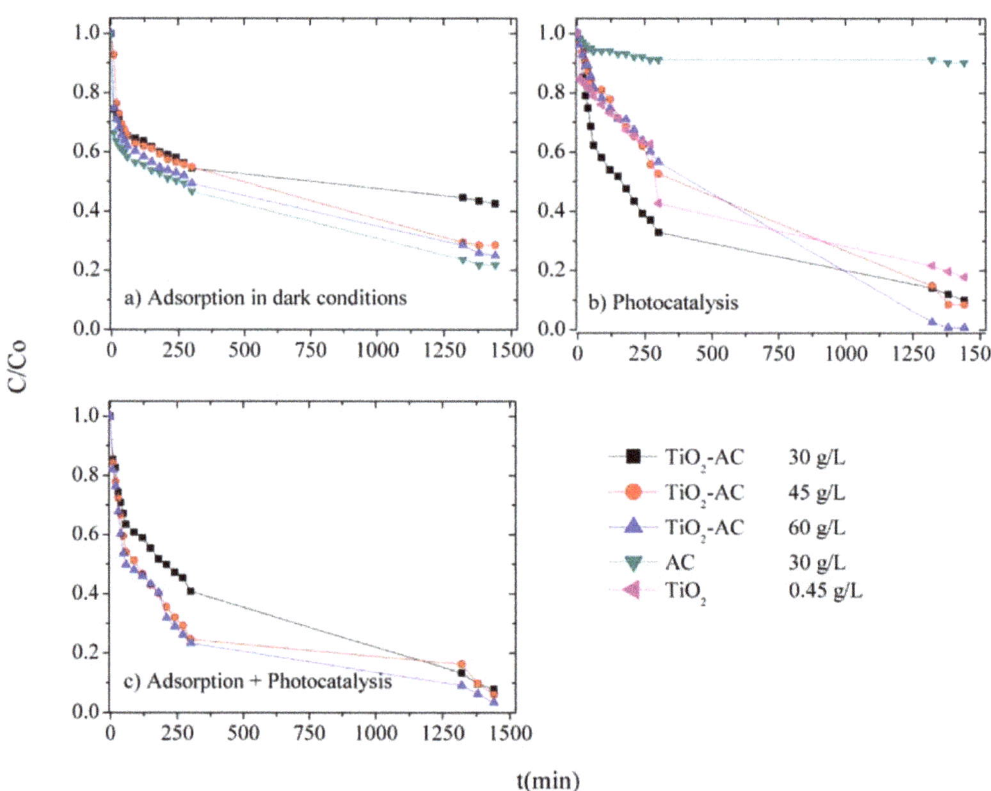

Figure 2. Cyanide ion degradation by (**a**) adsorption, (**b**) photocatalysis, and (**c**) simultaneous adsorption and photocatalysis.

2.4. Photocatalytic Cyanide Ion Degradation

Once the cyanide ion adsorption was performed during 24 h under dark conditions, the photocatalytic tests started by turning on the UV lamps. Tests were carried out at pH 10.5 with three concentrations 30, 45, and 60 g/L (tested in adsorption) during 24 h. More than 90% of cyanide ion was degraded in all assays using TiO$_2$-AC composites (Table 5). In order to compare the photocatalytic activity performance of materials used for composites on CN$^-$ degradation, tests with TiO$_2$ and AC were performed separately. Although we focused on the TiO$_2$-AC photocatalytic performance, AC yielded an unexpected 9.73% of cyanide ion degradation. Based on other investigations of AOPs with hydroxyl radicals, this result was explained by photocatalysis promoted by oxygen peroxide. This mechanism considers that H_2O_2 can be formed in an unstable and very fast chemical interaction of O_2, H^+, and e^- in an aqueous medium [29,30]. Moreover, the kinetics obtained for the system was pseudo-first order for all tests performed. TiO_2-AC 60 g/L composite assay aimed at 99.16% of cyanide ion degradation; meanwhile, 82.11% was obtained using TiO_2 0.45 g/L (equivalent mass of TiO_2 in the TiO_2-AC 60 g/L composite).

Table 5. Kinetic model results in cyanide ion photodegradation.

Parameter	AC 60 g/L	TiO$_2$ 0.45 g/L	TiO$_2$-AC 30 g/L	TiO$_2$-AC 45 g/L	TiO$_2$-AC 60 g/L
Individual photocatalytic degradation					
K_{app} (min^{-1})	4.02×10^{-5}	1.06×10^{-3}	1.34×10^{-3}	1.60×10^{-3}	4.90×10^{-2}
R^2	0.97	0.97	0.91	0.99	0.98
CN$^-$ Degradation (%)	9.73	82.11	90.08	91.39	99.16
Simultaneous adsorption and photocatalytic degradation					
K_{app} (min^{-1})	-	-	1.47×10^{-3}	1.39×10^{-3}	1.75×10^{-3}
R^2	-	-	0.83	0.98	0.98
CN$^-$ Degradation (%)	-	-	92.04	93.86	96.60

For comparison reasons, both processes adsorption and photocatalytic cyanide ion degradation were performed in simultaneous assays with the same TiO_2-AC concentrations established in this study. As shown in Figure 2, more than 90% of CN- was removed from the solution. In 24 h, 60 g/L of TiO_2-AC removed 97% of CN-under UV lights radiation in the continuous flow reactor. If this value is compared to the process that consisted of 24 h of adsorption (75%) and 24 h of photocatalysis (99%), a synergic effect between materials and phenomena appeared. The kinetics for simultaneous tests were pseudo-first order as well as an individual photocatalytic process, as shown in Table 5. Conversely, the kinetics of the adsorption process was not similar as simultaneous study kinetics.

3. Discussion

In order to test the adsorption and photocatalysis oxidation of cyanide ion in a continuous flow, a photo-reactor was fabricated based on previous investigations [19,30,31]. Scheme 1 summarizes the design of the photoreactor for CN^- degradation, which contributed to enhance the catalyst-fluid contact. Thus, we used a stable continuous flow of 6.60 mL/s of NaCN synthetic solution with an 8 mm liquid layer, inert material, and a maximum proximity between the UV source and material-fluid. These conditions indicated the advantage of AC and TiO_2 performance as a composite, because light can penetrate better in the composite. Since the photoreactor design considered a hollow box with a flow generated by a peristaltic pump through nine separated composite beds, O_2 from air can be introduced during all the processes, contributing to enhance the adsorption and photocatalytic oxidation of CN^-. Moreover, UV lamps radiation interacted directly with the fluid and the composite. On the other hand, granular AC (derived of coconut shell by steam activation) showed a high specific surface of 1336 m^2/g, which would favor the adsorption of contaminants such as CN^-. Indeed, the resistant properties of AC to thermal and mechanical processes and a 3.1 mm grain diameter allow building immobilized beds of the composite inside the photoreactor, avoiding post-recovery difficulties. The diameter of AC was smaller than TiO_2 grain size (170 nm), and SEM-EDX results showed that TiO_2 occupied the external surface of AC, allowing a major contact between UV light source and TiO_2 on the support [32]. The specific surface area of the TiO_2-AC composite decreased to 902 m^2/g due to the TiO_2 impregnation However, this result represents a high specific surface area available for the adsorption of cyanide ions. Based on the literature, the semiconductor would occupy the meso and macro porosity of AC; meanwhile, the microporosity is not affected during the wet impregnation of titania [33,34].

Adsorption and photocatalytic oxidation occurred simultaneously during the treatment of cyanide solutions. Thus, we performed two-stage tests to evaluate each process separately. Both processes were studied at 20 °C and pH = 10.5. During adsorption batch tests, 24 h were obtained as an equilibrium time (when adsorption does not continue). In addition, the Langmuir isotherm model was the better fit to the adsorption for AC and TiO_2-AC, where q_{max} was reduced from 155 to 52 mg CN^-/g TiO_2-AC once the impregnation technique was performed. The CN^- adsorption in AC was drastically reduced once titania was supported, because the TiO_2 added to the external surface of AC occupies large holes in the support. In the continuous flow reactor tests, we could estimate that the CN^- adsorption kinetic follows a pseudo-second-order reaction. This result showed that a chemisorption would take place as well as a physisorption of CN^- in AC and TiO_2-AC.

Results over 70% of CN^- oxidation in AC and composites revealed that in the process, in light absence, not only adsorption takes place but also an oxidation of cyanide ion could be carried out due to the –OH bonds of the AC surface and existing O_2 in aqueous pumping media [29,35].

As expected, granular TiO_2-AC composites showed a higher degradation of cyanide ion according to the amount introduced to the continuous flow system. Kinetics and CN^- degradation indicated the following order: TiO_2-AC 30 g/L < TiO_2-AC 45 g/L < TiO_2-AC 60 g/L < AC 60 g/L. Since pH 10.5 was performed in all tests, a negative charge of AC surface was expected, giving a repel effect with cyanide ion. Nonetheless, 75% and 78% of CN^- degradation were determined in light absence using TiO_2-AC 60 g/L and AC 60 g/L, respectively.

In photocatalytic tests, as an individual process study, more than 90% of CN^- removal was determined using TiO_2-AC composites. Thus, external TiO_2 on the AC surface contributed to degrade pollutants in a continuous flow system with composites immobilized with UV 15 W lamps irradiation. For comparison reasons, TiO_2 and AC were tested separately in weight amounts corresponding to 60 g/L TiO_2-AC. The results listed in Table 5 showed that titania aimed at 82% of CN^- removal; meanwhile, a surprising result of 10% CN^- degradation was obtained with AC. Although AC was not considered as a photo-catalyst, it showed CN^- removal due to the operational conditions and continuous flow system, and H_2O_2 could be formed rapidly during UV irradiation once the pair e^-/h^+ is formed [32,36]. Some investigations indicate that hydrogen peroxide can contribute to pollutant degradation in AOPs, since this compound can be formed in an electrochemical process associated to UV incidence and dissolved oxygen presence during the pumping process. These results ensure that TiO_2-AC composites enhanced the photocatalysis, increasing the synergic effect between AC and TiO_2. Kinetics and cyanide degradation were determined in order 60 g/L AC < 0.45 g/L TiO_2 < TiO_2-AC 30 g/L < TiO_2-AC 45 g/L < TiO_2-AC 60 g/L.

Simultaneous adsorption and photocatalysis tests were performed on UV radiation. More than 90% of CN^- was removed in all tests. However, in a comparison analysis of each concentration composite dosage, a numerical variation was detected. For 60 g/L of TiO_2-AC, 96% of CN^- removal was obtained during 24 h, where adsorption and photocatalysis took place at the same time. Whereas 74% and 99% of CN^- degradation were obtained in 24 h adsorption and 24 h of photocatalysis, respectively. Thus, a simultaneous process is suitable for cyanide ion degradation under operational continuous flow photorector design. Both photocatalysis and the simultaneous process (photocatalysis + adsorption) correspond to a pseudo-first-order reaction. This result is in concordance with the literature, which attributed the oxidation of pollutants by hydroxyl radical activity. The success of CN^- degradation was enhanced by TiO_2-AC presence and the design parameters of the photoreactor, because a stable system could be formed under a near interaction between the UV source and materials of this study.

4. Materials and Methods

4.1. TiO$_2$-AC Composites Preparation

TiO$_2$-AC composites were prepared by the wet impregnation of TiO$_2$ on activated carbon (AC). First, the AC was washed with distilled water under magnetic stirring, until a clarified washing solution was obtained. Then, 1 g of commercial TiO$_2$ (United States Pharmacopeia (USP) reference standard, 99% anatase) and 100 g of commercial activated carbon (CALGON GRC-20®) were added to 350 mL of distilled water under vigorous stirring for 2 h. The solvent was removed by evaporation, and composites were washed with distilled water several times. Finally, TiO$_2$-AC composites were dried at 110 °C for 2 h [37,38].

4.2. Physical and Chemical Characterization of GCR-20 Activated Carbon and TiO$_2$-AC Composite

Standardized sieves were used to determine the particle size distribution of AC by the Standard Test Method for Particle Size Distribution of Granular Activated Carbon ASTM-D2862. Moisture, volatile material, ash, and fixed carbon content were measured for the AC support according to ASTM-D3173 (Standard Test Method for Moisture in the Analysis Sample of Coal and Coke), ASTM-D3175 (Standard Test Method for Volatile Matter in the Analysis Sample of Coal and Coke), and ASTM-D3174 (Standard Test Method for Ash in the Analysis Sample of Coal and Coke from Coal).

The textural properties of the GCR-20 AC and TiO$_2$-AC composites were determined by N$_2$ physisorption isotherms in a Quantachrome Instruments Nova 4200e (Quantachrome Instruments, Boynton Beach, FL, USA). The BET model was used to determine the specific surface area.

A scanning electron microscopy analysis (SEM) was performed to analyze the TiO$_2$-AC composite texture with a Vega TESCAN (TESCAN, Brno, Czech Republic) microscope equipped with secondary electron (SE) and backscattered electron (BSE) detectors. The chemical mapping was determined by an energy-dispersive analysis of the X-ray (EDX) using Vega TESCAN scanning electron microscopy attached with a Bruker XFlash 5010 Detector (Bruker, Billerica, MA, USA), with an accelerating voltage of 20 kV under vacuum.

In order to determine the impregnation of TiO$_2$ on the activated carbon, all composites underwent microwave acid digestion with HNO$_3$, HF, and HCl. Titanium content in the acid solution was analyzed by inductively coupled plasma optical emission spectrometry (ICP-OES) (Perkin Elmer Optima 8000, Perkin Elmer Inc., Waltham, MA, USA).

4.3. Photo-Reactor Construction

A continuous photo-reactor was made of glass and covered with a wood box to isolate the fluid of the external conditions. Three 15 W UV lamps of 43.74 cm were attached to the lid of the reactor in parallel distribution in order to penetrate as far as possible the fluid and obtain the greatest proximity between the UV lamps, the photo-catalyst, and the fluid. The internal part of the photo-reactor made of glass is a rectangle of 135 cm × 50 cm with walls of 9 cm. Inside of this structure glass, plates of 46 cm × 4 cm were placed perpendicularly, as shown in Scheme 1. A 5-degree angle was adapted between the reactor and the horizontal level in order to maintain stable circulation of the fluid through the path formed inside the reactor.

Constant flow rate in the reactor, thickness of the liquid, residence time of the cyanide solution, and concentration of the catalyst/composite were measured to determine the optimal operating conditions.

4.4. Cyanide Adsorption Study

Sodium cyanide solutions with concentrations of 200, 400, 600, 800, and 1000 mg/L were prepared. The adsorption tests were carried out in a batch with 50 mL of sodium cyanide solution and 0.05 g of each AC and TiO$_2$-AC composite under continuous stirring for 24 h. The data of cyanide ion adsorption were studied using the mathematical model of the Langmuir isotherm. After adsorption-desorption equilibrium was achieved, the solid was filtered, and the cyanide ion concentration (in solution) was determined by titration with a 0.2256 N AgNO$_3$ solution [28].

The data were studied and modeled by the Langmuir isotherm according to the mathematical relationship described in Equation (7).

$$\frac{C_e}{q_e} = \frac{1}{q_{max}} \times C_e + \frac{1}{q_{max} \times b} \tag{7}$$

where C_e is the cyanide ion equilibrium concentration in the solution (mg/L), q_e is the equilibrium concentration of cyanide ion over the adsorbents (AC or TiO$_2$-AC) (mg/g), q_{max} is the maximum mass of cyanide ion adsorbed per 1 g of adsorbent (AC or TiO$_2$-AC) (mg/g), and b is the independent variable referring to the free energy of adsorption (L/mg).

4.5. Adsorption Study

4.5.1. Adsorption with AC

Adsorption tests were carried out in a photo-reactor of continuous flow under dark conditions at room temperature. A sodium cyanide solution of 500 mg/L was recirculated through a peristaltic pump (Geopump Inc., Medina, NY, USA) for 24 h. The pH of solution was adjusted at 10.5 throughout the adsorption experiments with the addition of NaOH. Concentrations of 30, 45, and 60 g/L of activated carbon were used for each test. Samples of 5 mL were taken each 10 min during the first hour and then each 30 min for three hours. The solution was recirculated overnight, and the next day, samples of 5 mL were taken each hour for 2 h. The cyanide ion concentration was determined by titration with a 0.2256 N AgNO$_3$ solution.

4.5.2. Adsorption with TiO$_2$-AC Composite and Photodegradation Process

Cyanide ion adsorption on AC and the photodegradation of cyanide ion occur simultaneously. Experiments of cyanide ion adsorption on the TiO$_2$-AC composite were carried out under the same conditions as in the adsorption with AC. At the end of the adsorption experiments, the photodegradation of cyanide ion was evaluated for 24 h under UV light. Samples of 5 mL were taken (in the time intervals explained in Section 4.5.1) to determine the cyanide ion concentration by titration with a 0.2256 N AgNO$_3$ solution.

4.6. Photocatalytic Cyanide Ion Degradation

Photocatalytic activity was performed in a continuous photo-reactor using an initial concentration of sodium cyanide of 500 mg/L and a composite concentration of 30, 45, and 60 g/L at pH 10.5 and room temperature. The system (Scheme 1) was maintained in dark conditions during 24 h to ensure adsorption-desorption equilibrium. Then, UV lamps (Philips, Amsterdam, The Netherlands, T8 G13 UV-C 15 W) were turned on to initiate the photodegradation. Aliquots of 5 mL were taken (in the time intervals explained in Section 4.5.1) to analyze the residual cyanide ion by titration with a 0.2256 N AgNO$_3$ solution.

Cyanide ion photodegradation with AC (saturated with cyanide) and TiO$_2$ of 60 g/L and 0.45 g/L (corresponding to the impregnation percentage) respectively were carried out in order to compare with the photodegradation results with the composite.

After the photodegradation test, samples of the remaining solution were taken to determine the concentration of titanium and sodium dissolved by atomic absorption spectroscopy.

5. Conclusions

Within this study, we built and performed a continuous flow photo-reactor during cyanide ion degradation using TiO$_2$-AC composites. Design operational conditions were established into the photoreactor where a minimum distance between UV source and fluid/composites and a stable continuous flow though immobilized composites were achieved. Therefore, inside of the reactor, adsorption and photocatalysis of CN$^-$ synthetic solutions were performed using TiO$_2$-AC in concentrations of 30, 45, and 60 g/L. When both processes, adsorption and photocatalysis, were studied separately, the adsorption of CN^- on AC in light absence decreased when TiO_2-AC was performed, achieving 75% of CN^- removal. Conversely, TiO_2-AC composites markedly enhance the photocatalytic process in comparison to individual TiO_2 and AC performances, achieving 99% of cyanide ion degradation. During 24 h of simultaneous process, 96% of CN^- was aimed. In the success operational conditions of the photoreactor, we determined that AC could contribute to photocatalysis cyanide degradation, since AC showed 10% of CN^- removal with UV irradiation. By other hand, in light absence, granular AC tested showed a considerable amount of cyanide degradation since –OH bonds on the AC surface and O_2 added to the system could contribute to cyanide removal. Thus, this reactor design and TiO_2-AC would represent an encouraging alternative of cyanide degradation in a continuous flow system due to their synergic effect in wastewaters remediation.

Supplementary Materials: The following are available online at https://www.mdpi.com/article/10.3390/catal11080924/s1, Figure S1. (a) SEM micrograph of AC, 783x; (b) SEM micrograph of TiO_2-AC composite, 927x; (c) EDX mapping performed on the micrograph of the TiO_2-AC composite, 927x, Figure S2. EDX mapping performed on a micrograph of the TiO_2-AC composite, 783x.

Author Contributions: Conceptualization, S.C. and D.E.; methodology, S.C., A.B.L. and D.E.; formal analysis, S.C., A.B.L.; investigation, S.C., A.B.L., E.d.l.T. and D.E.; resources, E.d.l.T. and D.E.; data curation, S.C., A.B.L.; writing—original draft preparation, S.C., A.B.L., L.E.M.-P.; writing—review and editing, S.C., A.B.L., L.E.M.-P. and D.E.; visualization, S.C., A.B.L., L.E.M.-P., E.d.l.T. and D.E.; supervision, L.E.M.-P., E.d.l.T. and D.E.; project administration, E.d.l.T. and D.E. All authors have read and agreed to the published version of the manuscript.

Funding: This research received no external funding.

Acknowledgments: The authors would like to thank the financial support provided by Escuela Politécnica Nacional through the project PII-DEMEX-20-01 belonging to the Master of Research in Metallurgy.

Conflicts of Interest: The authors declare no conflict of interest.

References

1. Lin, M.; Gu, Q.; Cui, X.; Liu, X. Cyanide Containing Wastewater Treatment by Ozone Enhanced Catalytic Oxidation over Diatomite Catalysts. *MATEC Web Conf.* **2018**, *142*, 4–10. [CrossRef]
2. Safavi, B.; Asadollahfardi, G.; Darban, A.K. Cyanide removal simulation from wastewater in the presence of Titanium dioxide nanoparticles. *Adv. Nano Res.* **2017**, *5*, 27–34. [CrossRef]
3. Mishra, J.; Pattanayak, D.S.; Das, A.A.; Mishra, D.K.; Rath, D.; Sahoo, N.K. Enhanced photocatalytic degradation of cyanide employing Fe-porphyrin sensitizer with hydroxyapatite palladium doped TiO_2 nano-composite system. *J. Mol. Liq.* **2019**, *287*, 110821. [CrossRef]
4. Koohestani, H.; Sadrnezhaad, S.K. Photocatalytic degradation of methyl orange and cyanide by using TiO_2/CuO composite, Desalin. *Water Treat.* **2016**, *57*, 22029–22038. [CrossRef]
5. Tarras-Wahlberg, N.H. Environmental management of small-scale and artisanal mining: The Portovelo-Zaruma goldmining area, southern Ecuador. *J. Environ. Manag.* **2002**, *65*, 165–179. [CrossRef] [PubMed]
6. Vangsnes, G.F. The meanings of mining: A perspective on the regulation of artisanal and small-scale gold mining in southern Ecuador. *Extr. Ind. Soc.* **2018**, *5*, 317–326. [CrossRef]
7. MAE (Ministerio del Ambiente). Texto Unificado Legislación Secundaria, Medio Ambiente, Libro VI. 2015, p. 187. Available online: http://www.ambiente.gob.ec/wp-content/uploads/downloads/2015/02/TEXTO-UNIFICADO-LEGISLACION-SECUNDARIA-MEDIO-AMBIENTE.pdf (accessed on 15 September 2019).
8. Pan, Y.; Zhang, Y.; Huang, Y.; Jia, Y.; Chen, L.; Cui, H. Synergistic effect of adsorptive photocatalytic oxidation and degradation mechanism of cyanides and Cu/Zn complexes over TiO_2/ZSM-5 in real wastewater. *J. Hazard. Mater.* **2021**, *416*, 125802. [CrossRef] [PubMed]
9. Guo, Y.; Wang, Y.; Zhao, S.; Liu, Z.; Chang, H.; Zhao, X. Photocatalytic oxidation of free cyanide over graphitic carbon nitride nanosheets under visible light. *Chem. Eng. J.* **2019**, *369*, 553–562. [CrossRef]
10. Pueyo, N.; Miguel, N.; Mosteo, R.; Ovelleiro, J.L.; Ormad, M.P. Synergistic effect of the presence of suspended and dissolved matter on the removal of cyanide from coking wastewater by TiO_2 photocatalysis. *J. Environ. Sci. Health Part A* **2017**, *52*, 182–188. [CrossRef]

11. Núñez-Salas, R.E.; Hernández-Ramírez, A.; Hinojosa-Reyes, L.; Guzmán-Mar, J.L.; Villanueva-Rodríguez, M.; Maya-Treviño, M.d.L. Cyanide degradation in aqueous solution by heterogeneous photocatalysis using boron-doped zinc oxide. *Catal. Today* **2019**, *328*, 202–209. [CrossRef]

12. Betancourt-Buitrago, L.A.; Hernandez-Ramirez, A.; Colina-Marquez, J.A.; Bustillo-Lecompte, C.F.; Rehmann, L.; Machuca-Martinez, F. Recent developments in the photocatalytic treatment of cyanide wastewater: An approach to remediation and recovery of metals. *Processes* **2019**, *7*, 225. [CrossRef]

13. Hernández-Alonso, M.D.; Coronado, J.M.; Maira, A.J.; Soria, J.; Loddo, V.; Augugliaro, V. Ozone enhanced activity of aqueous titanium dioxide suspensions for photocatalytic oxidation of free cyanide ions. *Appl. Catal. B Environ.* **2002**, *39*, 257–267. [CrossRef]

14. Ijadpanah-saravi, H.; Dehestaniathar, S.; Khodadadi-Darban, A.; Zolfaghari, M.; Saeedzadeh, S. Photocatalytic decomposition of cyanide in pure water by biphasic titanium dioxide nanoparticles. *Desalin. Water Treat.* **2015**, *57*, 20503–20510. [CrossRef]

15. Baeissa, E.S. Synthesis and characterization of sulfur-titanium dioxide nanocomposites for photocatalytic oxidation of cyanide using visible light irradiation. *Chin. J. Catal.* **2015**, *36*, 698–704. [CrossRef]

16. Chiang, K.; Amal, R.; Tran, T. Photocatalytic degradation of cyanide using titanium dioxide modified with copper oxide. *Adv. Environ. Res.* **2002**, *6*, 471–485. [CrossRef]

17. Omri, A.; Lambert, S.D.; Geens, J.; Bennour, F.; Benzina, M. Synthesis, Surface Characterization and Photocatalytic Activity of TiO_2 Supported on Almond Shell Activated Carbon. *J. Mater. Sci. Technol.* **2014**, *30*, 894–902. [CrossRef]

18. Harraz, F.A.; Abdel-Salam, O.E.; Mostafa, A.A.; Mohamed, R.M.; Hanafy, M. Rapid synthesis of titania-silica nanoparticles photocatalyst by a modified sol-gel method for cyanide degradation and heavy metals removal. *J. Alloys Compd.* **2013**, *551*, 1–7. [CrossRef]

19. Koohestani, H.; Hassanabadi, M.; Mansouri, H.; Pirmoradian, A. Investigation of photocatalytic performance of natural zeolite/TiO_2 composites. *Micro Nano Lett.* **2019**, *14*, 669–673. [CrossRef]

20. De la Torre Chauvin, E.H. Préparation de Charbon Actif à Partir de Coques de Noix de Palmier à Huile Pour la Récupération D'or et le Traitement D'effluents Cyanurés. Ph.D. Thesis, Université catholique de Louvain, Louvain-La-Neuve, Belgium, 2015.

21. De La Torre, E.; Adatty, M.; Gámez, S. Activated Carbon-Spinels Composites for Waste Water Treatment. *Metals* **2018**, *8*, 1070. [CrossRef]

22. Pilco, Y. Estudio de la Oxidación de Efluentes Cianurados en Presencia de Aire y Carbón Activado. Bachelor's Thesis, Escuela Politécnica Nacional, Quito, Ecuador, 2008.

23. Compagnoni, M.; Ramis, G.; Freyria, F.S.; Armandi, M.; Bonelli, B.; Rossetti, I. Innovative photoreactors for unconventional photocatalytic processes: The photoreduction of CO2 and the photo-oxidation of ammonia. *Rend. Lincei* **2017**, *28*, 151–158. [CrossRef]

24. Velasco, L.F.; Carmona, R.J.; Matos, J.; Ania, C.O. Performance of activated carbons in consecutive phenol photooxidation cycles. *Carbon* **2014**, *73*, 206–215. [CrossRef]

25. Marugán, J.; van Grieken, R.; Cassano, A.E.; Alfano, O.M. Intrinsic kinetic modeling with explicit radiation absorption effects of the photocatalytic oxidation of cyanide with TiO_2 and silica-supported TiO_2 suspensions. *Appl. Catal. B Environ.* **2008**, *85*, 48–60. [CrossRef]

26. Royaee, S.J.; Sohrabi, M.; Barjesteh, P.J. Performance evaluation of a continuous flow Photo-Impinging Streams Cyclone Reactor for phenol degradation. *Chem. Eng. Res. Des.* **2012**, *90*, 1923–1929. [CrossRef]

27. Motegh, M.; van Ommen, J.R.; Appel, P.W.; Kreutzer, M.T. Scale-Up Study of a Multiphase Photocatalytic Reactor—Degradation of Cyanide in Water over TiO_2. *Environ. Sci. Technol.* **2014**, *48*, 1574–1581. [CrossRef] [PubMed]

28. Matos, J.; Laine, J.; Herrmann, J.-M.; Uzcategui, D.; Brito, J.L. Influence of activated carbon upon titania on aqueous photocatalytic consecutive runs of phenol photodegradation. *Appl. Catal. B Environ.* **2007**, *70*, 461–469. [CrossRef]

29. Eskandari, P.; Farhadian, M.; Nazar, A.R.S.; Goshadrou, A. Cyanide adsorption on activated carbon impregnated with ZnO, Fe2O3, TiO_2 nanometal oxides: A comparative study. *Int. J. Environ. Sci. Technol.* **2020**, *18*, 297–316. [CrossRef]

30. Yang, W.; Zhou, M.; Cai, J.; Liang, L.; Ren, G.; Jiang, L. Ultrahigh yield of hydrogen peroxide on graphite felt cathode modified with electrochemically exfoliated graphene. *J. Mater. Chem. A* **2017**, *5*, 8070–8080. [CrossRef]

31. Moreira, J.; Lima, V.B.; Goulart, L.A.; Lanza, M.R.V. Electrosynthesis of hydrogen peroxide using modified gas diffusion electrodes (MGDE) for environmental applications: Quinones and azo compounds employed as redox modifiers. *Appl. Catal. B Environ.* **2019**, *248*, 95–107. [CrossRef]

32. Mashuri, S.I.S.; Ibrahim, M.L.; Kasim, M.F.; Mastuli, M.S.; Rashid, U.; Abdullah, A.H.; Islam, A.; Asikin-Mijan, N.; Tan, Y.H.; Mansir, N.; et al. Photocatalysis for organic wastewater treatment: From the basis to current challenges for society. *Catalysts* **2020**, *10*, 1260. [CrossRef]

33. Malato, S.; Blanco, J.; Vidal, A.; Fernández, P.; Cáceres, J.; Trincado, P.; Oliveira, J.C.; Vincent, M. New large solar photocatalytic plant: Set-up and preliminary results. *Chemosphere* **2002**, *47*, 235–240. [CrossRef]

34. Tanveer, M.; Guyer, G.T. Solar assisted photo degradation of wastewater by compound parabolic collectors: Review of design and operational parameters. *Renew. Sustain. Energy Rev.* **2013**, *24*, 534–543. [CrossRef]

35. Coronel, S.; Pauker, C.S.; Jentzsch, P.E.V.; de la Torre, E.; Endara, D.; Muñoz-Bisesti, F. Titanium dioxide/copper/carbon composites for the photocatalytic degradation of phenol. *Chem. Chem. Technol.* **2020**, *14*, 161–168. [CrossRef]

36. Kim, H.W.; Ross, M.B.; Kornienko, N.; Zhang, L.; Guo, J.; Yang, P.; McCloskey, B.D. Efficient hydrogen peroxide generation using reduced graphene oxide-based oxygen reduction electrocatalysts. *Nat. Catal.* **2018**, *1*, 282–290. [CrossRef]
37. Richald, M. Étude et Optimisation de Composites à Base de Charbon Actif et D'oxyde de Titane Pour L'oxydation Photocatalytique des ion Cyanures. Master's Thesis, Haute École Lucia De Brouckère, Bruselas, Bélgica, 2015.
38. Murillo, H. Obtención de un Compósito de Dióxido de Titanio y Carbón Activado Aplicado a la Oxidación Fotocatalítica del Ión Cianuro. Bachelor's Thesis, Escuela Politécnica Nacional, Quito, Ecuador, 2014.

Article

Enhanced SO$_2$ Absorption Capacity of Sodium Citrate Using Sodium Humate

Zhiguo Sun *, Yue Zhou, Shichao Jia, Yaru Wang, Dazhan Jiang and Li Zhang *

School of Environmental and Material Engineering, Shanghai Polytechnic University, Shanghai 201209, China; zhouyueyue626@163.com (Y.Z.); jiashichao2021@163.com (S.J.); wangyaru8468@163.com (Y.W.); jiangdazhan2021@163.com (D.J.)
* Correspondence: zgsun@sspu.edu.cn (Z.S.); zhangli@sspu.edu.cn (L.Z.); Tel.: +86-21-50211217 (Z.S.)

Abstract: A novel method of improving the SO$_2$ absorption performance of sodium citrate (Ci-Na) using sodium humate (HA–Na) as an additive was put forward. The influence of different Ci-Na concentration, inlet SO$_2$ concentration and gas flow rate on desulfurization performance were studied. The synergistic mechanism of SO$_2$ absorption by HA–Na and Ci-Na was also analyzed. The consequence shows that the efficiency of SO$_2$ absorption by Ci-Na is above 90% and the desulfurization time added with the Ci-Na concentration rising from 0.01 to 0.1 mol/L. Both the desulfurization efficiency and time may increase with the adding of HA–Na quality in Ci-Na solution. Due to adding HA–Na, the desulfurization efficiency of Ci-Na increased from 90% to 99% and the desulfurization time increased from 40 to 55 min. Under the optimum conditions, the desulfurization time of Ci-Na can exceed 70 min because of adding HA–Na, which is nearly doubled. The growth of inlet SO$_2$ concentration has little effect on the desulfurization efficiency. The SO$_2$ adsorption efficiency decreases with the increase of inlet flow gas. The presence of O$_2$ improves the SO$_2$ removal efficiency and prolongs the desulfurization time. Therefore, HA–Na plays a key role during SO$_2$ absorption and can dramatically enhance the SO$_2$ adsorption performance of Ci-Na solution.

Keywords: sodium citrate; sodium humate; SO$_2$; absorption

Citation: Sun, Z.; Zhou, Y.; Jia, S.; Wang, Y.; Jiang, D.; Zhang, L. Enhanced SO$_2$ Absorption Capacity of Sodium Citrate Using Sodium Humate. *Catalysts* **2021**, *11*, 865. https://doi.org/10.3390/catal11070865

Academic Editors: José Ignacio Lombraña, Héctor Valdés and Cristian Ferreiro

Received: 9 June 2021
Accepted: 18 July 2021
Published: 20 July 2021

Publisher's Note: MDPI stays neutral with regard to jurisdictional claims in published maps and institutional affiliations.

1. Introduction

It is well known that fossil fuels is mainly used to generate electrical energy in power plants and the combustion of fossil fuels generates SO$_2$, which is the major source of acid rain and a major air pollutant, which severely influences the atmosphere environment and human health if not controlled [1,2]. Controlling SO$_2$ is critical to improve air quality and has always caught people's eye in recent years due to the environmental issues [3–5]. Therefore, improving the desulfurization performance economically and effectively of existing desulfurization technology has become a research hotspot at home and abroad [6–8].

There are plenty of desulfurization processes developed on the laboratory scale, some of which are applied at industrial standards around the world [9]. In the traditional methods, limestone, sodium hydroxide solutions, calcium hydroxide and magnesium hydroxide and a number of organic solvents have been used as an adsorbent [10]. There are other desulfurization processes such as the citrate method. During the citrate process, SO$_2$ in the flue gas is absorbed by the sodium citrate (Ci-Na) solution [11,12]. According to the physical characteristics of Ci-Na [13], adopting the desulfurization technology of Ci-Na can meet the advantages of environmental protection, flexible operation, recyclable absorbents and recyclable SO$_2$ for resource utilization, also meeting the increasingly strict requirements of environmental protection, desulfurization, etc. [14–16].

Humic acid (HA) is a type of amorphous organic molecular compound, most of these extensively exist in nature. It can be obtained from lignite and peat [17,18]. Due to its "sponge-like" structure, HA produces a large surface area (330–340 m^2/g) and surface energy and has a strong adsorption capacity [19]. The adsorption capacity of HA is not only related to its surface area and surface energy, but also the swelling property of HA to water [20]. Sodium humate (HA–Na) is a water-soluble sodium salt of HA and a cost-effective absorbent, which reacts with H^+ to produce HA precipitate thus promoting the dissolution of SO_2 in the water [21]. They have been studied more broadly for biological breeding and pollution control due to their characteristics of adsorption, chelation and ion exchange [22]. HA–Na has higher swelling property than HA itself [23]. With the enhancement of the swelling property of HA, the active groups of HA can be more fully exposed in the aqueous solution and the probability of contact between HA and adsorbed ions is increased and then it improves the adsorption effect [24–26]. However, there have been few reports with regard to the addition of HA–Na to modify the Ci-Na solution to improve the adsorption capacity [27].

Sun et al. [28] investigated the desulfurization activity of the HA–Na/α-Al$_2$O$_3$ composite adsorbent on the fixed-bed quartz reactor. A series of characterization showed that coating α-Al$_2$O$_3$ fibers after being immersed in HA–Na solution can enhance the flue gas desulfurization performance of the α-Al$_2$O$_3$ carrier. The reason is that the HA–Na adsorbent has a stronger adsorption capacity for NH_4OH. The longer the conversion rate of SO_2 is maintained, the more NH_4OH will be adsorbed in the HA–Na/α-Al$_2$O$_3$ adsorbent. According to the previous study, it revealed that the HA–Na solution has good SO_2 absorption characteristics. The desulfurization products can be made into the HA compound fertilizer, which provides an economical and effective way to reduce SO_2 from flue gas [29].

This paper studies the absorption performance of HA–Na/Ci-Na and the desulfurization mechanism, which will lay the foundation for further research and popularization in the future.

2. The Enhancement Mechanism

HA–Na may bring an enhancement effect on SO_2 capture by the Ci-Na method. Table 1 shows the relevant reactions and the enhancement mechanism was put forward as follows: (1) The SO_2 absorption by Ci-Na mainly depend on the buffering properties of its absorbing solution. (2) After adding HA–Na, the carboxyl (COO–) and hydroxyl (OH–) of HA–Na reacts with H^+ rapidly and HA–Na is transferred to the HA sediment (Equation (8)). Due to these reactions, the reaction equilibrium of Equations (1)–(7) moves to the right and the amount of SO_2 dissolved into solution is increased. (3) HA–Na may reduce the rate of pH decline of Ci-Na solution since the HA–Na solution is also a kind of acidic buffer solution, which also may enhance SO_2 absorption.

Table 1. The reaction equation of CO_2 capture.

Reaction Equation	Number
$SO_2(g) \leftrightarrow SO_2(aq)$	(1)
$SO_2(aq) + H_2O \leftrightarrow 2H^+(aq) + SO_3^{2-}(aq)$	(2)
$C_i^{3-}(aq) + H^+(aq) \leftrightarrow HC_i^{2-}(aq)$	(3)
$HC_i^{2-}(aq) + H^+(aq) \leftrightarrow H_2C_i^-(aq)$	(4)
$H_2C_i^-(aq) + H^+(aq) \leftrightarrow H_3C_i(aq)$	(5)
$H_2O \leftrightarrow H^+(aq) + OH^-(aq)$	(6)
$R - (COONa)_n(aq) \leftrightarrow R - (COO^-)_n(aq) + nNa^+(aq)$	(7)
$R - (COO^-)_n + nH^+(aq) \leftrightarrow R - (COOH)_n(s)$	(8)

R–(COONa)$_n$ is the structural formula of HA–Na and R–(COOH)$_n$ is the structural formula of HA.

3. Results and Discussion

3.1. Desulfurization Performance of Only Ci-Na

The influence of different concentrations of Ci-Na on the removal rate of SO_2 was analyzed as can be seen in Figure 1. It shows the relationship of Ci-Na concentration and SO_2 desulfurization efficiency [30]. The SO_2 absorption by different concentrations of

Ci-Na all shows higher efficiency and the SO_2 absorption efficiency had no obvious change and basically maintained above 90% with the increase of Ci-Na concentration. The duration of high efficiency desulfurization also added with the increasing of Ci-Na concentration. When the concentration of Ci-Na added from 0.01 to 0.1 mol/L, the desulfurization time was extended from 20 to 80 min, which was increased by 4 times.

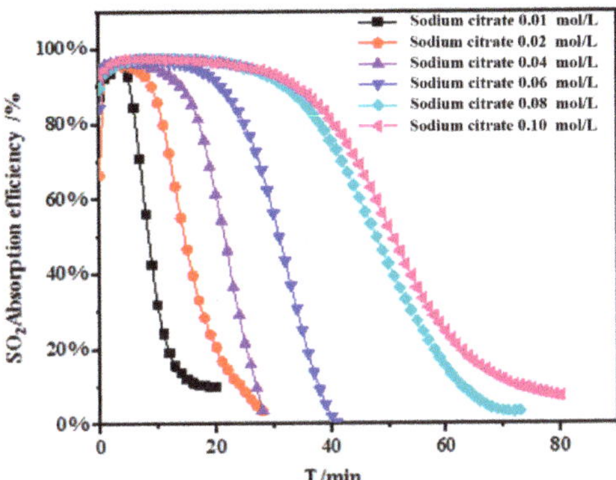

Figure 1. Effect of Ci-Na concentration on desulfurization efficiency. SO_2 = 2300 ppm, gas flow = 1.68 L/min and absorption solution = 60 mL and 25 °C.

Figure 2 shows the relationship between desulfurization time and Ci-Na concentration. The desulfurization time experienced two rapid growth phases with the growth of Ci-Na concentration and it tended to be flat after 0.08 mol/L. The Ci-Na solution is weakly alkaline and the citrate ion has good buffering capacity. In this experiment, the concentration of 0.06 mol/L Ci-Na was selected as the optimum condition and the SO_2 absorption efficiency was 96.4% and the duration was 40 min.

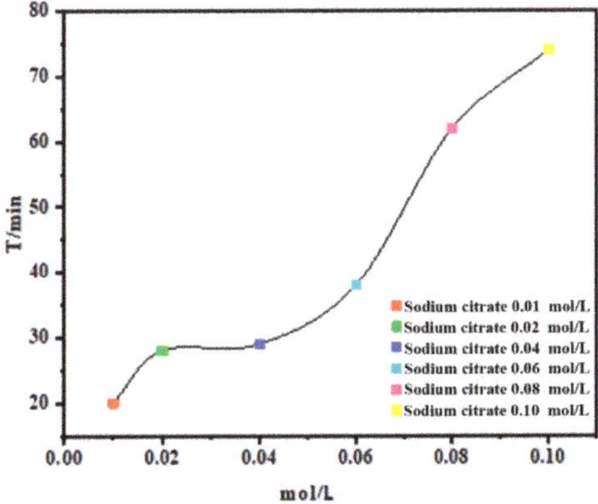

Figure 2. The effect of Ci-Na concentration on desulfurization time. SO_2 = 2300 ppm, gas flow = 1.68 L/min and absorption solution = 60 mL and 25 °C.

3.2. Effect of HA–Na Concentration

The different quantity of HA–Na was a significant factor on the reduction of SO_2 concentration, hence a series of experiments were carried out to study the effect of quantity on desulfurization efficiency [31]. The desulfurization efficiency using only HA–Na solution was shown in Figure 3. With the quantity increasing of HA–Na, the SO_2 absorption efficiency increased from 82% to 98%, which also had a certain impact on the break-through time. In addition, as the amount of HA–Na quantity increased (from 0.05 to 2.4 g), the desulfurization time also was enhanced and almost remained above 40 min when HA–Na mass was 2.4 g. The reason is as follows: the HA–Na solution is alkaline (generally the PH value is 10), and the hydroxide (OH^-) in the solution is rapidly neutralized with the generated H^+. Moreover, a large number of acid ions ionized by HA–Na (such as COO- and OH^-), which will interact with a large number of H^+. The H^+ combines with HA–Na to generate HA precipitation, which moves the dissolution balance to the right and promotes the dissolution of more SO_2 into the HA–Na solution.

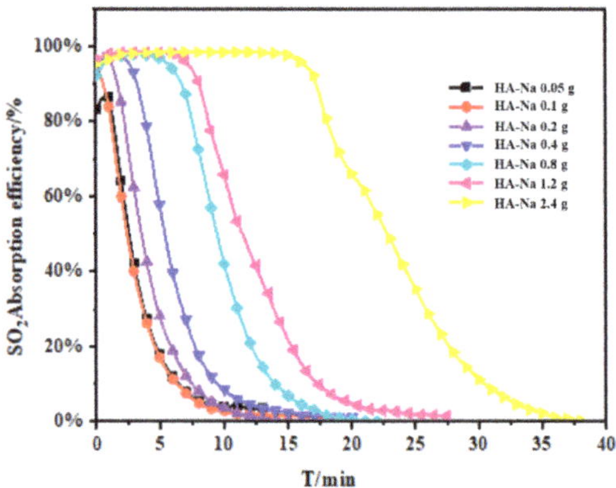

Figure 3. The effect of different quality HA–Na on desulfurization efficiency. SO_2 = 2300 ppm, gas flow = 1.68 L/min and absorption solution = 60 mL and 25 °C.

3.3. Effect of the Additive Amount of HA–Na on the Desulfurization Performance of Ci-Na

The addition of a different quantity of HA–Na may be one of the factors affecting the desulfurization efficiency of Ci-Na [32]. HA–Na was added into Ci-Na solution as an additive, such as 0.05 g, 0.1 g, 0.2 g, 0.4 g, 0.8 g, 1.2 g and 2.4 g, respectively, and the desulfurization effect was shown in Figure 4. The SO_2 absorption efficiency increased as the adding amount of HA–Na, and the saturation time also was improved, from 40 to 70 min. The reason may be that the addition of HA–Na increases the hydroxide ion (OH^-) in the solution, promoting more SO_2 absorption. Ci-Na and HA–Na had a synergistic effect for SO_2 absorption. This is more clearly confirmed in Figure 5.

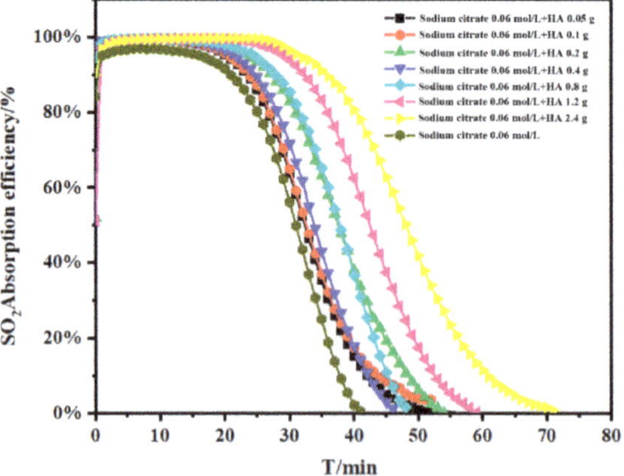

Figure 4. Effect of HA–Na additive on the desulfurization effect. SO_2 = 2300 ppm, gas flow = 1.68 L/min, absorption solution = 60 mL and Ci-Na = 0.06 mol/L and 25 °C.

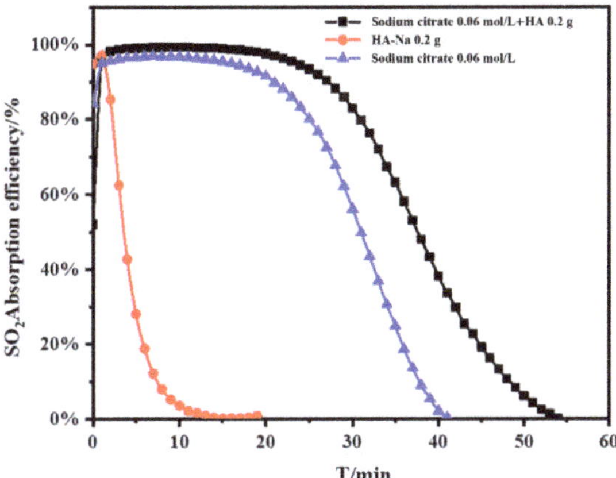

Figure 5. Comparison of HA–Na and Ci-Na solution. SO_2 = 2300 ppm, gas flow = 1.68 L/min, absorption solution = 60 mL and Ci-Na = 0.06 mol/L and 25 °C.

It is evident from Figure 5 that the addition of HA–Na can remarkably enhance the desulfurization efficiency and saturation time of Ci-Na. It was also found that the SO_2 absorption efficiency was close to 0% at 40 min when there was no HA–Na added, but it was still about 50% after adding HA–Na. Moreover, the desulfurization time increased by 15 min. The causes of this phenomenon are various [33]. In addition to the hydrolysis of HA–Na to generate hydroxide ions, it can also ionize the acid radical ions (carboxylate), thus consuming H^+ to move the dissolution balance to the right and cooperating with Ci-Na to absorb more SO_2.

3.4. Effect of the Inlet SO_2 Concentration

The concentration of SO_2 is different in the actual industrial flue gas. Hence, it might be necessary to research the influence of SO_2 concentration on SO_2 absorption efficiency. The influence of different SO_2 concentrations on the SO_2 removal efficiency as illustrated in Figure 6. Simulated flue gas with the SO_2 concentrations of 1000 ppm, 2300 ppm and 3000 ppm were used for the desulfurization experiment. The results are represented in Figure 6 that with the increase of SO_2 concentration, the desulfurization time of reaching saturation decreased from 92 to 40 min and diminished by 2.3 times. Moreover, the desulfurization time decreased significantly at 40 min, only about 5% under the high SO_2 concentration condition, while the desulfurization time was still close to 100% under the condition of low SO_2 concentration. The result shows that the inlet SO_2 concentration had a certain influence on the removal efficiency.

The main reason is that the driving force of mass transfer increased with the increasing of SO_2 concentration, which is beneficial to the absorption reaction [34]. However, the SO_2 capacity per unit volume of the solution was constant. As the inlet SO_2 concentration increased, the mass transfer rate was heightened while the time of SO_2 absorption saturation was shortened. So, the SO_2 absorption rate will be accelerated and the desulfurization time will be reduced.

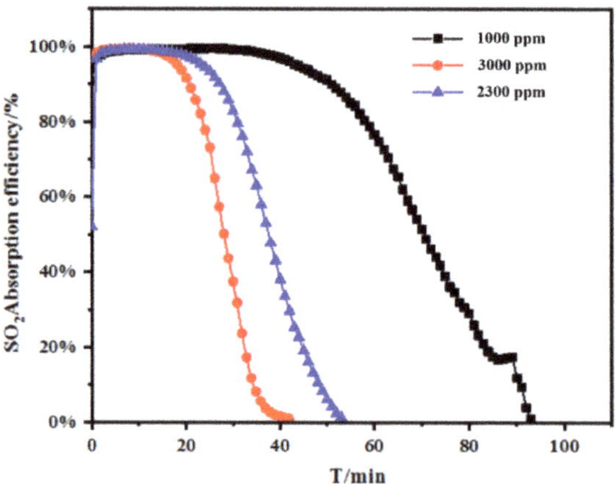

Figure 6. Effect of SO_2 concentration on desulfurization efficiency. Gas flow = 1.68 L/min, absorption solution = 60 mL, Ci–Na = 0.06 mol/L and HA–Na = 0.2 g and 25 °C.

3.5. Effect of the Gas Flow Rate

Most of the experiments were discussed as the influence of the gas flow rate on SO_2 removal efficiency. The initial inlet gas flow rate were respectively set as 1.0 L/min, 1.3 L/min and 1.6 L/min in the experiment. The results was presented in Figure 7. It is proved by the experiment that the removal efficiency of SO_2 increased as the initial inlet gas flow decreased. The increasing of the gas flow rate reduced the driving force of the absorption reaction, which is unfavorable for the desulfurization reaction. At the same time, the gas flow rate increased and the gas–liquid reaction time was reduced. A part of SO_2 was released before the reaction, which affected the absorption efficiency. In general, increasing gas flow had only some negative consequences.

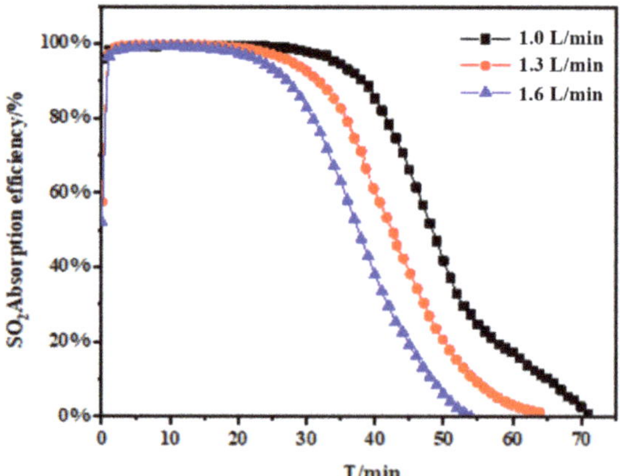

Figure 7. Effect of the gas flow rate. SO_2 = 2300 ppm, absorption solution = 60 mL, Ci-Na = 0.06 mol/L and HA–Na = 0.2 g and 25 °C.

3.6. Effect of O_2

The actual industrial flue gas contains a variety of ingredients, such as O_2. For instance, the flue gas of coal-fired power plant typically contains about 5–15 vol% O_2 [35]. Therefore, the existence of O_2 in the simulated flue gas was also explored. Figure 8 shows the effect of the presence of O_2 on the SO_2 removal efficiency. The experimental results indicated that SO_2 absorption efficiency at 50 min was improved significantly from 5% to 95% by the addition of 15% O_2. It can be seen that the existence of O_2 increased the desulfurization efficiency noticeably and also prolonged the desulfurization time from 55 to 80 min. The possible reasons were as follows [36]. The presence of O_2 could be more effective in improving O_2 dissolving into water so that the centration of dissolved O_2 into the solution was far higher than before, which is conducive to the oxidation of sulfite. According to Equation (9), it can infer that dissolved O_2 could accelerate the oxidation of sulfate. This reduces the concentration of HSO_3^- in the liquid phase and makes Equation (10) shift to the right. The liquid phase mass transfer coefficient was reduced.

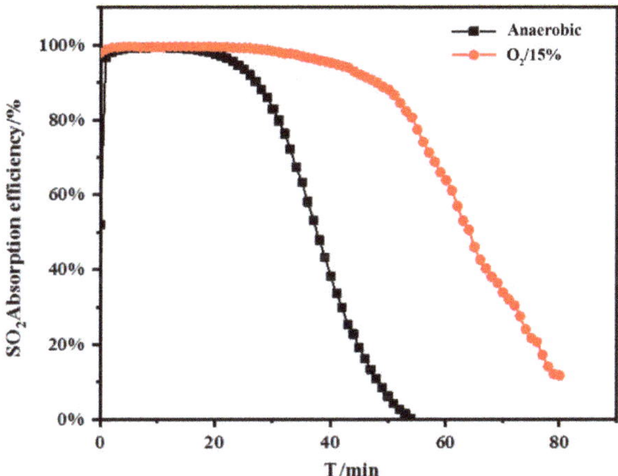

Figure 8. Effect of O_2. SO_2 = 2300 ppm, gas flow = 1.68 L/min, absorption solution = 60 mL, Ci-Na = 0.06 mol/L and HA–Na = 0.2 g and 25 °C.

After the addition of O_2, the desulfurization process will be accompanied by the following reactions.

$$2SO_3{}^{2-}(aq) + O_2(g) \rightarrow 2SO_4{}^{2-}(aq) \tag{9}$$

$$HSO_3{}^-(aq) \leftrightarrow H^+(aq) + SO_3{}^{2-}(aq) \tag{10}$$

It could be deduced that more O_2 in the solution participated in the desulfurization reaction and both the amount of SO_2 absorbed and desulfurization time were increased.

4. Materials and Methods

4.1. Sample Preparation

Ci-Na, sodium hydroxide, sodium acetate, acetic acid solution, anhydrous ethanol and sodium carbonate were from Sino pharm Chemical Reagent Co., Ltd., in Shanghai, China. HA–Na was from Shanghai Jincheng Biochemical Co., Ltd, in Shanghai, China. Deionized water was made in the laboratory.

4.2. Desulfurization Test

A principle diagram of the experimental devices are represented in Figure 9 below. Absorption experiments of the SO_2 in the laboratory consisted of SO_2, O_2 and balance N_2 as simulated flue gas. The SO_2, O_2 and N_2 gases were provided by cylinders. The experiment adopted SO_2 with a concentration range of 1000–3000 ppm. The total flow rate of the simulated flue gas was controlled with a mass flow controller (MFC). The flue gas analyzer was used to monitor the change of SO_2 concentration at the inlet and outlet of the reactor (KANE-9506, UK) [29].

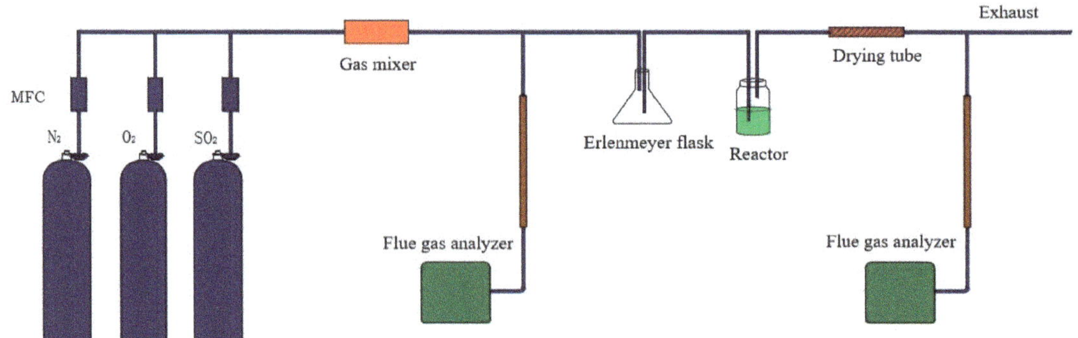

Figure 9. Schematic diagram of the experimental apparatus.

The absorption efficiency SO_2 can be obtained by the following formula:

$$\eta = \frac{(C_{in} - C_{out}) \times 100\%}{C_{in}} \tag{11}$$

where η is the SO_2 absorption efficiency and C_{an} and C_{oot} are the inlet and outlet of the SO_2 concentration, respectively.

5. Conclusions

The new desulfurization method with Ci-Na/HA–Na solution was put forward. The influence of different Ci-Na concentration, inlet SO_2 concentration, flow rate and other elements on the desulfurization performance were studied. The mechanism of HA–Na as an addition agent to improve the desulfurization performance of Ci-Na was discussed. For the absorption process, the higher the Ci-Na concentration and the lower the inlet flue gas flow, the more conducive to the SO_2 absorption. The presence of O_2 had a slight influence on the desulfurization efficiency. HA–Na played a key role during SO_2 absorption by the Ci-Na solution and can improve obviously the desulfurization performance of the Ci-Na solution.

Author Contributions: Conceptualization, Z.S.; methodology, S.J.; validation, Z.S.; investigation, Y.W. and D.J.; resources, Z.S.; writing—original draft preparation, Y.Z.; writing—review and editing, Z.S. and L.Z.; project administration, Z.S. All authors have read and agreed to the published version of the manuscript.

Funding: The authors gratefully acknowledge financial support by National Natural Science Foundation of China (No. 21806101), Natural Science Foundation of Shanghai (No.16ZR1412600), Research Center of Resource Recycling Science and Engineering, Shanghai Polytechnic University and Gaoyuan Discipline of Shanghai—Environmental Science and Engineering (Resource Recycling Science and Engineering), Cultivate discipline fund of Shanghai Polytechnic University (No.XXKPY1601).

Data Availability Statement: No data associated with this publication to be link. All the data associated in presented in this paper.

Conflicts of Interest: The authors declare no conflict of interest.

References

1. Mohd, M.; Hayyiratul-Fatimah, Z.; Chong, K.; Khairulazhar, J.; Lim, Y.; Sarrthesvaarni, R. Futuristic advance and perspective of deep eutectic solvent for extractive desulfurization of fuel oil: A review. *J. Mol. Liq.* **2020**, *306*, 112870.
2. Akyalçın, L.; Kaytakoğlu, S. Flue gas desulfurization by citrate process and optimization of working parameters. *Chem. Eng. Process.* **2009**, *49*, 199–204. [CrossRef]
3. Antony, R.; Tian, C.; Hong, F.; Zhi, Y.; Jie, F.; Wen, L. A comprehensive review on oxidative desulfurization catalysts targeting clean energy and environment. *J. Mater. Chem. A* **2020**, *8*, 2246.

4. Yong, Y.; Zhong, M.; Jin, L.; Ru, D.; Jia, L. Simultaneous removal of SO_2, NO_x and Hg^0 by O_3 oxidation integrated with bio-charcoal adsorption. *Fuel Chem. Technol.* **2020**, *48*, 1452–1460.
5. Guo, H.; Zhi, S.; Han, G. Novel Process of Simultaneous Removal of SO_2 and NO_2 by sodium humate solution. *Environ. Sci. Technol.* **2010**, *44*, 6712–6717.
6. Wei, S.; Jia, Z.; Bao, W.; Shu, L.; Jun, H. New insight into investigation of reduction of desulfurization ash by pyrite for clean generation SO_2. *J. Clean. Prod.* **2020**, *253*, 120026.
7. Jia, L.; Zhi, Y.; Si, L.; Qi, J.; Jian, Z. Review on oxidative desulfurization of fuel by supported heteropolyacid catalysts. *J. Ind. Eng. Chem.* **2020**, *82*, 1–16.
8. Lu, C.; Zhi, S.; Jin, X.; Meng, W.; Jia, F.; Li, Z. Reactivity Improvement of Ca-Based CO_2 Absorbent Modified with Sodium Humate in Cyclic Calcination/Carbonation. *ACS Omega* **2020**, *5*, 8867–8874.
9. Yi, Z.; Run, H.; Tian, W.; Chun, Y. Follow-up research for integrative process of pre-oxidation and post-absorption cleaning flue gas: Absorption of NO_2, NO and SO_2. *Chem. Eng. J.* **2015**, *273*, 55–65.
10. Castillo-Villalón, P.; Jorge, R.; Vargas-Luciano, J.A. Analysis of the role of citric acid in the preparation of highly active HDS catalysts. *J. Catal.* **2014**, *320*, 127–136. [CrossRef]
11. Budukva, S.V.; Klimov, O.V.; Noskov, A.S. Effect of citric acid and triethylene glycol addition on the reactivation of CoMo/γ-Al_2O_3 hydrotreating catalysts. *Catal. Today* **2019**, *329*, 35–43. [CrossRef]
12. Gao, L.; Yuan, L.; Jun, H. Intensifying effects of zinc oxide wet flue gas desulfurization process with citric acid. *J. Environ. Chem. Eng.* **2018**, *7*, 102831.
13. Xiu, J.; You, L.; Meiduo, G.U. Absorption of Sulphur Dioxide with Sodium Citrate Buffer Solution in a Rotating Packed Bed. *Chin. J. Chem. Eng.* **2011**, *19*, 687–692.
14. Kazakova, M.O.; Kazakova, M.A.; Vatutina, Y.V.; Larina, T.V.; Noskov, A.S. Comparative study of MWCNT and alumina supported CoMo hydrotreating catalysts prepared with citric acid as chelating agent. *Catal. Today* **2020**, *357*, 221–230. [CrossRef]
15. Zhang, W.; Yu, H.; Wei, W.; Shu, R.; Kai, Z. Efficient Removal of Sulfuric Acid from Sodium Lactate Aqueous Solution Based on Common-Ion Effect for the Absorption of SO_2 of Flue Gas. *Energy Fuels* **2019**, *33*, 4395–4400.
16. Fateme, R.; Rownaghi, A.; Saman, M.; Ryan, P.; Christopher, W. SO_x/NO_x Removal from Flue Gas Streams by Solid Adsorbents: A Review of Current Challenges and Future Directions. *Energy Fuels* **2015**, *29*, 5467–5486.
17. Zheng, X.; Deng, L.; Fei, W.; Zhi, S.; Zheng, L. Simultaneous removal of NO and SO_2 with a new recycling micro-nano bubble oxidation-absorption process based on HA-Na. *Sep. Purif. Technol.* **2020**, *242*, 116788.
18. Motta, F.L.; Santana, A. Production of humic acids from oil palm empty fruit bunch by submerged fermentation with Trichoderma viride: Cellulosic substratesand nitrogen sources. *Biotechnol. Prog.* **2013**, *29*, 631–637. [CrossRef] [PubMed]
19. Hong, D.; Tong, S.; Ting, H.; Xiang, L.; Ze, G.; Xing, W.; Yong, C. Interactions between cerium dioxide nanoparticles and humic acid: Influence of light intensities and molecular weight fractions. *Environ. Res.* **2021**, *195*, 110861.
20. Shu, X.; Ya, X.; Guo, W.; Zhen, M. Adsorption of heavy metals in water by modifying Fe_3O_4 nanoparticles with oxidized humic acid. *Colloid Surf. A* **2021**, *616*, 126333.
21. Kun, Y.; Xin, Y.; Jia, X.; Ling, J.; Wen, W. Sorption of organic compounds by pyrolyzed humic acids. *Sci. Total Environ.* **2021**, *781*, 146646.
22. Trckova, M.; Lorencova, A.; Hazova, K.; Sramkova, Z. Prophylaxis of post-weaning diarrhea in piglets by zinc oxide and sodium humate. *Vet. Med.* **2015**, *63*, 351–360.
23. Hashish, K.I.; Fatma, E.M.; Azza, M.M. Influence of potassium humate on growth and chemical constituents of Jatropha Curcus L. *Int. J. Chemtech. Res.* **2015**, *8*, 279–283.
24. Fábio, O.; Natália, A.; Raul-Castro, C.R.; Luciano, C. Substrate biofortification in combination with foliar sprays of plant growth promoting bacteria and humic substances boosts production of organic tomatoes. *Sci. Hortic.* **2015**, *183*, 100–108.
25. Wen, C.; Zhong, L.; Yan, L.; Xiao, X.; Qiang, L.; Xun, Z. Improved electricity generation, coulombic efficiency and microbial community structure of microbial fuel cells using sodium citrate as an effective additive. *J. Power Sources* **2021**, *482*, 228947.
26. Yan, Z.; Xue, Z.; Yong, S.; Jiang, W. Enhanced performance of calcium-enriched coal ash for the removal of humic acids from aqueous solution. *Fuel* **2015**, *141*, 93–98.
27. Guo, D.; Zhi, J. Preparation of Sodium Humate-Modified Biochar Absorbents for Water Treatment. *ACS Omega* **2019**, *4*, 536–542.
28. Zhi, S.; Han, G.; Guo, H.; Yan, L. Preparation of Sodium Humate/α-Aluminum Oxide Adsorbents for Flue Gas Desulfurization. *Environ. Eng. Sci.* **2009**, *26*, 1249–1255.
29. Zhi, S.; Yu, Z.; Han, G.; Guo, H. Removal of SO_2 from Flue Gas by Sodium Humate Solution. *Energy Fuels* **2010**, *24*, 1013–1019.
30. Biswa, B.; Sung, J. Oxidative desulfurization and denitrogenation of fuels using metal-organic framework-based/-derived catalysts. *Appl. Catal. B* **2019**, *259*, 118021.
31. Guo, D.; Zhi, J. Sodium humate as an effective inhibitor of low-temperature coal oxidation. *Thermochim. Acta* **2019**, *673*, 53–59.
32. Kentaro, K.; Masatoshi, N. Active sites of sulfided NiMo/Al_2O_3 catalysts for 4, 6-dimethyldibenzothiophene hydrodesulfurization-effects of Ni and Mo components, sulfidation, citric acid and phosphate addition. *Catal. Today* **2017**, *292*, 74–83.
33. Kutus, B.; Dudás, C.; Friesen, S.; Peintler, G.; Pálinkó, I.; Sipos, P.; Buchner, R. Equilibria and Dynamics of Sodium Citrate Aqueous Solutions: The Hydration of Citrate and Formation of the Na_3Cit^0 Ion Aggregate. *J. Phys. Chem.* **2020**, *124*, 9604–9614. [CrossRef] [PubMed]

34. Nadeesha, H.K.; Patricio, X.P.; Dionysios, D.D.; Al-Abed, S.R. Recent advances in flue gas desulfurization gypsum processes and applications—A review. *J. Environ. Manag.* **2019**, *251*, 109572.
35. Seong-Pil, K.; Huen, L. Recovery of CO_2 from Flue Gas Using Gas Hydrate: Thermodynamic Verifification through Phase Equilibrium Measurements. *Environ. Sci. Technol.* **2000**, *34*, 4397–4400.
36. You, L.; Jia, L.; Tian, Y. Surface modification of SiC powder with HA-Na: Adsorption kinetics, equilibrium and mechanism. *Langmuir* **2018**, *34*, 9645–9653.

 *catalysts* 

Article

Characterization of Anaerobic Biofilms Growing on Carbon Felt Bioanodes Exposed to Air

Raúl M. Alonso [1], Guillermo Pelaz [1], María Isabel San-Martín [1], Antonio Morán [1] and Adrián Escapa [1,2,*]

[1] Chemical and Environmental Bioprocess Engineering Group, Natural Resources Institute (IRENA), Universidad de León, Avda. de Portugal 41, E-24009 Leon, Spain; ralog@unileon.es (R.M.A.); gpelg@unileon.es (G.P.); msanb@unileon.es (M.I.S.-M.); amorp@unileon.es (A.M.)
[2] Department of Electrical Engineering and Automatic Systems, Campus de Vegazana s/n, Universidad de León, E-24071 León, Spain
* Correspondence: aescg@unileon.es

Received: 14 October 2020; Accepted: 16 November 2020; Published: 18 November 2020

Abstract: The role of oxygen in anodic biofilms is still a matter of debate. In this study, we tried to elucidate the structure and performance of an electrogenic biofilm that develops on air-exposed, carbon felt electrodes, commonly used in bioelectrochemical systems. By simultaneously recording the current density produced by the bioanode and dissolved oxygen concentration, both inside and in the vicinity of the biofilm, it was possible to demonstrate the influence of a protective aerobic layer present in the biofilm (mainly formed by *Pseudomonas* genus bacteria) that prevents electrogenic bacteria (such as *Geobacter* sp.) from hazardous exposure to oxygen during its normal operation. Once this protective barrier was deactivated for a long period of time, the catalytic capacity of the biofilm was severely affected. In addition, our results highlighted the importance of the material's porous structure for oxygen penetration in the electrode.

Keywords: exoelectrogen; biocatalyst; microenvironment; porous electrode; anaerobic

---

1. Introduction

The term microbial electrochemical technologies (METs) covers a group of bio-based electrochemical devices that hold great potential for practical applications in the fields of bioremediation and energy production/management [1]. The anodic processes of many METs (typically microbial fuel cells and microbial electrochemical cells) rely on the biocatalytic conversion of organic matter into electricity, requiring the presence of certain types of microorganisms, frequently referred to as electrogens, electricigens, or electrogenic microorganisms [2], which are capable of performing extracellular electron transfer to a solid electrode.

Mimicking natural processes, most of these electrogenic microorganisms can carry out their life cycle forming biofilms on the surface of the electrodes [3,4], where they benefit from favourable physical–chemical (micro)environments and useful trophic relationships. These microenvironments are caused by diffusive and/or reactive gradients of different substances, such as electron acceptors or donors that do not exist outside these multi-organism scaffolds [3]. Normally, anodic biofilms are colonised by electrogens, and it is very common that, depending on the source of the inoculum, carbon source, physicochemical parameters, and operational conditions, very complex microbial communities thrive on the surface of bioanodes [4]. In this regard, the presence of dissolved oxygen (DO) in an anodic medium can determine, to a significant extent, the structure of anodic biofilms, as many microorganisms involved in their functioning are anaerobic or facultative anaerobic [5]. In addition, the role and impact of oxygen in the performance of bioanodes has aroused some controversy [6]. On one hand, some authors claim that oxygen can have a beneficial impact on the degradation of complex organic compounds [6] and even improve the overall performance under some conditions [7]. On the other hand, several authors maintain that oxygen can promote the proliferation of aerobic microorganisms that compete with electrogens for organic matter, decreasing the coulombic efficiency (i.e., reducing the percentage of electrons initially available in the organic matter that are converted to electrical current in the external circuit), which explains why the bioanodes of METs frequently operate under anaerobic conditions [8].

Microsensors can provide an invaluable tool for understanding the nature, structure, and metabolic interactions within biofilms. They were initially applied to the study biofilms in the area of human diseases [9], and today, their use is widespread in microbial ecology for both natural and engineered systems [10], including METs [11]. For METs in particular, microsensors have been used to monitor chemical parameters (e.g., pH or redox potential) or the concentration of substances, such as hydrogen [12], acetate [13], nitrite [14], or oxygen [15] within biofilms. Focusing on oxygen, some publications are dedicated to revealing the local distribution of DO concentrations in specific biofilms, extracting significant conclusions [16,17], such as the importance of pH control in aerobic biocathodes [16] or the role of cathodic biofilms as oxygen diffusion barriers in single-chamber microbial fuel cells [17]. However, these investigations are mainly carried out on flat surfaces and rarely on porous surfaces, such as carbon felt, which is frequently used in METs and is also a good candidate for potential real-life applications [18]. The characterization of DO profiles inside biofilms growing on porous electrodes could provide valuable information on the implications of oxygen diffusion inside these elements and its consequences on biofilm development and functioning.

The objective of this research is to study the distribution of DO in electrogenic biofilms and their surroundings when these microbial aggregates grow, exposed to air, in carbon felt electrodes. It also tries to understand the effect of DO on the catalytic activity of an anodic biofilm fed with acetate.

2. Results and Discussion

After inoculation, the bioanode was operated in batch mode (electrode potential: +0.1 V vs. Ag/AgCl) with the anodic compartment open to air, so oxygen could freely diffuse into the anodic medium. The end of every batch cycle was determined by a drop in the current density (CD), below 10% of the maximum value achieved in that particular cycle. This resulted in an average cycle duration of approximately five days. Total organic carbon (TOC) was monitored daily, and at the end of each batch cycle (TOC was <10 mg·L^{-1}), the reactor was replenished with fresh medium and carbon source, providing an acetate concentration of 10 mM. After the first cycle (six days of operation), a biofilm was clearly visible on the electrode surface. To ensure pseudo-steady state conditions, the anode was operated in this manner for 42 additional days before the experimental period began.

2.1. The Interplay between DO and CD

To understand the behaviour of the exoelectrogenic biofilm when oxygen is present in the anodic medium, both the CD and the DO concentration, at a fixed height of 100 μm above the apparent surface of the biofilm, were simultaneously recorded for 15 days (Figure 1). Like the acclimation period, when the current fell below 10% of the maximum value, the TOC was checked to be less than 10 mg·L^{-1}, and then, the cell medium was spiked (the medium was not replaced) with sodium acetate to reach a final (theoretical) acetate concentration of 10 mM. After the addition of sodium acetate (blue stars in Figure 1), the CD immediately increased, while the oxygen concentration in the liquid phase declined abruptly, suggesting the existence of at least two competing processes: electrogenic and aerobic metabolisms. When the substrate was available again, the metabolism of aerobic and electrogenic microorganisms was reactivated. The role of microorganisms present in the planktonic phase can be seen as marginal due to the replacement of the medium in each feeding cycle.

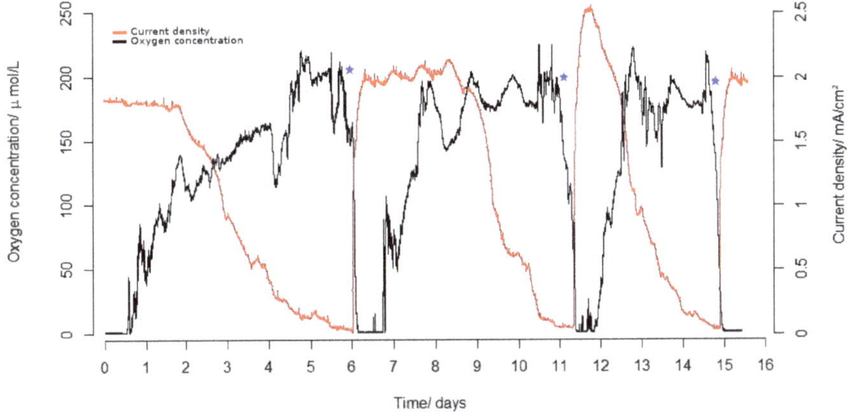

Figure 1. Biofilm current density (CD) and dissolved oxygen (DO) concentration profiles at a constant depth of 100 µm on the apparent surface of the biofilm during three consecutive feeding cycles. The blue stars denote spiking with sodium acetate (to a theoretical concentration of 10 mM).

Using a coculture of *Escherichia coli* and *Geobacter Sulfurreducens*, Qu et al. [19] also reported oxygen depletion in the bulk of the anolyte of a microbial fuel cell (MFC), showing the importance of microorganisms other than exoelectrogens in the maintenance of anoxic conditions in anodes exposed to oxygen. In the referred study, exoelectrogenesis was inhibited without the presence of *E. Coli* (the main oxygen scavenger), demonstrating its essential role in this anodic microbial community. In our experiments, the near-zero DO concentration was maintained for 12–18 h and then increased, agreeing with an expected decrease in the carbon source, deduced from the current profile (Figure 1). This fact indicates that the acetate concentration is the limiting factor in oxygen consumption. Figure 1 also shows that the biofilm is capable of maintaining exoelectrogenic activity, despite the presence of DO in the bulk. In the following experimental section, we will go into the biofilm and will try to clarify, among other things, if DO ever penetrates the biofilm during substrate oxidation cycles.

2.2. Oxygen Concentration Profile within the Biofilm

By studying the DO profile from the bulk to the interior of the electrode, functional aspects of the assembly liquid phase/biofilm/porous electrode can be understood. The experimental procedure leading to the achievement of the DO profiles proved to be challenging, since it is necessary to keep the cell in stationary conditions for an extended period of time. As Supplementary Materials, two prematurely interrupted DO profiles, not used in the subsequent analysis, were included. The oxygen probe was initially placed in the bulk within the vicinity of the electrode at an apparent distance of 500 µm from the biofilm surface. This height was taken as the reference position with a value of 0 µm (Figure 2) with the help of a camera (Figure 1). This initial position (0 µm) was the height where the profiling process began, and from here, the DO profile was periodically recorded at different depths (Figure 2). The DO concentration in the bulk did not change during the experiment. The reader can observe how the height reference and experimental procedure were different from the reference system used in Section 3.1. After all, in this section, the microelectrode moved, and in the previous experiment, it remained stationary. At a depth of about 355 µm, the DO concentration started to decline, almost linearly (slope -0.353 µM·µm^{-1}), which is indicative that the probe is entering the diffusion boundary layer (DBL). In the biofilm vicinity (from 290 to 570 µm), which has been amplified in Figure 3, we found at least three sub-regions that can be distinguished by changes to the shape of the DO concentration profiles. As mentioned before, the first sub-region could be associated with the DBL, which is in direct contact with the biofilm. From Figure 2, we can estimate its thickness as approximately 75 µm, a value which is in agreement with the thickness of the DBL found by other authors for aerobic biofilms composed of bacteria of the genus *Pseudomonas* [20]. Interestingly, this genus was found in a high relative abundance in our biofilm; for more details, see Section 3.4. The second sub-region, with a quite similar slope to the previous one, was slightly less steep ($-0.360\ \mu M\cdot\mu m^{-1}$), suggesting that the probe was entering a new phase, most probably the biofilm. The small difference between the slopes of the first and second sub-regions reinforces the idea that the diffusion coefficient in the outer layers of the biofilm is close to that of oxygen in water [21]. The oxygen flux (J) in this zone, calculated from Equation (6), was $7.39 \times 10^{-7}\ mol\cdot m^{-2}\cdot s^{-1}$, a value which is on the same order of magnitude as others reported in the literature for aerobic biofilms [22,23].

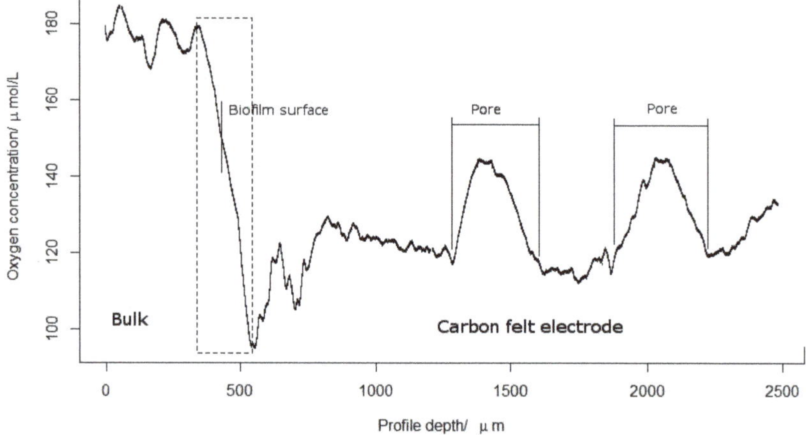

Figure 2. DO concentration profile in a carbon felt electrode colonised by an exoelectrogenic biofilm.

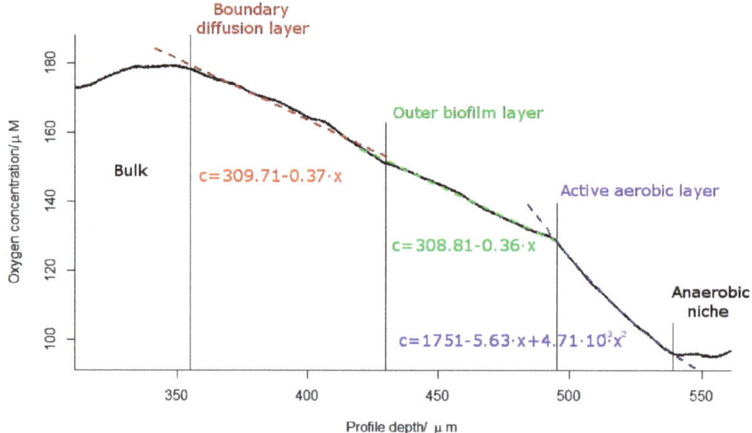

Figure 3. Detailed DO concentration profile in the vicinity of the biofilm. In the colours corresponding to each zone, fitted regression models are provided.

A clearer change in the DO profile came at a depth of ~495 μm, which can be attributed to significant oxygen consumption (Equation (1)) and (in the absence of another apparent oxygen sink) to the presence of an aerobic metabolism. This oxygen consumption was distributed over a 45 μm-thick region, in which a concave profile (positive second spatial derivative) could be observed. The reactive consumption criterion in biofilms and sediments has been extensively detailed by Berg et al. [24]. The fitting of a second-degree polynomic model allowed the oxygen consumption rate (OCR) to be estimated, following Equation (5). The resulting value was 9.42×10^{-3} mol·s^{-1}·m^{-3}, which is on the same order of magnitude as other OCRs reported in literature for aerobic biofilms [25,26]. Below this presumable aerobic layer, the DO concentration stabilised for ~20 μm. The resulting, almost flat profile, reveals the existence of another biofilm region, in which oxygen is no longer being consumed. This led us to hypothesise that this zone is mainly inhabited by microorganisms other than aerobes, which thrive because of the "oxygen-diffusion barrier" provided by the upper layers of the biofilm. This narrow microenvironment seems to be the logical candidate to host the *Geobacter* population, which is the most important exoelectrogen found in our biofilm (see Section 3.4). Several authors maintain that these low DO conditions are optimal for the growth and development of *Geobacter*; however, this is still an issue subject to some controversy, as discussed in a recent review by Reguera and Kashefi [27]. This review summarises research that concludes a certain tolerance of *Geobacter* to oxygen, as reflected in its capacity to encode proteins of the oxidative stress response and to adapt to microaerophilic conditions.

The relatively small thickness of this "exoelectrogen" layer does not contradict the high CD observed in our set-up (~2 mA·cm^{-2}), which is in agreement with previous studies that show that the exoelectrogenic activity of *Geobacter* is due to "active" layers of little thickness relative to the total width of the biofilm [28,29].

When the microsensor tip was placed at a depth of 550 μm (leaving the biofilm vicinity and entering the bulk of the carbon felt electrode), the oxygen concentration began to slightly increase again, and over a very wide region (spanning depths from ~800 to 2500 μm), the average DO was about 130 μM. In this zone, the DO concentration profile presented two very pronounced "hills" that can be attributed to the existence of large cavities within the carbon felt (Figure 1). These hills showed similar slope values to those found in the DBL of the external surface of the electrode, thus, suggesting a common subjacent phenomenon. Likely, these cavities are acting as "channels" that allow the transport of DO through the electrode, revealing a complex structure that has been extensively studied in electrochemical energy-storage systems [30–32]. A relevant reason for the lack of colonisation by exoelectrogens of carbon felt electrodes in deep layers may be due to its porous structure that may be channelling DO to the inner domain. Although, the size of the pores (~300 μm, see Figure 2) also suggests that the limited diffusion of the buffering substances is a plausible hypothesis for this lack of colonisation, as has been pointed out in recent research [33]. In this investigation, for cavities in bioanodes between 100 and 500 μm, the limiting factor for the development of the biofilm was determined to be the accumulation of metabolites that triggers local acidification. As a consequence of these results, this same structure could be relevant to carry or remove other substances related to the microbial metabolism (e.g., substrate, inhibitors, protons), highlighting the importance of the optimization of the volume/active area electrode relationship in different applications of METs, an aspect which has been remarked on in other studies [33–36]. The behaviour of carbon felt anodes without attached biofilm has not been analysed in this study, since there is an extensive literature on the subject and the results obtained have been acceptably consistent with the expected cavity distributions [37,38].

2.3. The Biofilm under Substrate Limited Conditions

In the final phase of the experiment, the anodic biofilm was kept under starving conditions for seven days (once CD declined below 10% of its maximum value, no acetate was fed to the anode). After this period, when acetate was fed again, the oxygen concentration decreased, as observed previously (results not shown, the behaviour was similar to that depicted in Figure 1), revealing a hypothetic reactivation of the aerobic metabolism. However, CD profiles did not resume previous values and remained below 5% of the maximum current value, which suggests that electrogenic microorganisms were affected by either the lack of a carbon source, the prolonged exposure to oxygen, or both. However, the lack of carbon source can be ruled out, since previous work has shown that electrogenic biofilms can remain active for long periods of time in the absence of a carbon source [39,40].

In contrast, the exposure to oxygen seems a more plausible explanation. Indeed, when acetate is present and available to the aerobic microorganisms, these are capable of depleting oxygen from the bulk (Figure 1), thus, protecting the electrogenic microorganisms in contact with the electrode from DO. In the absence of a carbon source, the aerobic microorganisms are no longer capable of consuming oxygen, and the "barrier effect" is lost, which increases the local DO concentration and exposure times in deep layers of the biofilm. However, this is a hypothesis that needs to be supported by further experiments.

2.4. Biofilm Composition

The microbial composition of the biofilm was analysed by pyrosequencing after 90 days of operation. The total number of pyrosequencing reads were 41,523, and the valid reads were 26,002 (62.6%). The number of operational taxonomic units (OTUs) found in the sample was 126, a number that indicates the high specialization of this community.

The taxonomic composition at the genus level (Figure 4) of the anodic community shows a great relative abundance of the exoelectrogen *Geobacter* (33.2%). The presence of this anaerobic bacteria in an air-exposed surface is not surprising in light of other research that has shown the presence of *Geobacter* in both natural [41] and engineered aerobic biofilms [19,42]. The stratification of DO through the carbon felt (see Figures 3 and 4) probably promoted the appearance of "low oxygen" microenvironments, coherent with the tolerance of *Geobacter* to the presence of DO [27]. The most abundant genus on the biofilm was *Pseudomonas* (45.5%), which is considered a facultative anaerobe that could be favoured as a result of microaerophilic conditions [43] and has been recognised as electroactive. However, *Pseudomonas* can be discarded as the main exoelectrogen in our study since the reported current densities for this genus are lower than expected for *Geobacter*-based electrogenic biofilms [7,44], which are in accordance with those observed in this study (~2 mA·cm^{-2}). Species assigned to this taxon, such as *P. aeruginosa*, are capable of oxidising organic matter using oxygen as a terminal electron acceptor, inducing steep gradients in DO concentrations in pure culture biofilms [7,20]. Another less prevalent genus in the sample was *Clostridium* (6.9%), a strict anaerobic that can ferment a variety of substrates. The genera *Flavobacterium* (4.5%) and *Bacillus* (4.1%) can generally be considered aerobic. However, their role in this community is not fully understood. *Flavobacterium* has also been identified in cathodic biofilms exposed to oxygen [17].

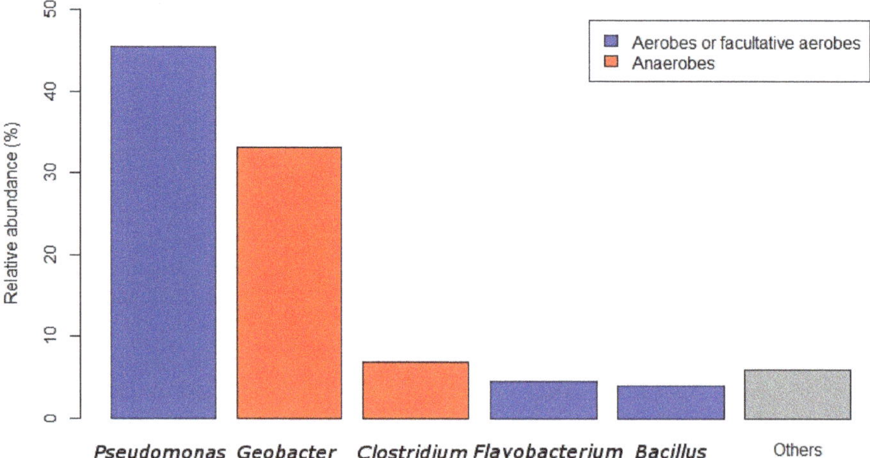

Figure 4. Microbial community composition of the anodic biofilm at the genus level.

3. Experimental

3.1. Reactor Set-Up and Biofilm Growth

Tests were carried out on a biofilm growing in a porous electrode in a specially designed single-chamber microbial electrolysis cell under aerobic conditions for all experimental procedures (Figure 1). The methacrylate reactor allowed for the easy handling of the microelectrode, while using a three-electrode electrical arrangement, establishing a working electrode (WE) potential of +0.1 V with respect to the reference electrode (RE).

A carbon-felt WE (Sigracell® GFA 6, SGL, Wiesbaden, Germany), a commercial RE microelectrode (Ag/AgCl, +0.197 vs. SHE, Unisense, Aarhus, Denmark), and a platinum mesh (Goodfellow, London, UK) counter electrode (CE) was used. The dimensions of the WE and CE were 8 × 2 cm and 2 × 2 cm, respectively. The inoculation procedure consisted of mixing fluvial sediment, freshly obtained with growth medium, in a 1:5 volume ratio. The growth medium composition per litre was 0.87 g of K_2HPO_4, 0.68 g of KH_2PO_4, 0.25 g of NH_4CL, 0.453 g of $MgCl_2 \cdot 6H_2O$, 0.1 g of KCl, 0.04 g of $CaCl_2 \cdot 2H_2O$, and 10 mL of mineral solution. The mineral solution composition was detailed by Marshall et al. [45]. Sodium acetate was added to the mixture as a carbon source to reach a final concentration of 6.1 mM in the start-up cycle. The reactor was kept at room temperature (22 ± 2 °C). Feeding was done in batch mode (i.e., by replacing the entire medium at the beginning of every new cycle). Sodium acetate was added to achieve a 10 mM concentration in the medium (except in the start-up cycle) when the current dropped below 10% of the maximum value. The biofilm was maintained for 60 days in the described growing conditions (acclimation period) before starting the analyses of this study.

3.2. DO Concentration Measurement

DO concentration in the exoelectrogenic biofilm was measured using a microelectrode (OX-10, UNISENSE, Aarhus, Denmark) that works as a Clark-type electrode [46]. The tip diameter chosen was approximately 10 μm, which ensures an adequate relationship with the estimated thickness of the biofilm (less than 100 μm). The microelectrode was introduced into the thin film cell using a programmable micropositioning system. The step size in the z axis (axis in which the microelectrode passes through the electrode vertically to its surface, Figure 5) was 1 μm ± 20 nm. The permanence time of the microsensor in each position was 3 s, since the response time to reach 90% of the signal, as detailed by the manufacturer, is 1–3 s. The current signal was acquired and registered using a specific amplifier (Microsensor Multimeter, UNISENSE, Aarhus, Denmark) and the associated software (UNISENSE logger). The sensor was calibrated with a two-point procedure. First, it was placed in an air-saturated medium, and then, the zero level was obtained by generously bubbled nitrogen in the same medium. In order to place the microsensor tip at the desired height over the biofilm surface, a video microscope system (Model VCAM3) was used. The analysis of the DO concentration profiles was based on the one-dimensional form of Fick's first law, assuming the biofilm can be considered a flat slab (Equation (1)):

$$\frac{\partial c}{\partial t} = D \cdot \frac{\partial^2 c}{\partial^2 x} - OCR \tag{1}$$

where c is the concentration (mol·m^{-3}), D is the diffusion coefficient (m^2·s^{-1}), and OCR (mol·m^{-3}·s^{-1}) is the oxygen consumption rate (OCR is assumed to be constant during the experiment). The D value for oxygen in water at 20 °C (2.27×10^{-9} m^2·s) was obtained from de Beer et al. [22]. For the biofilm, D was corrected using a relative effective diffusion coefficient (Equation (2)):

$$D_{bio} = D_{aq} \cdot \gamma \tag{2}$$

where D_{bio} is the oxygen diffusion coefficient in the biofilm, D_{aq} is the oxygen diffusion coefficient in water, and γ is the effective diffusion coefficient, with a value taken as 0.88, according to

Lewandowski et al. [21], and it is consistent with other studies [47]. The steady-state solutions to Equation (1) yields Equations (3)–(5):

$$\frac{\partial^2 c}{\partial^2 x} = \frac{OCR}{D_{bio}} \tag{3}$$

$$\frac{\partial c}{\partial x} = \frac{OCR}{D_{bio}} x + A \tag{4}$$

$$c = \frac{OCR}{D_{bio}} x^2 + Ax + B \tag{5}$$

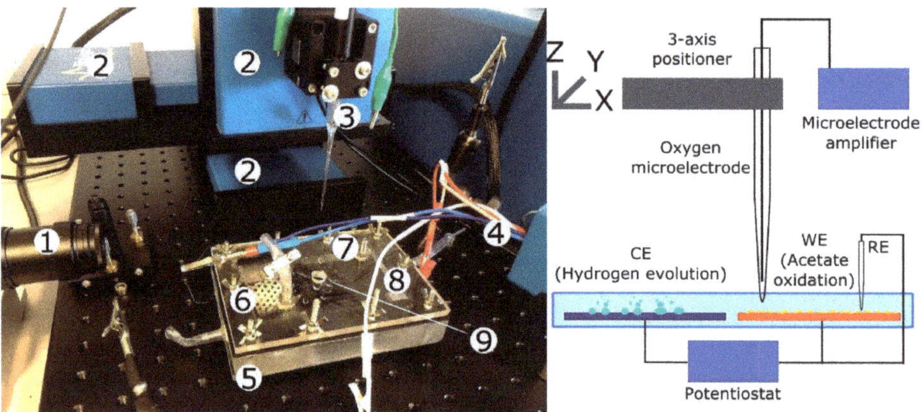

Figure 5. Experimental set-up: camera (**1**); three-axis micropositioner (**2**); microelectrode (**3**); connections to the potentiostat (**4**); methacrylate cell (**5**); counter electrode (CE) (**6**); reference electrode (RE) adapter (**7**); working electrode (WE) (**8**); and microelectrode sampling hole (**9**).

Equation (5), where A and B are constants, allows us to detect the presence of net oxygen consumption and calculate the OCR in the biofilm if the quadratic term is present (the sign of the second spatial derivative of the DO concentration profile can be used to discern between oxygen consumption or production [24]). In this sense, if the quadratic term is not present, it is possible to estimate the oxygen flux (J), following Equation (4) with $OCR = 0$ (Equation (6)):

$$J = D_{bio} \frac{\partial c}{\partial x} \tag{6}$$

where J is in mol·m^{-2}·s^{-1}. The first or second order models have been used to calculate the coefficients in Equation (3), according to convenience, in each zone of the profile, using R software [48] to perform the least squares regression. R has also been used to elaborate the graphs.

3.3. Experimental Procedure

DO concentration profiles were recorded at various stages during the substrate degradation cycles, once the start-up and stabilization period was completed. The most stable profiles, like those analysed in this article, were produced during the constant part of the current cycle, which occurs between the feeding and final decay. To perform the DO concentration profiles, the microelectrode tip was placed 500 μm from the apparent surface of the biofilm and then moved 5 mm through the carbon felt electrode. The position of the microelectrode was changed every 3 s for profile recording, assuming that the nominal response time was less than this interval. For a constant height measurement, the microelectrode was held outside the diffusion boundary layer (DBL), 100 μm above the apparent surface of the biofilm. Total organic carbon (TOC) was determined by a TOC multi N/C 3100 (Analytikjena, Jena, Germany).

3.4. Microbial Community Analysis

Once the experiment was completed, a sample was taken for the analysis of microbial communities. For this purpose, a 1×1 cm portion of carbon felt was cut using sterilised scissors. Genomic DNA from the electrodic sample was extracted with the PowerSoil® DNA Isolation Kit (MoBio Laboratories Inc., Carlsbad, CA, USA), following the manufacturer's instructions. The entire DNA extract was used for the pyrosequencing of the 16S-rRNA gene-based massive library, targeting the eubacterial region V1–V3 16S-rRNA, and performed at MR DNA (www.mrdnalab.com, Shallowater, TX, USA), utilising MiSeq equipment (Illumina, San Diego, CT, USA). The primer set used was 27Fmod (5′-AGRGTTTGATCMTGGCTCAG-3′) /519 R modBio (5′-GTNTTACNGCGGCKGCTG-3′). Diluted DNA extracts were used as a template for polymerase chain reaction (PCR) reactions. The obtained DNA reads were compiled in FASTq files for further bioinformatics processing. The trimming of the 16S-rRNA bar-coded sequences into libraries was carried out using QIIME software, version 1.8.018 [49]. Quality filtering of the reads was performed at Q25 quality prior to grouping into operational taxonomic units (OTUs) at a 97% sequence homology cut-off. The following steps were performed using QIIME, a denoising procedure using a denoiser algorithm [50]. Final OTUs were taxonomically classified using BLASTn against a database derived from the Ribosomal Database Project II (RDPII, http://rdp.cme.msu.edu) and the National Centre for Biotechnology Information (NCBI, www.ncbi.nlm.nih.gov).

4. Conclusions

The proliferation and biocatalytic activity of *Geobacter* in air-exposed anodes depends on the anaerobic microenvironment provided by the aerobic microorganisms present in outer layers of the biofilm, which consume DO. The porous structure of the carbon felt causes the channelling of DO in internal zones. This study confirms *Geobacter*'s tolerance to sustained oxygen concentrations below 90 µM. During periods in which the biofilm is maintained without substrate, oxygen diffuses through the biofilm height. If these periods without substrate are longer than six days, the inner electroactive layers seem to be irreversibly affected, and the exoelectrogenic activity does not resume again after feeding.

Supplementary Materials: The following are available online at http://www.mdpi.com/2073-4344/10/11/1341/s1, Figure S1: Dissolved oxygen profile. Rejected because the data acquisition system was accidentally interrupted at the end of the experiment; Figure S2: Another dissolved oxygen profile. Rejected because the data acquisition system was accidentally interrupted at the end of the experiment.

Author Contributions: R.M.A. contributed to the conceptual design of the experiment, carrying out the experimental procedures, data analysis, results interpretation and drafting the manuscript. G.P.: performed the microbiological analysis and provided assistance in the laboratory. M.I.S.-M. also provided assistance in the laboratory and contributed to results' interpretation. A.M.: contributed to results interpretation, being responsible for project administration and funding acquisition. A.E.: contributed to the conceptualization, results interpretation, project supervision and writing/reviewing the manuscript. All authors have read and agreed to the published version of the manuscript.

Funding: This research was possible thanks to the financial support by 'Consejería de Educación de la Junta de Castilla y León' (ref: LE320P18), a project co-financed by FEDER funds. R. M. Alonso thanks the University of León for his predoctoral contract.

Conflicts of Interest: The authors declare that the research was conducted in the absence of any commercial or financial relationships that could be construed as a potential conflict of interest.

References

1. Escapa, A.; Mateos, R.; Martínez, E.J.; Blanes, J. Microbial electrolysis cells: An emerging technology for wastewater treatment and energy recovery. From laboratory to pilot plant and beyond. *Renew. Sustain. Energy Rev.* **2016**, *55*, 942–956. [CrossRef]
2. Logan, B.E. Exoelectrogenic bacteria that power microbial fuel cells. *Nat. Rev. Microbiol.* **2009**, *7*, 375–381. [CrossRef]
3. Flemming, H.-C.; Wingender, J.; Szewzyk, U.; Steinberg, P.; Rice, S.A.; Kjelleberg, S. Biofilms: An emergent form of bacterial life. *Nat. Rev. Microbiol.* **2016**, *14*, 563–575. [CrossRef] [PubMed]
4. Koch, C.; Harnisch, F. Is there a specific ecological niche for electroactive microorganisms? *ChemElectroChem* **2016**, *3*, 1282–1295. [CrossRef]
5. Wang, C.; Shen, J.; Chen, Q.; Ma, D.; Zhang, G.; Cui, C.; Xin, Y.; Zhao, Y.; Hu, C. The inhibiting effect of oxygen diffusion on the electricity generation of three-chamber microbial fuel cells. *J. Power Sources* **2020**, *453*, 227889. [CrossRef]
6. Yang, L.-H.; Zhu, T.-T.; Cai, W.-W.; Haider, M.R.; Wang, H.-C.; Cheng, H.-Y.; Wang, A.-J. Micro-oxygen bioanode: An efficient strategy for enhancement of phenol degradation and current generation in mix-cultured MFCs. *Bioresour. Technol.* **2018**, *268*, 176–182. [CrossRef]
7. Yong, X.-Y.; Yan, Z.-Y.; Shen, H.-B.; Zhou, J.; Wu, X.-Y.; Zhang, L.-J.; Zheng, T.; Jiang, M.; Wei, P.; Jia, H.-H.; et al. An integrated aerobic-anaerobic strategy for performance enhancement of Pseudomonas aeruginosa-inoculated microbial fuel cell. *Bioresour. Technol.* **2017**, *241*, 1191–1196. [CrossRef]
8. Zhang, X.; Cheng, S.; Wang, X.; Huang, X.; Logan, B.E. Separator characteristics for increasing performance of microbial fuel cells. *Environ. Sci. Technol.* **2009**, *43*, 8456–8461. [CrossRef] [PubMed]
9. Saxena, P.; Joshi, Y.; Rawat, K.; Bisht, R. Biofilms: Architecture, resistance, quorum sensing and control mechanisms. *Indian J. Microbiol.* **2019**, *59*, 3–12. [CrossRef] [PubMed]
10. Lewandowski, Z.; Beyenal, H. *Fundamentals of Biofilm Research*; CRC Press: Boca Raton, FL, USA, 2013.
11. Beyenal, H.; Babauta, J.T. *Biofilms in Bioelectrochemical Systems: From Laboratory Practice to Data Interpretation*; John Wiley & Sons: Hoboken, NJ, USA, 2015.
12. Maegaard, K.; Garcia-Robledo, E.; Kofoed, M.V.W.; Agneessens, L.M.; de Jonge, N.; Nielsen, J.L.; Ottosen, L.D.M.; Nielsen, L.P.; Revsbech, N.P. Biogas upgrading with hydrogenotrophic methanogenic biofilms. *Bioresour. Technol.* **2019**, *287*, 121422. [CrossRef]
13. Atci, E.; Babauta, J.T.; Sultana, S.T.; Beyenal, H. Microbiosensor for the detection of acetate in electrode-respiring biofilms. *Biosens. Bioelectron.* **2016**, *81*, 517–523. [CrossRef] [PubMed]
14. Lee, W.H.; Wahman, D.G.; Pressman, J.G. Amperometric carbon fiber nitrite microsensor for in situ biofilm monitoring. *Sens. Actuators B Chem.* **2013**, *188*, 1263–1269. [CrossRef]
15. Zhou, L.; Yan, X.; Yan, Y.; Li, T.; An, J.; Liao, C.; Li, N.; Wang, X. Electrode potential regulates phenol degradation pathways in oxygen-diffused microbial electrochemical system. *Chem. Eng. J.* **2020**, *381*, 122663. [CrossRef]
16. Wang, Z.; Deng, H.; Chen, L.; Xiao, Y.; Zhao, F. In situ measurements of dissolved oxygen, pH and redox potential of biocathode microenvironments using microelectrodes. *Bioresour. Technol.* **2013**, *132*, 387–390. [CrossRef] [PubMed]
17. Montpart, N.; Rago, L.; Baeza, J.A.; Guisasola, A. Oxygen barrier and catalytic effect of the cathodic biofilm in single chamber microbial fuel cells. *J. Chem. Technol. Biotechnol.* **2018**, *93*, 2199–2207. [CrossRef]
18. San-Martín, M.I. *Bioelectrochemical Systems for Energy Valorization of Waste Streams*; Leicester, D.D., Ed.; IntechOpen: Rijeka, Yugoslavia, 2018; Chapter 8; ISBN 978-1-78923-711-5.
19. Qu, Y.; Feng, Y.; Wang, X.; Logan, B.E. Use of a coculture to enable current production by geobacter sulfurreducens. *Appl. Environ. Microbiol.* **2012**, *78*, 3484–3487. [CrossRef]
20. Kragh, K.N.; Hutchison, J.B.; Melaugh, G.; Rodesney, C.; Roberts, A.E.L.; Irie, Y.; Jensen, P.Ø.; Diggle, S.P.; Allen, R.J.; Gordon, V.; et al. Role of multicellular aggregates in biofilm formation. *MBio* **2016**, *7*, e00237-16. [CrossRef]
21. Lewandowski, Z.; Walser, G.; Characklis, W.G. Reaction kinetics in biofilms. *Biotechnol. Bioeng.* **1991**, *38*, 877–882. [CrossRef]
22. De Beer, D.; Stoodley, P.; Roe, F.; Lewandowski, Z. Effects of biofilm structures on oxygen distribution and mass transport. *Biotechnol. Bioeng.* **1994**, *43*, 1131–1138. [CrossRef]

23. Mclamore, E.S.; Zhang, W.; Porterfield, D.M.; Banks, M.K. Membrane-aerated biofilm proton and oxygen flux during chemical toxin exposure. *Environ. Sci. Technol.* **2010**, *44*, 7050–7057. [CrossRef]

24. Berg, P.; Risgaard-Petersen, N.; Rysgaard, S. Interpretation of measured concentration profiles in sediment pore water. *Limnol. Oceanogr.* **1998**, *43*, 1500–1510. [CrossRef]

25. Ning, Y.-F.; Chen, Y.-P.; Shen, Y.; Zeng, N.; Liu, S.-Y.; Guo, J.-S.; Fang, F. A new approach for estimating aerobic–anaerobic biofilm structure in wastewater treatment via dissolved oxygen microdistribution. *Chem. Eng. J.* **2014**, *255*, 171–177. [CrossRef]

26. Stewart, P.S. Diffusion in biofilms. *J. Bacteriol.* **2003**, *185*, 1485–1491. [CrossRef] [PubMed]

27. Reguera, G.; Kashefi, K. The electrifying physiology of Geobacter bacteria, 30 years on. *Adv. Microb. Physiol.* **2019**, *74*, 1–96.

28. Sun, D.; Chen, J.; Huang, H.; Liu, W.; Ye, Y.; Cheng, S. The effect of biofilm thickness on electrochemical activity of geobacter sulfurreducens. *Int. J. Hydrogen Energy* **2016**, *41*. [CrossRef]

29. Sun, D.; Cheng, S.; Wang, A.; Li, F.; Logan, B.E.; Cen, K. Temporal-spatial changes in viabilities and electrochemical properties of anode biofilms. *Environ. Sci. Technol.* **2015**, *49*, 5227–5235. [CrossRef]

30. Banerjee, R.; Bevilacqua, N.; Eifert, L.; Zeis, R. Characterization of carbon felt electrodes for vanadium redox flow batteries—A pore network modeling approach. *J. Energy Storage* **2019**, *21*, 163–171. [CrossRef]

31. Döner, A.; Karcı, İ.; Kardaş, G. Effect of C-felt supported Ni, Co and NiCo catalysts to produce hydrogen. *Int. J. Hydrogen Energy* **2012**, *37*, 9470–9476. [CrossRef]

32. González, Z.; Sánchez, A.; Blanco, C.; Granda, M.; Menéndez, R.; Santamaría, R. Enhanced performance of a Bi-modified graphite felt as the positive electrode of a vanadium redox flow battery. *Electrochem. Commun.* **2011**, *13*, 1379–1382. [CrossRef]

33. Chong, P.; Erable, B.; Bergel, A. Effect of pore size on the current produced by 3-dimensional porous microbial anodes: A critical review. *Bioresour. Technol.* **2019**, *289*, 121641. [CrossRef]

34. Blanchet, E.; Erable, B.; De Solan, M.-L.; Bergel, A. Two-dimensional carbon cloth and three-dimensional carbon felt perform similarly to form bioanode fed with food waste. *Electrochem. Commun.* **2016**. [CrossRef]

35. Brunschweiger, S.; Ojong, E.T.; Weisser, J.; Schwaferts, C.; Elsner, M.; Ivleva, N.P.; Haseneder, R.; Hofmann, T.; Glas, K. The effect of clogging on the long-term stability of different carbon fiber brushes in microbial fuel cells for brewery wastewater treatment. *Bioresour. Technol. Rep.* **2020**, *11*, 100420. [CrossRef]

36. Flexer, V.; Jourdin, L. Purposely designed hierarchical porous electrodes for high rate microbial electrosynthesis of acetate from carbon dioxide. *Acc. Chem. Res.* **2020**, *53*, 311–321. [CrossRef]

37. Ma, P.; Ma, H.; Galia, A.; Sabatino, S.; Scialdone, O. Reduction of oxygen to H2O2 at carbon felt cathode in undivided cells. Effect of the ratio between the anode and the cathode surfaces and of other operative parameters. *Sep. Purif. Technol.* **2019**, *208*, 116–122. [CrossRef]

38. Smith, R.E.G.; Davies, T.J.; Baynes, N.d.B.; Nichols, R.J. The electrochemical characterisation of graphite felts. *J. Electroanal. Chem.* **2015**, *747*, 29–38. [CrossRef]

39. Marozava, S.; Röling, W.F.M.; Seifert, J.; Küffner, R.; von Bergen, M.; Meckenstock, R.U. Physiology of geobacter metallireducens under excess and limitation of electron donors. Part II. Mimicking environmental conditions during cultivation in retentostats. *Syst. Appl. Microbiol.* **2014**, *37*, 287–295. [CrossRef]

40. Lin, B.; Westerhoff, H.V.; Röling, W.F.M. How geobacteraceae may dominate subsurface biodegradation: Physiology of geobacter metallireducens in slow-growth habitat-simulating retentostats. *Environ. Microbiol.* **2009**, *11*, 2425–2433. [CrossRef]

41. Commault, A.S.; Barrière, F.; Lapinsonnière, L.; Lear, G.; Bouvier, S.; Weld, R.J. Influence of inoculum and anode surface properties on the selection of Geobacter-dominated biofilms. *Bioresour. Technol.* **2015**, *195*, 265–272. [CrossRef]

42. Nevin, K.P.; Zhang, P.; Franks, A.E.; Woodard, T.L.; Lovley, D.R. Anaerobes unleashed: Aerobic fuel cells of Geobacter sulfurreducens. *J. Power Sources* **2011**, *196*, 7514–7518. [CrossRef]

43. Peix, A.; Ramírez-Bahena, M.-H.; Velázquez, E. Historical evolution and current status of the taxonomy of genus Pseudomonas. *Infect. Genet. Evol.* **2009**, *9*, 1132–1147. [CrossRef]

44. Su, W.; Zhang, L.; Li, D.; Zhan, G.; Qian, J.; Tao, Y. Dissimilatory nitrate reduction by Pseudomonas alcaliphila with an electrode as the sole electron donor. *Biotechnol. Bioeng.* **2012**, *109*, 2904–2910. [CrossRef] [PubMed]

45. Marshall, C.W.; Ross, D.E.; Fichot, E.B.; Norman, R.S.; May, H.D. Electrosynthesis of commodity chemicals by an autotrophic microbial community. *Appl. Environ. Microbiol.* **2012**, *78*, 8412–8420. [CrossRef] [PubMed]

46. Revsbech, N.P. An oxygen microsensor with a guard cathode. *Limnol. Oceanogr.* **1989**, *34*, 474–478. [CrossRef]

47. Stewart, P.S. A review of experimental measurements of effective diffusive permeabilities and effective diffusion coefficients in biofilms. *Biotechnol. Bioeng.* **1998**, *59*, 261–272. [CrossRef]
48. R Core Team. *R: A Language and Environment for Statistical Computing*; R Foundation for Statistical Computing: Vienna, Austria, 2015.
49. Caporaso, J.G.; Kuczynski, J.; Stombaugh, J.; Bittinger, K.; Bushman, F.D.; Costello, E.K.; Fierer, N.; Pena, A.G.; Goodrich, J.K.; Gordon, J.I.; et al. QIIME allows analysis of high-throughput community sequencing data. *Nat. Methods* **2010**, *7*, 335–336. [CrossRef] [PubMed]
50. Reeder, J.; Knight, R. Rapidly denoising pyrosequencing amplicon reads by exploiting rank-abundance distributions. *Nat. Methods* **2010**, *7*, 668–669. [CrossRef]

Publisher's Note: MDPI stays neutral with regard to jurisdictional claims in published maps and institutional affiliations.

Review

Metal–Organic Frameworks (MOFs) and Materials Derived from MOFs as Catalysts for the Development of Green Processes

Gonzalo Valdebenito, Marco Gonzaléz-Carvajal, Luis Santibañez and Patricio Cancino *

Facultad de Ciencias Químicas y Farmacéuticas, Universidad de Chile, Santiago 8380000, Chile;
gonvaldebenito@ug.uchile.cl (G.V.); mgonzalezcar96@gmail.com (M.G.-C.); santibanezcamposl@gmail.com (L.S.)
* Correspondence: pcancino@ciq.uchile.cl

Abstract: This review will be centered around the work that has been reported on the development of metal–organic frameworks (MOFs) serving as catalysts for the conversion of carbon dioxide into short-chain hydrocarbons and the generation of clean energies starting from biomass. MOFs have mainly been used as support for catalysts or to prepare catalysts derived from MOFs (as sacrifice template), obtaining interesting results in the hydrogenation or oxidation of biomass. They have presented a good performance in the hydrogenation of CO_2 into light hydrocarbon fuels. The common patterns to be considered in the performance of the catalysts are the acidity of MOFs, metal nodes, surface area and the dispersion of the active sites, and these parameters will be discussed in this review.

Keywords: metal–organic frameworks; heterogeneous catalysis; carbon dioxide; biomass; hydrogenation; oxidation; Fisher-Tropsch

Citation: Valdebenito, G.; Gonzaléz-Carvajal, M.; Santibañez, L.; Cancino, P. Metal–Organic Frameworks (MOFs) and Materials Derived from MOFs as Catalysts for the Development of Green Processes. *Catalysts* **2022**, *12*, 136. https://doi.org/10.3390/catal12020136

Academic Editors: José Ignacio Lombraña, Héctor Valdés and Cristian Ferreiro

Received: 31 December 2021
Accepted: 19 January 2022
Published: 22 January 2022

Publisher's Note: MDPI stays neutral with regard to jurisdictional claims in published maps and institutional affiliations.

1. Introduction

In recent years, the world has seen a significant increase in temperature due to different pollutants present in the atmosphere. These pollutants are responsible for the famous greenhouse gas effect [1]. CO_2 is one of the pollutants found in greater concentration in the atmosphere and its emission is mainly due to the use of fossil fuels (heating, automobiles, industrial chimneys) [1,2]. Today there are many proposals to generate a cleaner and healthier environment for living, but one of the proposals that attract the most attention is the conversion of CO_2 to hydrocarbon molecules, which can later be used as fuel in different areas [3–5]. This proposal largely solves the problem of atmospheric pollution on the planet and opens up new areas for the development and research of renewable energies [4].

One of the main problems found in the conversion of CO_2 into hydrocarbon is the need for high-energy processes [5]. Different chemical processes have been investigated that can help to implement a fast and cheap process for the synthesis of renewable energies from CO_2, such as the use of different types of raw material catalysts [6].

In the last few decades, the valorisation of biomass has garnered great interest. This is due to the fact that by using biomass, one is able to obtain valuable chemical products and sustainable biofuel. The production of this kind of product aims to complement conventional energy sources, especially fossil fuel used in most types of transport [4]. Biomass can be classified into two groups: edible, which is rather scarce, and the much more abundant non-edible biomass. Among the edible kind, we find starch and oils, which are created by processing oilseeds such as rapeseed, miscanthus or soy, and can be transformed into biofuel, also known as biodiesel [7]. This fuel can also be produced from recycled oils obtained from food and gastronomic industries. Fat acids in the oils from both sources, fresh and recycled, are converted into biodiesel by going through a transesterification process that, along with the esters corresponding to the biodiesel, yields glycerol, a compound that is produced in quantities exceeding its industrial demand [8]. Because of this, the refining and upgrading of glycerol are desirable in order to prevent this by-product from turning into waste and harnessing the biomass completely [8]. Glycerol is considered a platform compound, that is, a chemical species from which other species with added value can be formed, such as diols or glycerol carbonate, which are used in various industries. However, the chemical reactions involved in the conversion of glycerol usually need to be carried out in the presence of a catalyst. It is common to use homogeneous catalysts for these reactions [9].

With respect to the non-edible biomass, the most abundant is lignocellulose. It is part of the structural basis of every plant, so it can be obtained from crops, farming waste or forest waste [10]. Lignocellulosic biomass is divided into three parts: cellulose, hemicellulose and lignin, and for their separation, different methods have been employed, such as thermal methods, pyrolysis or gasification, all of which require high amounts of energy and are not very selective [10]. Fermentation is a more selective method, but it is not efficient in terms of yield. The catalytic conversion of biomass is a better method since the chemical specificity of catalysts enhances the selectivity and, when used in optimal conditions, can also enhance the yielding rate while reducing the required energy [11].

Heterogeneous catalysts are the most used for the conversion of lignocellulosic biomass, particularly supported catalysts. This kind of material consists of a catalytically active molecule, mostly transition metal species, hosted into the structure of a different and chemically stable material known as the support [12,13]. Different compounds have been used as support, such as alumina, zeolites or activated carbon. This provides supported catalysts with an augmented surface area, which improves the substrate/catalyst interaction [14,15].

Another kind of catalyst is that of coordination polymers, among which metal–organic frameworks (MOFs) are the most used. MOFs are classified within the group of porous materials. These compounds are formed by two important parts that define the structural base: the inorganic moiety and the organic linker [16]. The inorganic part corresponds to the metallic center, which presents simultaneous coordinated bonds with different ligands of organic nature. The structure of these species presents porosity, which is the fundamental characteristic of metal–organic frameworks. The existence of pores is mainly due to the geometrical network that is formed between the bonds of the organic ligands and the metallic centres. The length of the organic ligands and the extensive bonds formed with the metallic centres contribute greatly to the size and diameter of the pores present in the MOF, thus determining the catalytic activity [16,17]. On the other hand, the volume of the pores and the specific surface area that these ligands provide in the network characterize MOFs as excellent immobilizers for molecular catalysts.

The physical and chemical characteristics of these compounds can be modified, as different types of MOF can be formed by using different metallic centres and organic ligands, thus generating greater selectivity towards a certain chemical reaction. All these properties allow the synthesis of a wide range of compounds with different characteristics, which will help to choose the most suitable type of MOF for various reactions [3–5]. Regarding the organic ligands used for the synthesis of MOFs, the most common are polyamines, phosphonates, crown ethers and carboxylates (Figure 1) [16].

The use of these compounds is quite varied because they can become very versatile or specific, depending on the area in which their use is required. Some of the applications can be as adsorbent materials in aqueous solutions, due to their properties and structural characteristics, in addition to the low cost [18]. Within the novel research and technologies related to the use of new types of energy from renewable sources, the use of MOFs as electrochemical capacitors and in fuel electrodes for rechargeable batteries has been developed [17].

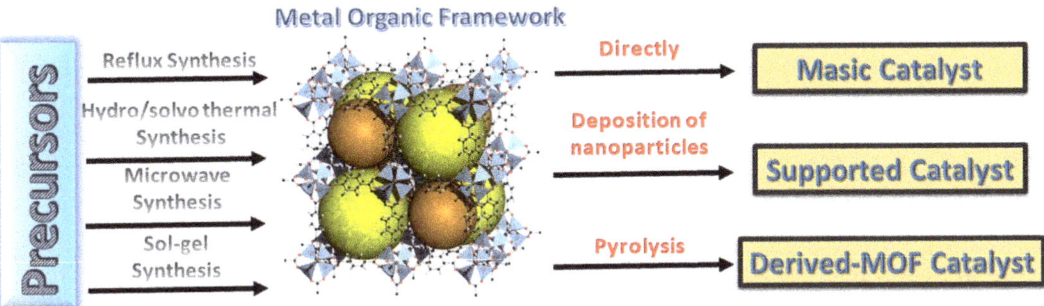

Figure 1. Summary of main routes to obtain catalysts based on MOFs.

Besides, the structure and pores of MOFs can be modified by post-synthetic procedures that allow the material to be more selective, determining greater efficiency in the CO_2 adsorption process; these modifications lead to an efficient catalytic process [18–21]. The use of MOFs in heterogeneous catalysis has been of great importance due to their physical and chemical properties; these greatly improve the parameters necessary to be considered for a useful catalyst [18]. The properties of MOFs that stand out are the high thermal stability, the high porosity that serves as an active site for chemical reactions and the enormous regenerative capacity that allows the reuse of these compounds, which finally lead to a lower total cost for the catalytic reactions [19]. Altering the structure of a MOF is not such a complex process and it facilitates the addition of different ligands to the structure, thus making the network more useful in different scenarios, such as the solution medium in which the reaction is taking place [20,21]. These processes allow MOFs to be used in many areas, such as electrocatalysis, photocatalysis and biocatalysis [22–24].

Different ways of covering the need to create and use various types of renewable energies, such as fuels and biofuels, by the use of CO_2 and biomass as the main raw material will be highlighted in the following sections. For the synthesis and chemical conversion of CO_2, catalysts will be mentioned based on the properties of metal–organic frameworks, which affect different variables that significantly reduce the reaction parameters. In this way, MOFs are shown to decrease the monetary and energy costs of the processes, which are involved in the transformation of CO_2 and biomass into valuable products.

2. MOFs for Conversion of CO_2 into Light Hydrocarbons Fuels

Carbon dioxide is a very stable molecule, comparable to water ($\Delta G°_f = -396$ kJ/mol). Due to this fact, the use of CO_2 as a precursor to obtaining valuable products requires high energy; especially CO_2 conversion to light hydrocarbon fuels (LHCF) [25,26].

An excellent alternative to producing renewable fuels starting with CO_2 is using catalysts, thus reducing the carbon footprint of the processes. The heterogeneous catalysts offer an alternative to decreasing the high activation energies associated with CO_2 activation to produce renewable fuels. Zeolites, oxides, nanoparticles and supported catalysts have shown promising results in the preparation of LHCF by CO_2 reduction. This type of catalyst is capable of controlling the selectivity to methane, C_2–C_8 paraffins and olefins, with a detriment in the production of CO [27–35].

MOFs have been shown to be very active catalysts; they are useful in different chemical reactions to produce valuable products. [36–39]. However, in this particular field, they have not been used very often, despite having many attractive characteristics for the conversion, for example, of CO_2 into LHCF. These characteristics include a high surface area, high pore size, Brönsted acid sites and a well-defined position of the active sites in the crystalline network.

Several researchers have worked using MOFs in CO_2 conversion into LHCF [40–43]. Generally, in these reactions, MOFs have been used as supporting material or to manufacture MOF-derived catalysts by pyrolysis. Particularly, the generation of MOF-derived catalysts by pyrolysis offers interesting characteristics when compared to the classical supported catalysts, that is, high dispersion of the active sites, thermal stability and surface area.

The performance, benefits and future perspective of the use of MOFs or MOF-derived catalysts to produce light hydrocarbon fuels starting with CO_2 will be presented and commented on in the following section.

2.1. MOFs Used as Supporting Materials for Catalysts

As stated above, MOFs have been used as supports for catalysts, which are then utilized to convert CO_2 into LHCF. One of the most studied catalytic systems is metallic nanoparticles (NPs) supported on MOFs since the porous nature of MOFs permits them to accommodate the NPs.

Hu et al. [44] prepared five catalysts based on MIL-53(Al) and ZIF-8(Zn), using different modifications in the syntheses in order to obtain a variety of morphologies. All catalysts were charged with α-Fe_2O_3 NPs. The most remarkable point was that, despite the conversion for all catalysts being lower than 30%, the selectivity to hydrocarbons was very high, between 75–80%. Additionally, the performance of the MOFs as the supporting material was compared against γ-Al_2O_3 and α-Fe_2O_3 (without support) as control reactions. The results showed clear differences associated with user support; the conversion and the selectivity to hydrocarbons were higher when using MOFs as support than that of the controls. Moreover, when comparing the two MOFs, it was possible to observe some differences relative to the produced amount of methane and olefins. It became evident that the selectivity was sensible to the acidity of the MOF used in each catalytic reaction. Moreover, the conversion was also modulated by the morphology of the corresponding MOF and particle size of the catalyst and surface area. The surface area was 1535 $m^2\,g^{-1}$ for MIL-53(Al) and 1918 $m^2\,g^{-1}$ for ZIF-8(Zn), with ZIF-8(Zn) being more active than MIL-53(Al) as a catalyst.

Tarasov et al. [45] used MIL-53(Al) as support, but they embedded the material with Co NPs as active metal sites. The obtained results were similar to those obtained by Hu et al. [44]; the conversion was between 25–38% and the selectivity to hydrocarbons was 74–91%. The study focused on the effect of the reaction conditions, mainly the temperature, which ranged from 260 to 340 °C. As the temperature was raised, the conversion and the selectivity increased. The amount of methane increased with higher temperatures, while the obtained amount of C_2–C_5 and C_{5+} was reduced. The Brönsted acid sites present in the catalyst permitted the regulation of the relationship between paraffins and olefins. The proposed mechanism related to this catalytic reaction assumed that the Co NPs generated interactions with CO_2, thus reducing it to CO. This step was followed by the Fisher-Tropsch reaction (FTR), which produced light hydrocarbons (Figure 2).

The hydrogenation of CO_2 was studied by Zhao et al. [46] using a Zr-based metal–organic framework, $Zr_6O_4(OH)_4(BDC)_6$ (UiO-66) (BDC = 1,4-benzenedicarboxylate), as support for Ni NPs generated by an in situ reduction, obtaining highly monodispersed ultrasmall nanoparticles (1.6–2.6 nm) (Ni@UiO-66). The catalysts were prepared by modifying the load of Ni NPs from 5 to 30 wt%. The dependence of the conversion with temperature was studied for the hydrogenation of CO_2 to CH_4; the highest conversions were achieved when the loadings were 20, 25 and 30 wt% at 340 °C. The selectivity was tested for a reaction time of 100 h at 300 °C, using a load of 20% (20Ni@UiO-66), obtaining 58% of conversion and 100% of selectivity to CH_4 [46].

A similar study by Zhen et al. [47] reported the synthesis of Ni@MIL-101 with the same loadings of Ni NPs. The catalyst had a high surface area of 3297 $m^2\,g^{-1}$, which showed an excellent performance to methanation of CO_2, achieving 100% of the conversion with a 100% selectivity to CH_4 at 300 °C. On the other hand, Zhen et al. [48] also used MOF-5 as support for Ni NPs. The catalyst exhibited a surface area of 2961 $m^2\,g^{-1}$ at 300 °C; the conversion was 57%, obtaining exclusively CH_4 (Table 1).

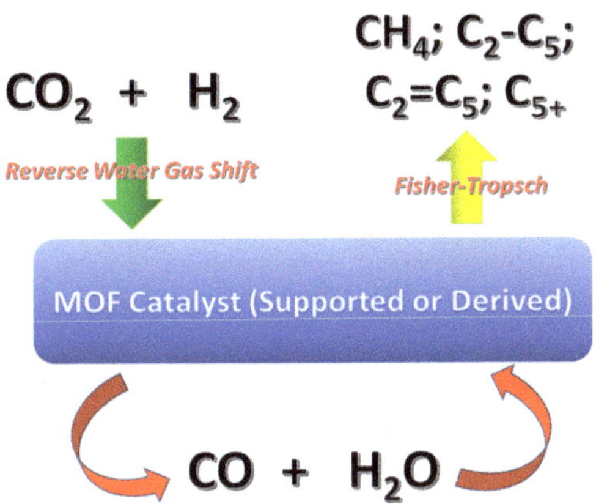

Figure 2. Scheme of the reactions to convert CO_2 in LHCF using MOFs as catalysts.

Table 1. Summary of catalytic results for the hydrogenation of CO_2 with different supported catalysts.

Catalyst	CO_2 Conversion (%)	Mass of Catalysts (g)	Temperature (°C)	Selectivity (%)					Reference
				CO	CH$_4$	C$_2$–C$_4$$^-$	C$_2$–C$_4$$^=$	C$_5$$^+$	
MIL-53(Al)/Fe$_2$O$_3$	20	1	300	20	41	34	0	5	[44]
ZIF-8(a)(Zn)/Fe$_2$O$_3$	23	1	300	23	21	23	21	12	[44]
ZIF-8(b)(Zn)/Fe$_2$O$_3$	25	1	300	21	22	30	15	12	[44]
ZIF-8(c)(Zn)/Fe$_2$O$_3$	30	1	300	17	25	40	8	10	[44]
10%Co/MIL-53(Al)	25	0.8	260	26	26	22	0	26	[45]
10%Co/MIL-53(Al)	30	0.8	300	19	28	24	0	28	[45]
10%Co/MIL-53(Al)	38	0.8	340	9	32	27	0	32	[45]
5Ni@UiO-66	8	0.1	300	#	#	#	#	#	[46]
10Ni@UiO-66	13	0.1	300	#	#	#	#	#	[46]
15Ni@UiO-66	25	0.1	300	#	#	#	#	#	[46]
20Ni@UiO-66	58	0.1	300	100	0	0	0	0	[46]
25Ni@UiO-66	50	0.1	300	#	#	#	#	#	[46]
30Ni@UiO-66	41	0.1	300	#	#	#	#	#	[46]
20Ni@MIL-101	100	0.2	300	100	0	0	0	0	[47]
10Ni@MOF-5	57	0.2	300	100	0	0	0	0	[48]

The authors do not give information about the selectivity.

The hydrogenation of CO_2 can also produce MeOH. The generation of MeOH, starting from CO_2, was tested by Somorjai et al. [49] using Cu NPs (18 nm) over two different MOFs as support, UiO-66, MIL-101 and ZIF-8. The reactions were carried out at 175 °C and 0.4 g of the catalyst. The results were interesting since only CuUiO-66 was able to catalyze the reaction, obtaining 5% conversion and 100% selectivity. The X-ray photoelectron spectroscopy data obtained on the surface of the catalyst showed that the presence of Zr oxide clusters or Zn oxides generated a strong interaction with the Cu NPs. This interaction allowed the coexistence of different oxidation states of copper, which play different roles in the mechanism of the production of methanol.

In a similar way, Lin et al. [50] developed Cu/ZnOx NPs supported on UiO-bpy in order to take advantage of the synergistic interaction that occurs between copper and zinc. The reaction was conducted at 250 °C and 0.1–0.3 g of catalyst; the conversion was 17.4% and the selectivity to MeOH was 85.6%. The importance of the Zn was demonstrated using only Cu in the reaction; the performance of the catalyst decreased. On the other hand, a TPD of CO_2 study revealed that the vacancies or unsaturated Zr sites of the support (UiO-bpy) interact with the CO_2 in the catalytic cycle. These open sites can adsorb CO_2 and accept dissociated hydrogen atoms to form Zr-H, likely playing a key role in catalytic CO_2 hydrogenation.

2.2. Catalysts Obtained by Pyrolysis of MOFs

Another synthetic route to obtaining heterogeneous catalysts for CO_2 activation is the preparation of these by pyrolysis of MOFs. Several papers can be found in the literature reporting this technique, and these will be summarized in this section.

Liu et al. [51] synthesized Fe-based catalysts by pyrolysis of Fe-MIL-88. The catalysts contained a high load of Fe, in comparison with classical supported catalysts. The authors controlled the temperature and the time of the syntheses of these catalysts, with the loading of Fe in the obtained catalysts resulting between 34–51%. The used conditions in the preparation of these catalysts affected the mean particle size and dispersion. The authors remarked that the dispersion of the active sites was more homogeneous than in the classical supported catalysts. The results were promising; the best performance was achieved with a 46% conversion and 82% selectivity to light hydrocarbons, with C_2–C_4 paraffins (34%) being the main products. Compared with the aluminum oxide supported catalysts, these MOF-derived catalysts were capable of obtaining higher CO_2 conversion and better selectivity of valuable products. The MOF-derived catalysts showed larger stability than aluminum oxide supported catalysts due to the high dispersion of metal particles in the matrix.

Another Fe-based catalyst was synthesized by pyrolysis of (Fe-BDC) MOF (BDC = 1,4-benzenedicarboxylate). The hydrogenation of CO_2 after 12 h of reaction was promoted by sodium, together with the MOF-derived catalyst (Na-Fe@C). The catalyst exhibited interesting results, showing a conversion close to 40% and a selectivity of 80% to hydrocarbons, with C_{5+} as the main product [52]. The authors also proposed a combination of hierarchical zeolite H-ZSM-5 with the prepared MOF-derived catalyst, expecting to obtain aromatics as the main products. The acid environment contributed by the zeolite did not produce significant changes in the conversion; however, the new reaction environment permitted the isolation of aromatic compounds. Moreover, the aromatic compounds were the main products of the catalytic reaction. Different concentrations of zeolite were tested and compared, and the most active one used 0.2 M of H-ZSM-5 together with Na-Fe@C, with selectivities close to 30% for C_{5+} and 42% for aromatic compounds.

A similar process was tested by Martin, Dusselier et al. [53] by combining a chabazite-type zeolite (CHA) with In-ZrMOF, starting from (Zr)UiO-67-bipy-In, which was calcined at 580 °C for 5 h. The new material presented good dispersion and strong acidity, which allowed for the couple activation of CO_2. The catalytic reaction was carried out at 375 °C, and the products were obtained with space–time yields of 0.1 mol of CO_2 converted to light olefins per gram of In per hour. However, the selectivity to CO was high at 70%, and the remaining 30% was distributed between olefins (25%), with several minor products also distributed among paraffins (2%), methane (2%) and oxygenated products (1%). In this case, the acidity of the material allowed the breaking of the C-O bond, thus activating CO_2. Nevertheless, the following step corresponding to the hydrogenation of CO_2 was not successful due to the high amount of reported CO.

Ramirez et al. [54] reported the preparation of several catalysts by pyrolysis of Basolite F300 (Fe-MOF). These catalysts contained a loading of ca. 41% of Fe. The set of catalysts was prepared by the addition of 0.75% of fourteen different promotors and was used to evaluate the improvement of the reverse water gas shift (RWGS) activity (Fe, Cu, Mo, Rh, Pt, Ni) [55–60] or olefin selectivity (Li, Na, K, Mg, Mn, Ca, Zn, Co) [61–65]. The results permitted us to conclude (Table 2) that the use of the different metals as promotors could activate or inhibit the reaction. For example, K, Pt and Cu activated the reaction, increasing the conversion in comparison with the used control catalysts. In contrast to the catalysts without promotors, with Mo, Ca and Mn, the conversion decreased.

Table 2. Summary of catalytic results of Fe/C catalysts using different metals as promotors at 320 °C.

Catalyst	Conv. (%)	Selectivity (%)				
		CO	C_1	C_2–C_6	C_2=C_6	C_{7+}
Fe/C	24	39	39	12	1	9
Fe/C + Fe	25	40	38	12	1	9
Fe/C + Cu	29	23	53	15	2	7
Fe/C + Mo	22	50	30	9	1	10
Fe/C + Li	26	38	40	12	2	8
Fe/C + Na	27	38	32	15	4	11
Fe/C + K	35	17	18	7	40	18
Fe/C + Mg	22	48	39	9	1	3
Fe/C + Ca	24	40	43	13	1	3
Fe/C + Zn	23	41	45	13	1	0
Fe/C + Ni	26	34	52	13	1	0
Fe/C + Co	26	32	51	14	2	1
Fe/C + Mn	23	45	38	8	1	8
Fe/C + Pt	30	22	51	19	3	5
Fe/C + Rh	25	17	63	19	1	0

The best promotors were K for the olefin selectivity and Rh for the RWGS reaction, while the worst was Mo for this same reaction. The metals that resulted as promotors considerably enhanced the selectivity towards hydrocarbons, varying it between 83% and 50%. Thus, the most active catalyst was that which contained K as promotor (conversion 35% and selectivity 83%). Furthermore, when this catalyst was used, a higher amount of olefins (C_2–C_6) was obtained. The reaction conditions using this catalyst were investigated, modulating the loading of K, the temperature of the reaction and time. The reaction time did not present a significant effect and the loading of K did not change the conversion when the loading was above 0.75%, while a lower loading of K decreased the conversion and selectivity to HC. Finally, the optimal temperature resulted at 375 °C, achieving 40% of conversion. Data corresponding to the catalytic reactions carried out at 320 °C are summarized in Table 2.

Pyrolyzed MOFs have also been used in methanation reactions. From pyrolyzed ZIF-67 (Co), Co-xH_2/Ar was obtained with a surface area of 125 m^2 g_{Co}^{-1}. [66]. The methanation reaction was conducted at 300 °C; the conversion achieved a 48.5% with a 97% of selectivity to CH_4. In the same way, Ni@C was prepared by pyrolysis of Ni-MOF, and the surface area was shown to be 177 m^2 g^{-1}. The temperature dependence of the reaction was studied, and the conversion achieved 100% when the temperature of the reaction exceeded 300 °C, with 100% selectivity [67]. Additionally, to understand the reaction mechanism, in situ DRIFTS spectra were recorded, in order to show the catalyst co-interacting with CO_2 and H_2 at different temperatures (between 50 to 400 °C). From the results, it was possible to conclude that an intermediary species, Ni–CO_2, was formed due to the interaction between the catalyst and CO_2, followed by the reaction of dissociated H to form Ni–C–$(OH)_2$ intermediates. These intermediates continued reacting with dissociated H to produce CH_4 and H_2O.

2.3. Perspectives and Conclusions

The generation of renewable fuels from CO_2 is a field intensely explored in recent years. However, the direct use of MOFs as catalysts for this reaction is scarce and opens up a great perspective of possibilities, mainly in the generation of light hydrocarbon fuels from CO_2. The recent reports discussed in this work show the first steps that have been taken in this area, with interesting and promising results. Two strategies have been used to convert CO_2 in LHCF: MOFs used as supporting material for catalysts, and catalysts obtained by pyrolysis of MOFs.

When MOFs are used as support, the intrinsic characteristics of MOFs are more determinant in the result, modulating the conversion and the selectivity by surface area, dispersion or/and acidity. Moreover, the acidity of the support modulates the selectivity, for example, the ratio between the obtained olefins and the paraffins. It is important to consider that, at the time of designing a MOF as support, the use of a metal ion that will provide an acidic environment has to be considered in order to increase the production of hydrocarbons over CO. Additionally, the utilization of specific organic linkers, which will increase the porosity and the surface area and directly affect the conversion in these processes, has to be also evaluated.

On the other hand, the use of MOFs to prepare catalysts by pyrolysis will completely change the original properties of the MOF. Moreover, the obtained new catalyst will present high dispersion, low particle size and high surface area. The main point in the design of the MOF is the selection of the active metal ion. The cited authors have used Fe or Co to prepare this type of catalyst due to the fact that these metals are active in the RWGS and the Fisher-Tropsch reaction.

An unexplored way to face these reactions is to directly test MOFs as catalysts, combining the acidity of an active metal ion and the organic linkers that will generate high surface areas. The combination of metals ions to produce bimetallic MOFs may also be a promising area of focus to enhance the conversion and the selectivity to LHCF. Moreover, these bimetallic MOFs may also be used in pyrolysis processes in order to obtain new bimetallic nanoparticles with high dispersion, activity and selectivity.

In summary, MOFs have shown promising results, presenting high selectivity to LHCF. Nevertheless, they have not been extensively used in this field and, therefore, remain an open field of research. The use of catalysts is very important to generate sustainable processes to convert CO_2 into hydrocarbons, decreasing the carbon footprint and helping to mitigate the greenhouse effect. The use of MOFs in these catalytic reactions still presents many challenges, thus opening a number of new areas of research.

3. MOFs and Their Application in Hydrogenation and Oxidation Reactions for Biomass Valorisation

This section will discuss the structural and chemical properties of MOFs, which must be considered when choosing them as support or catalysts for biomass upgrading reactions.

3.1. Hydrogenation

Platform compounds found in biomass can undergo diverse transformations depending on the type of reaction in which they are involved. Hydrogenation or reduction reactions are useful to obtain alcohol from carbonyl compounds [68] or saturated hydrocarbons by cleavage of C–O bonds [69,70]. Herein, we present the properties of MOFs, which directly influence the behaviour of hydrogenation reactions of biomass platform compounds (Figure 3).

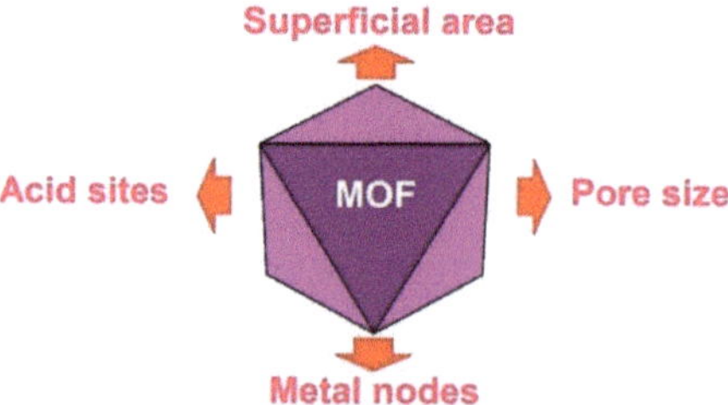

Figure 3. Relevant parameters to be considered in the preparation of MOFs, which will be used as catalysts in biomass transformation.

3.1.1. Surface Area and Pore Size

Ruiqi et al. [68] reported the use of MOFs designated as UiO-66 and H-UiO-66 as support for different transition metal nanoparticles (Au, Pt, Cu and Pd). These materials were used as heterogeneous catalysts in the hydrogenation of furfural, using molecular hydrogen to obtain furfuryl alcohol. The best results were achieved in relatively mild conditions (60 °C and 0.5 Mpa) using water as solvent and Pd/H-UiO-66 as the catalyst, yielding a complete conversion and >99% selectivity to furfuryl alcohol with a TOF of 66.7 h^{-1}. These good results are attributable to the high dispersion and uniform distribution of the nanoparticles (NPs) over the highly porous MOF. This allows reducing the mass diffusion resistance and grants great accessibility for the substrate to interact with the active sites of the catalyst (Pd NPs). ATR-IR spectroscopy showed that the principal interaction between the substrate and the catalyst occurred through the C=O and C=C bonds (stretching bands at 1670 cm^{-1} and 1540 cm^{-1}, respectively), which explains the high selectivity of the catalytic system [71]. Since the hydrogenation of the carbonyl group of furfurals is necessary for alcohol formation, the interaction between carbonyl groups from the substrate and the Pd NPs from the catalyst is fundamental for high conversion and selectivity [72].

Yuan et al. [71] also studied MOF UiO-66 as a support for several metallic NPs. The obtained results indicated that the best catalyst was Ru/UiO-66, which showed a furfural conversion of 94.9% with 100% selectivity to furfuryl alcohol. The Ru/UiO-66 catalyst presented the highest surface area, as compared to the other catalysts studied in this work. Another interesting feature was that the Ru NPs in the catalyst were oxidized in contact with air. However, the obtained RuOx species were again reduced under the reaction conditions, giving rise to the active sites for the hydrogenation of the substrate. The authors suggest that the good results obtained were attributed to the small size of the Ru NPs (less than 3 nm), the interaction of the Ru NPs with the support and the redox properties of the RuOx species.

Jian et al. [73] used Pd@UiO-66, Pd@UiO-66–NH_2 and Pd/UiO-66–NH_2 in the catalytic hydrogenation of levulinic acid (LA) to obtain γ-valerolactone (GVL). Results showed that a good dispersion of Pd NPs throughout the MOF improves the conversion. The worst dispersion was found in Pd/UiO-66–NH_2, where Pd NPs were located along the edges of the MOF instead of inside the MOF. This, because the traditional impregnation technique yields Pd NPs with a 5.8 nm diameter, was too big to penetrate the MOF structure. On the other hand, Pd@UiO-66 and Pd@UiO-66–NH_2 were synthesized with a double solvent method, using water and n-hexane [74], which produced Pd NPs with a mean particle size of 1.1 nm, thus enabling them to enter the MOF structure. This procedure enhanced the dispersion of the NPs. It is worth noting that the NH_2 group had a notorious influence on the loading of Pd contained within the MOF [75]. The percentage of Pd present in the catalyst was determined by ICP-AES, showing a 0.57% for Pd@UiO-66, and 0.94% Pd@UiO-66–NH_2 (with an expected theoretical value of 1%). These results were explained by the coordinating capacity of the –NH_2 group, which retained the Pd NPs inside the MOF. Under the selected conditions for the catalytic studies, Pd@UiO-66–NH_2 presented the best results both in structural properties, including Pd loading and dispersion, and in catalytic activity, converting 98.2% of the substrate with a selectivity of 100% towards the GVL after 2 h of reaction (hydrogen pressure, 2 MPa; temperature, 140 °C).

The proposed mechanism for the transformation of lactic acid to GVL sets the necessity of sites that allow the activation of hydrogen and acid sites, including both Lewis and Brönsted, which promote the LA dehydration. The Pd NPs take care of the first step, activating the hydrogen when it disperses and adsorbs on them, while the Zr_6 clusters generate the Lewis and Brönsted acid sites that interact with the –OH group from the LA [76–78]. When both molecular hydrogen and LA have interacted with the catalyst, the first hydrogenation step occurs, producing 4-hydroxipentanoic acid adsorbed in the NP@MOF system, followed by acid catalysed dehydration allowing the intramolecular esterification thus yielding GVL [73]. This proposed mechanism remarks the importance of good NPs dispersion along the MOF structure since it is a determinant for the hydrogenation step. Another relevant factor is the existence of the acid sites present in the MOF structure, which makes possible not only the activation of LA but also the final dehydration step, which is a common important factor in biomass upgrading reactions.

3.1.2. Acidity

As mentioned, acid sites play a very important role in the properties of a catalyst. For example, Xiang et al. [79] studied the chemical transformation of furfuryl alcohol into 4-hydroxi-2-cyclopenthenol (HCP) and 2-cyclopenthenone (CPE), using FeZn, FeZn-P, Cr-MIL-101 and Fe-MIL-100 as catalysts. The Lewis acid sites are necessary for this reaction since the substrate–catalyst interaction occurs on these sites to promote dehydration, aldol condensation and hydrogenation of the substrate. In the FeZn and FeZn-P catalysts (double-metal cyanide kind of MOFs or DMC), the Zn^{II} ions correspond to the acid sites, while for Cr-MIL-101 and Fe-MIL-100, the Cr^{III} and Fe^{III} are the acid sites, respectively. The catalysts acidity was determined by FT-IR, measuring the absorption bands of pyridine chemisorbed on the Lewis (1450 cm^{-1}) and Brönsted (1540 cm^{-1}) acid sites [80]. This study determined that both FeZn and FeZn-P catalysts presented a Lewis/Brönsted acid site density ratio of 10.2 and 12.8, respectively, while Cr-MIL-101 and Fe-MIL-100 only presented the Lewis acidity with acid site densities of 2790 and 3072 $mmol \cdot g^{-1}$, respectively [79].

For the transformation of furfuryl alcohol, Brönsted acids such as H_2SO_4, HCl, *p*-toluensulfonic acid, H-ZSM-5 (an acidified commercial zeolite), hyaluronic acid or Amberlyst-15 (a strongly acid exchange resin) readily catalyse the hydrolysis reaction, favoring the production of levulinic acid [81,82]. For example, when using Amberlyst-15, only a 16.5% selectivity towards the HCP is achieved, yielding mostly levulinic acid as the product. On the other hand, all the catalysts used by this research group presented Lewis acid sites and exhibited higher yields for the HCP product. Cr-MIL-101 and Fe-MIL-100 presented a relatively lower conversion, being below 85%, with selectivity values ca. 70% versus 88.8% of conversion (approximately 86% selectivity towards HCP) achieved with FeZn, and 95.2% conversion (96.2% selectivity towards HCP) achieved with FeZn-P. This behaviour was attributed to the weaker acidity of Cr^{III} and Fe^{III} ions as compared to that of the Zn^{II} ions.

When comparing the two DMC catalysts, the better results obtained for FeZn-P are attributed, according to the authors, to the higher surface area (42.6 $m^2 \cdot g^{-1}$) and acid density (1349 $mmol \cdot g^{-1}$). However, it is worth noting that while Fe- and Cr-based catalysts present surface areas 20 and 70 times higher than that of FeZn-P and higher Lewis acid densities as well, their conversions and selectivity still remain lower than FeZn-P. This may be related to the absence of Bronsted acidity in Cr-MIL-101 and Fe-MIL-100, which would have a significant impact on the catalytic activity of both materials in the furfuryl-alcohol conversion.

The two DMC catalysts were used to study the conversion of furfuryl alcohol to CPE, using from 2 to 6 MPa, augmenting the CPE yield from 20.4% to 61.5%, the highest achieved conversion. This indicates that the DMC materials can be used as acid catalysts and hydrogenation catalysts. The authors point out that DMC catalysts have been primarily used as acid catalysts, so, by being active in hydrogenation catalysis, they become bifunctional catalysts for both types of reactions. This was corroborated with the study of the furfuryl alcohol conversion to CPE using other catalysts in the presence of molecular hydrogen. The acid catalysts Amberlyst-15 and Al-MCM-41 were not able to catalyse this reaction, while the bifunctional DMC and MIL catalysts reported in this work displayed a considerable CPE yield with the used experimental conditions, thus confirming the necessity of both acid and hydrogenation catalytic active sites.

A point worth considering is the fact that the acidity of a MOF can be modified with post-synthetic methods, such as the dispersion of Brönsted acid molecules (for example, sulfuric or phosphoric acid) throughout the MOF. Phosphoric acid, for example, is able to host inside the pores of MIL-101, enhancing its activity in the catalytic hydrodeoxygenation of biomass oleic acid to yield saturated hydrocarbons [68]. In this work, the concentrations of phosphoric acid dispersed in the Pt@MIL-101 supported system were varied, resulting in a dramatic enhancement of the oleic acid conversion from 15% to 57% when the concentration of phosphoric acid was raised from 30 mM to 60 mM. Further increase of the concentration of phosphoric acid produced more discrete augmentations (conversions of 68% and 86.5% for 120 mM and 240 mM, respectively). A higher content of phosphoric acid hosted inside the MIL-101 pores increased the Pt particle size distribution values supported on the MOF. Moreover, the presence of this Brönsted acid had a direct influence on the reaction mechanism, switching from a decarbonylation/hydrogenation system to a hydrodeoxygenation and favouring the production of heptadecane over octadecane.

Another example is presented by Zhang et al. [83], who reported the use of MIL-101–SO_3H(x) as catalysts (x = 25, 50, 80 and 100) that have different amounts of –SO_3H groups. These catalysts were studied in the GVL hydrogenation reaction to obtain ethyl valerate, with 98% conversion and 83% selectivity using H_2 at 250 °C for 10 h. The authors mention that the incorporation of a sulfonic acid functional group (–SO_3H) in the MOF is favorable due to the properties that this group confers to the catalyst, such as strong acidity, favorable textural properties and high hydrothermal stability. The results indicated that a greater number of –SO_3H acid groups produced a greater conversion of GVL. On the other hand, the obtained good results were attributed to the fact that the palladium supported on the acid materials showed bifunctional catalytic properties, allowing an increase in the density of acid sites. This resulted in increased activity in the opening of the GVL ring and the subsequent hydrodeoxygenation of the C=O bond, allowing the intermediate ethyl pentenoate to be obtained, which was then hydrogenated by the Pd particles to form ethyl valerate.

3.1.3. Metal Nodes

The work by Anil et al. [84] indicates that metal node coordination plays an important role in the catalytic activity of a MOF. The catalytic hydrogenation of furfural (FUR) was studied, using Zr-based MOFs (Zr-MOFs) such as UiO-66, UiO-67 and DUT-52 as catalysts, and 2-propanol (2P) as the hydrogen source. The reaction, in absence of the catalyst, presents a conversion under 2%, and from 2% to 5% when carried out in the presence of these Zr-MOFs. The values of conversion do not vary much, whether the reaction is catalysed or not, nor even when the used catalysts are varied, despite the fact that these have different structural properties [85]. While UiO-67 has the highest surface area and pore size, its catalytic activity was not significantly higher than that of UiO-66 or DUT-52, thus indicating that these parameters were not determinant for the catalytic system. However, the number of metal nodes coordinated with carboxylic groups in the unit cell of different MOFs appears to be relevant. When using catalysts with fewer nodes, the conversion is improved. DUT-52 (12 Zr nodes) presented 2.1% conversion, while DUT-67 (8 Zr nodes) had 16.4% and MOF-808 (6 Zr nodes) showed a dramatic increase with 81.3% of conversion for the furfural hydrogenation. This relation between the lower number of nodes and the higher conversion values is explained by the fact that fewer nodes mean more unused space around the metal ions, allowing for better accessibility of the substrate molecules to the active Zr sites, thus yielding higher conversions.

The number of acid and basic sites is another parameter influenced by the number of metal nodes [86,87], and it is determined by TPD-NH_3 and TPD-CO_2, with MOF-808 presenting the highest values for available acid (0.85 mmol·g^{-1}) and basic (0.15 mmol·g^{-1}) sites. This is consistent with the fact that MOF-808 is the most active of all the studied Zr-MOFs.

The activation of MOF-808 with methanol modifies the catalyst metal nodes, altering its structure without affecting the crystalline or porous properties of the MOF. This treatment allows the removal of coordinated formate groups, µ3–OH and physisorbed DMF, and generates the coordination of methoxi and terminal –OH groups, in addition to forming the coordinatively unsaturated Zr sites, which facilitate the furfural and 2-propanol adsorption on the metal nodes and enhance the catalytic activity. This has been proven by comparing the results obtained for the same experimental conditions (T: 40 °C, t: 24 h, Cat/FUR/2P:

0.1/0.5/12.5) for the hydrogenation of furfural, using the as-synthesized and the methanol activated forms of MOF-808 as catalysts. The activated form surpassed the as-synthesized conversion by over 60% (27.5% versus 96.5%), corroborating the effect of activating the catalysts prior to its use [84].

The methanol activation technique was previously reported by Yang et al. [88], who studied the activation of UiO-66 and Nu-1000, corroborating the formation of methoxi groups on the metal nodes, which modify the surface chemistry of the MOFs. The same author with a different research group [89] made use of this technique for the catalytic dehydration of tert-butyl alcohol. The activated MOF was characterized by FTIR and compared with the as-synthesized material. Two characteristic ν_{C-H} absorption bands at 2931 cm^{-1} and 2866 cm^{-1}, corresponding to formate groups coordinated to the Zr nodes (Zr–OOCH), disappeared and were replaced by absorption bands at 2929 cm^{-1} and 2827 cm^{-1}, corresponding to the ν_{C-H} vibrations from methoxi groups coordinated with the Zr nodes (Zr–OCH$_3$) [90]. Thus, the substitution of the formate groups by methoxi groups necessary for the activation was confirmed. When analysing the spectrum of the MOF activated with deuterated methanol, it is possible to observe that the methoxi bands are shifted to lower energies (2203 cm^{-1} and 2060 cm^{-1}), as expected for the ν_{C-D} vibration of the deuterated methoxi coordinated group. Using spectral data, it was possible to prove that the generation of these metoxi coordinated groups is in fact produced by the methanol activation method.

The activated M-MOF-808 also proved to be an efficient catalyst in the hydrogenation reaction of carbonylic biomass derivatives to produce alcohols. Among the studied substrates, 5-hydroximethylfurfural (5-HMF), ethyl levulinate, 5-methylfurfural, benzaldehyde, 4-chlorobenzaldehyde and acetophenone were hydrogenated to their respective alcohols in mild conditions (T: 82 °C, S/Cat: 0.106), yielding conversions and selectivities above 90%. The summary of the reactions and its catalytic results are shown in the Figure 4 and Table 3.

Table 3. Different catalysts and catalytic conditions in hydrogenation reactions for biomass valorisation.

	Catalyst	Catalyst (mg)	Sudstrate	Substrate (mmol)	Hydrogen Sours	T (°C)	t (h)	Conv. (%)	Selectivity (%)	Product	Ref.
a	Pd/H-UiO-66	0.3 mol%	FUR	5.0	0.5 MPa H$_2$	60	3	100	>99	FOL	[68]
b	Ru/UiO-66	100	FUR	0.1 mL FUR in 9.9 mL H$_2$O	0.5 MPa H$_2$	20	4	94.9	100	FOL	[71]
c	Pd@UiO-66–NH$_2$	150	LA	5	2 MPa H$_2$	140	2	98	100	GVL	[73]
d	FeZn-P	100	FOL	10.2	-	150	6	95	88.2	HCP	[79]
d	FeZn-P	100	FOL	10.2	4 MPa H$_2$	150	6	95	61.5	CPE	[79]
e	Pd/MIL-101–SO$_3$H	100	GVL	10	3 MPa H$_2$	250	10	98	83	Ethyl valerate	
f [a]	Pt/P@MIL	50	Oleic acid	0.0047 (mmol/s·g$_{cat}$)	2 MPa H$_2$	300	2	75	-	HDO products	[70]
g	M-MOF-808	100	FUR	10	IPA [b] 416 mmol	40	24	96.5	88.6	FOL	[84]

[a] Catalysis in gas phase. [b] IPA: 2-propanol.

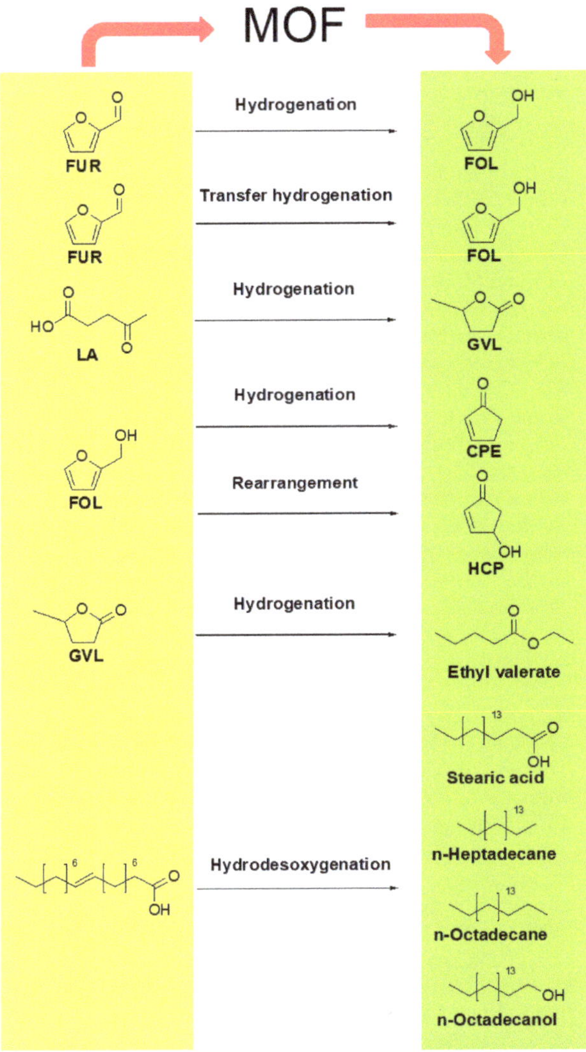

Figure 4. Main products starting from different biomass precursors.

3.2. Oxidation

MOFs as Sacrifice Templates (MOF Derived Catalysts)

As already mentioned, the use of MOFs as heterogeneous catalysts is based on the adequate structural properties of the material, such as the ordered metal ions distribution and the porosity of the MOF, which allows a good substrate diffusion in the catalytic process [91–93]. However, the use of MOFs as sacrifice templates to obtain metal nanoparticles (NPs) or supported metal oxides has become of great interest [94–96]. This method usually consists of the generation of a MOF-support or MOF-nanoparticle system, which is then calcinated or pyrolyzed [97], yielding a material with the reduced metal ions from the MOF highly dispersed along the support or as nanoparticles. This allows the combination of the porous properties of the support with the ordered dispersion of the catalytically active species, producing a metal-support interface with interesting properties.

Yu-Te Liao et al. [98] reported the preparation of the cage shaped catalysts, $Au_xPd@Co_3O_4$, used in the oxidation of 5-hydroximethylfurfural (HMF). These materials are alloys of Au-Pd nanoparticles supported on cobalt oxide and were obtained using a De novo process, reported previously by the same group [99,100]. This method mixes gold and palladium ions with the linkers and Co ions before producing ZIF-67. This yields the Au-Pd-ZIF-67 system, which is then calcinated to form the $Au_xPd@Co_3O_4$ catalysts.

The catalysts were prepared with different Au/Pd ratios, i.e., 0.1:0, 0.1:0.1, 0.05:0.1, 0.1:0.2, 0.05:0.2, 0.025:0.2, 0:0.2. The surface area and the pore size varied from 11 to 75 $m^2 \, g^{-1}$ and 10 to 55 nm, respectively. The highest performing catalyst, $Au_{0.5}Pd@Co_3O_4$ (Au/Pd ratio 0.1:0.1), presents a specific surface area of 64 $m^2 \, g^{-1}$, much lower than the original MOF specific surface area, 722 $m^2 \, g^{-1}$, due to the decomposition of the organic framework. Even when it is known that a high specific surface area enhances the catalytic activity of a MOF [97], in this case, the original structure is sacrificed to prioritize a better interaction with Co^{II} ions. The thermal treatment performed on the precursor (Au-Pd-ZIF-67) is relevant, because this method allows a uniform distribution of the metallic NPs in the Co_3O_4 cages, corroborated by TEM-EDX. Moreover, XPS analysis confirmed the presence of highly active Au^I ions in the interface between the NPs alloy and the Co_3O_4 support. The Au^I ion formation has been observed in systems where Au is embedded within reducible oxide supports [101,102]. To clarify the contribution of the MOF synthesis method upon the presence of Au^I ions, the authors compared $Au_{0.5}Pd/Co_3O_4$, obtained with the Sol-immobilization method, with $Au_{0.5}Pd@Co_3O_4$, obtained with the De novo method, showing Au^+/Au^0 ratios of 0 and 0.23, respectively. This study served to verify that the method used to synthesize the MOF affects the presence of Au^I ions. In the De novo method, the metallic NPs are surrounded by the ZIF-67 crystalline lattice that contains uniformly distributed Co^{II} ions, while in the Sol-immobilization method, the NPs are attached to the Co_3O_4 affecting the interface between the Au and the Co_3O_4. This was also proved by the Co^{III}/Co^{II} ratio in the cages of $Au_{0.5}Pd/Co_3O_4$ y $Au_{0.5}Pd@Co_3O_4$ (2.52 and 2.13, respectively), showing that there is a higher interaction between the metallic NPs and the Co_3O_4 in the $Au_{0.5}Pd@Co_3O_4$ system. A lower Co^{III}/Co^{II} ratio represents a higher interaction because it has been demonstrated that Au^0 and/or Pd^0 interact with Co^{III}, forming $Au^{+\delta}$, Pd^{II} y Co^{II} ions and decreasing the Co^{III}/Co^{II} ratio [103].

The catalytic results indicate that $Au_xPd@Co_3O_4$ catalysts present rather high HMF conversion values (from 57% to 100%) to different products, i.e., 5-hydroxymethyl-2-furancarboxylic acid (HMFCA), 2,5-diformylfuran (DFF), 5-formylfurancarboxylic acid (FFCA) and 2,5-furandicarboxylic acid (FDCA). The best results were achieved with the $Au_{0.5}Pd@Co_3O_4$, which presented a 100% conversion and 95% selectivity towards the desired product FDCA at 90 °C, using NaOH and Na_2CO_3 as basic media and 10% H_2O_2 as the oxidizing agent. The good results are attributed to the cage-like structure of $Au_{0.5}Pd@Co_3O_4$ that has a homogeneous distribution of the Au-Pd NPs covered by the cobalt ions, thus increasing the Au/Co_3O_4 interface. These structural features are necessary for the catalytic process to occur since Pd forms part of the NPs alloy, allowing an increase of Au^I ions. The presence of Au and Au^I ions produces hemiacetal and alcoxide intermediates in the catalytic process, which then interact with hydroperoxide radicals and oxygen produced in situ along the Au-Co_3O_4 interface. This allows the generation of HMFA, FFCA and FDCA [101,104,105]. The high performance of $Au_{0.5}Pd@Co_3O_4$ in the oxidation of HMF to FDCA is mainly attributed to the Au-Co_3O_4 interface that facilitates the HMFCA oxidation to FFCA. This is a slow step in the reaction, and it is relevant in the production of FDCA [106,107]. On the other hand, there is an important contribution from the hydroperoxide radicals produced from the H_2O_2, which activates the Lewis acid sites on the Co_3O_4 and Au^I and accelerates the oxidation reaction.

Feng et al. [108] also used ZIF-67 as a sacrifice template to obtain cobalt-based N-doped carbon materials (Co@C–N), using these as heterogeneous catalysts in the oxidative esterification of HMF to produce dimethyl furandicarboxilic acid (DMFDCA). Unlike the work of Yu-Te Liao et al., this group pyrolyzed ZIF-67 alone to generate Co@C–N derivatives. The thermogravimetric analysis determined the pyrolysis temperature; this is when the organic linkers decompose over 500 °C. Thus, pyrolysis temperatures of 600, 700, 800 and 900 °C were used. It was determined that the transformation of ZIF-67 into metallic Co implies a mass loss of 73.3%. However, for this system, the mass loss for the temperatures mentioned previously was between 28.3% and 46.6%. This indicates the formation of a carbon-containing material instead of a material composed of only Co NPs. To better understand the structural aspects of the catalysts, XRD analyses were performed on the Co@C–N materials. The results showed the presence of graphite carbon sheets and metallic Co. The intensity of the diffraction peaks suggests the formation of a Co phase with a high crystallization degree. On the other hand, the Raman spectra of the Co@C–N samples revealed G and D carbon bands characteristic for sp^2 graphite carbon and disoriented carbon, respectively [109]. The intensity ratio of the I^G/I^D bands for the Co@C–N (600), Co@C–N (700), Co@C–N (800) and Co@C–N (900) materials was 0.86, 0.97, 1.29 and 1.97, respectively, which reveals that the crystalline degree of graphite carbon could be improved with higher temperatures. These results make clear that it is possible to obtain carbon-based materials containing Co NPs.

As mentioned above, the thermal treatment used on the MOF precursor is key to achieving an adequate distribution of metallic NPs over the material structure, since it derives in good dispersion of the catalytically active sites. SEM analysis proved that this was achieved since the pyrolyzed NPs roughly kept the polyhedric form of ZIF-67 with a slight contraction. Moreover, the SEM analysis made evident that the Co NPs were highly dispersed on the material, which was mainly due to the order of the organic linkers from the ZIF-67 precursor.

The presence of nitrogen and its influence in the catalytic process was investigated as well. The N content for Co@C–N (600), Co@C–N (700), Co@C–N (800) y Co@C–N (900) was found to be 9.7, 5.8, 2.1 and 0.6%, respectively, showing a decrease at higher temperatures. Besides, two kinds of nitrogen species were found by XPS technique in the Co@C–N catalysts by two high-resolution N1s peaks that suggest the presence of pyridine nitrogen (c.a. 399 eV) and graphite nitrogen (c.a. 401 eV) [110]. These results indicate that the relative content of graphite N increased with the pyrolysis temperature, while the relative pyridine N content decreased as the used temperature was increased. This implies a higher graphitization at higher temperatures, which is beneficial for better electronic mobility. Reports indicate that a strong interaction between metallic NPs and graphite walls can enhance the activity of nanocarbon catalysts' activity due to the electron transfer from the metal atoms to the carbon shells; a process that occurs until the Fermi level reaches equilibrium [111,112]. Moreover, the pyridine N doping produces structural flaws in the C–N compounds when forming the O_2 adsorption sites and basic sites, which are necessary for the formation of oxygen radicals, as well as the substrate interaction with the surface of the catalysts. These structural features had a clear repercussion in the catalytic results obtained for the oxidative esterification of HMF to produce DMFDCA.

The catalytic reactions were performed at 100 °C, using sodium carbonate as the base, methanol as the solvent and 2 MPa of molecular oxygen as the oxidant. The most active catalyst turned out to be Co@C–N (800) since it showed an appropriate nitrogen content in its structure and a relatively high graphitic-N/C ratio, which contributed to the balance between the nitrogen contents and nitrogen species. This balance can result in higher electron mobility and more active sites, which helps in the activation and reduction of O_2. The strong electronic interaction between the Co atoms and N-doping carbon is prone to generate strong electron acceptors (Co nanoparticles) and electron donors (C–N composites), leading to the high catalytic activity of Co@C–N (800) in the oxidative esterification reaction.

Another example of a MOF as a sacrifice template was reported by Bao et al. [113] who synthetized an active catalyst for the aerobic oxidation of HMF to FDCA with holey 2D Mn_2O_3 nanoflakes (M400). These were obtained by the calcination of the MnTPA-MOF at 400 °C. A remarkable detail is that the thermal treatment used on the MnTPA precursor generates a change in the MOF structure. This was observed in the calcination at 250, 300, 350 and 400 °C of MnTPA to obtain the M250, M300, M350 and M400 materials, respectively. The PXRD, FESEM and N_2 adsorption/desorption characterization analysis demonstrated these structural changes caused by the used calcination temperature. The PXRD diffractograms showed that the thermal treatment at 250 and 300 °C (M250 and M300) resulted in the decreasing and widening of the diffraction peaks compared to the MnTPA precursor, which implies the solvent elimination and partial amorphization of the original MOF. When the temperature rises to 350 °C, the diffraction peaks of the MnTPA precursor completely disappear, generating peaks that can be attributed to Mn_2O_3 (JCPDS 41-1442) and Mn_3O_4 (JCPDS 01-1127). This suggests the decomposition of the crystalline MnTPA, forming metal oxides. The data collected from the FESEM analysis supports this theory as a flake-like structure is observed with a smooth surface and a thickness of 60–130 nm for the pristine MnTPA, which is preserved in the M250 and M300 material with few differences in surface and thickness. For the M350 material, the surface of the flakes became rough, and pores began to appear as the thickness decreased to 30 nm. When the calcination temperature increased to 400 °C, the thickness of the flakes was ca. 20 nm, and more holes were observed. This, combined with the FTIR spectra of all materials, showed that the structure changes due to the decomposition of the organic linkers losing the crystalline structure to yield metal oxides. It is worth mentioning that the morphological changes aim for a higher surface area of the catalyst, which, as we have seen before, is a very important factor for the catalytic activity of any heterogeneous catalyst. N_2 adsorption/desorption studies were performed, and the results corroborate a higher porosity of M400 over the rest of the materials. The study showed that M400 presents a surface area of 61.5 m^2 g^{-1} while the pore size distribution showed micropores of 2.4 nm and mesopores of 22.7 nm, attributed to the nanoflakes pores and the interflake pores, respectively, proving that the M400 material presented the most adequate characteristics to be used as the catalyst.

Another important characteristic in a catalyst is the presence of metallic species that have an adequate redox couple to carry out the catalytic process. XPS spectra showed that M400 presented Mn^{II}, Mn^{III} and Mn^{IV} ions, which allowed the presence of Mn^{II}/Mn^{IV} and Mn^{III}/Mn^{IV} redox couples (peak intensities of 0.36 and 0.77, respectively). These data indicate a relatively high proportion of Mn^{IV} with respect to the other oxidation states.

M350 and M400 were selected as catalysts for the aerobic oxidation of HMF to FDCA due to the structural properties of these materials. The reactions were made at 100 °C and 1.4 Mpa of O_2 in the presence of $NaHCO_3$ (three equivalents of $NaHCO_3$ for each equivalent of HMF) for 24 h. For comparison, the same reaction conditions were used in three other reactions, i.e., in the absence of the catalyst, using activated Mn_2O_3 as the catalyst and using activated MnO_2 as the catalyst. The results showed that in the absence of any catalyst, the total yield of HMFCA and FDCA was below 4%. For the activated Mn_2O_3, the total yield for HMFCA, FFCA and FDCA did not surpass 13%, and for the activated MnO_2, only a 12% yield was observed for the desired FDCA, with a 91.1% of HMF conversion, revealing a poor selectivity. The M350 material showed an almost total HMF conversion with an FDCA total yield of 76.1%, but this was lower than the 99.5% FDCA yield of M400 with a 100% HMF conversion. This difference was attributed to the presence of Mn_3O_4 in the M350 structure, while the good results obtained with M400 as the catalyst were attributed to the abundant surface pores present in the M400 material since they allow a better substrate diffusion and increased the catalytically active sites available on the catalyst surface.

Up to this point, it has become evident that the pyrolysis temperature is important to obtain a heterogeneous catalyst with adequate structural properties such as pore size, surface area, metal dispersion and metal-support interface. Fang et al. [114] remark on the importance of thermal control in the synthesis process. In this work, the ZIF-67 MOF was encapsulated in the mesoporous silica KIT-6, which acted as support to obtain the ZIF-67@KIT-6 precursor. This precursor was then calcined at 250 °C to produce a catalyst with highly dispersed cobalt oxide NPs that maintained the support structure (Co@KIT-6). The authors indicated that the control of certain parameters, such as heating rate, temperature and calcination time, was crucial to achieving ultrafine metal oxide NPs. In this work, a sufficient calcination time was critical for the complete oxidation of metal ions and the elimination of organic linkers. The calcination temperature and the heating rate were carefully set to 250 °C (above the MOF decomposition temperature in an O_2 atmosphere [115]) and 1 °C/min respectively. According to the Kirkendall effect, the dispersed and ordered metal ions tend to aggregate to reduce their surface-free energy under thermal treatment. At higher temperatures (e.g., 275 °C or 300 °C), the metal ions are more prone to aggregate in bigger particles (>5 nm). On the other hand, faster heating rates (2 and 3 °C/min) accelerate the decomposition process, producing larger aggregates (c.a. 5 nm). Hence, the key elements to achieve ultrafine Co_3O_4 NPs were slow heating rates, low calcination temperatures and sufficient thermolysis time.

Co@KIT-6 was used as a catalyst in the aerobic oxidation of HMF to obtain FDCA at 80 °C while using water as solvent. The authors made a comparative study using KIT-6 and Co/KIT-6. The original KIT-6 did not present conversion under the reaction conditions. Besides, the Co/KIT-6 prepared with a typical impregnation method exhibited a poor activity, achieving a 10% HMF conversion and total FDCA yield of 9.5%. The results obtained with Co@KIT-6 indicated a 100% HMF conversion with a 99% selectivity towards the desired FDCA. These good results were attributed to the presence of highly oxidant ultrafine Co_3O_4 NPs confined in the KIT-6 mesopores, which were effective for the oxidation of HMF to FDCA, following the Mars van Krevelen reaction mechanism. According to this mechanism, the lattice oxygen in Co_3O_4 facilitates HMF oxidation to FDCA, being subsequently replenished by the oxygen present in the reaction atmosphere.

Fang et al. [116] have also reported the synthesis of FeCo/C bimetallic catalysts, used in the aerobic oxidation reaction of HMF to obtain DFF. In this case, the FeCo/C catalysts were obtained from MIL-45b MOF by thermolysis at 500, 600, 700 and 800 °C, with a heating rate of 1 °C min^{-1} for 6 h. These were named FeCo/C(T), where T represents the calcination temperature. The studies indicate that the best catalyst was FeCo/C(500), with which conversion and selectivity of >99% to the DFF product were obtained. The good results were attributed to the fact that the hollow structure of the FeCo/C(500) catalyst promotes the adsorption of HMF, as well as the rapid desorption of the formed DFF, leading to a higher product yield. The catalysts obtained at higher temperatures, i.e., 600, 700 and 800, did not present the same properties since the pores collapsed due to the effect of the temperature. This highlights that thermal control in the synthesis of the catalyst is a relevant parameter. The summary of the reactions and its catalytic results are shown in the Figure 5 and Table 4.

Table 4. Catalytic conditions for the different catalysts of Figure 5 in the HMF oxidation.

	Catalyst (mg)	Substrate (mmol)	Base	Oxidant	Temperature (°C)	Time (h)	Conv. HMF (%)	Ref.
a	10	0.3	NaOH	H_2O_2	90	1	100	[98]
b	10	0.3	NaOH/Na_2CO_3	H_2O_2	90	1	100	[98]
c	100	0.5	Na_2CO_3	O_2	100	5	99	[108]
d	100	0.5	Na_2CO_3	O_2	100	5	99	[108]
e	150	50 mM	$NaHCO_3$	O_2	100	24	>99	[113]
f	150	50 mM	$NaHCO_3$	O_2	100	24	>99	[113]
g	1 mol%	0.1	-	Air	80	2	100	[114]
h	20 mol%	1	Na_2CO_3	O_2	100 °C	6	>99	[116]

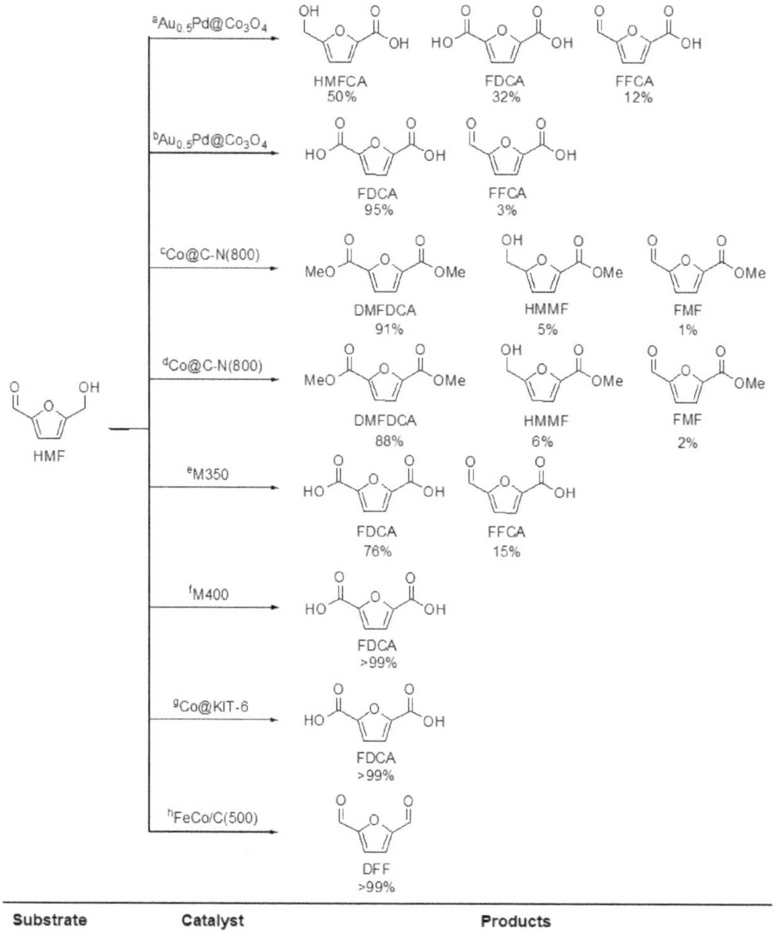

Figure 5. HMF oxidation with different catalysts, synthesized using MOFs as sacrifice templates.

3.3. Perspectives and Conclusions

It is evident that MOFs are a great alternative to be used in the synthesis of heterogeneous catalysts due to their high number of properties and versatility. Due to this, it is important that this type of compound be studied in catalytic reactions that involve the use of substrates from biomass. In this way, the properties of MOFs that were described in this section are necessary to achieve good results in chemical transformations, which involve platform compounds such as furfural or levulinic acid, since their derivatives are important in the current chemical industry.

In this chapter, four properties were highlighted that are considered relevant when designing a heterogeneous catalyst—based on MOFs for hydrogenation reactions—and that can finally be extrapolated to other catalytic reactions. The first important characteristics are the pore size and the dispersion of the nanoparticles within the MOF. Both are important because they are related to the accessibility of the substrate to the active sites of the catalyst, and the availability of these sites to carry out the catalytic process. In both cases, the choice of the MOF and the catalyst synthesis method are important. Second, the type of acidic sites that the catalyst may have must be considered, these being the Brönsted and Lewis acidic sites. The acidic sites are essential to generate the possibilities of interaction of the substrate with the active metallic sites of the catalyst; in addition, being able to modulate the selectivity of the reaction, depending on the type of acid site. Finally, the connectivity of the metallic nodes with the ligands must be taken into account in such a way that there is an optimal ratio of connected nodes that allows maintaining an adequate pore size and high availability of active acid sites. It is possible to note that all the aforementioned characteristics point to two central properties, the availability of the active sites of the catalyst and the accessibility of the substrate to these active sites. Therefore, we consider these parameters and the relationship between them to be of vital importance in designing MOFs that may have interesting potential in the transformation of biomass.

Funding: This research was funded by PROYECTO FONDECYT INICIACION 1119042.

Data Availability Statement: Not applicable.

Conflicts of Interest: The authors declare no conflict of interest.

References

1. Yu, D.; He, L. Introduction to CO_2 utilization. *Green Chem.* **2021**, *23*, 3499–3501. [CrossRef]
2. Garba, M.; Usman, M.; Khan, S. CO_2 towards fuels: A review of catalytic conversion of carbon dioxide to hydrocarbons. *J. Environ. Chem. Eng.* **2021**, *9*, 104756. [CrossRef]
3. Chen, Z.; Du, S.; Zhang, J.; Wu, X. From 'Gift' to gift: Producing organic solvents from CO_2. *Green Chem.* **2020**, *22*, 8169–8182. [CrossRef]
4. Hwang, S.-M.; Zhang, C.; Han, J.H.; Park, H.-G.; Kim, Y.T.; Yang, S.; Jun, K.-W.; Kim, S.K. Mesoporous carbon as an effective support for Fe catalyst for CO_2 hydrogenation to liquid hydrocarbons. *J. CO_2 Util.* **2020**, *37*, 65–73. [CrossRef]
5. Wang, J.; Al Qahtani, M.; Wang, X. One-step plasma-enabled catalytic carbon dioxide hydrogenation to higher hydrocarbons: Significance of catalyst-bed configuration. *Green Chem.* **2021**, *23*, 1642–1647. [CrossRef]
6. Wang, Y.; Winter, L.; Chen, J.; Yan, B. CO_2 hydrogenation over heterogeneous catalysts at atmospheric pressure: From electronic properties to product selectivity. *Green Chem.* **2021**, *23*, 249–267. [CrossRef]
7. Li, X.; Luo, X.; Jin, Y. Heterogeneous sulfur-free hydrodeoxygenation catalysts for selectively upgrading the renewable bio-oils to second generation biofuels. *Renew. Sustain. Energy Rev.* **2018**, *82*, 3762–3797. [CrossRef]
8. Santillan-Jimenez, E.; Crocker, M. Catalytic deoxygenation of fatty acids and their derivatives to hydrocarbon fuels via decarboxylation/decarbonylation. *J. Chem. Technol. Biotechnol.* **2012**, *87*, 1041–1050. [CrossRef]
9. Davis, K.; Yoo, S.; Shuler, E.; Sherman, B.; Lee, S.; Leem, G. Photocatalytic hydrogen evolution from biomass conversion. *Nano Converg.* **2021**, *8*, 6. [CrossRef]
10. Rajeswari, G.; Jacob, S.; Chandel, A.; Kumar, V. Unlocking the potential of insect and ruminant host symbionts for recycling of lignocellulosic carbon with a biorefinery approach: A review. *Microb. Cell. Fact* **2021**, *20*, 107. [CrossRef]
11. Serrano-Ruiz, J.C.; West, R.M.; Dumesic, J.A. Catalytic Conversion of Renewable Biomass Resources to Fuels and Chemicals. *Annu. Rev. Chem. Biomolecular Eng.* **2010**, *1*, 79–100. [CrossRef] [PubMed]
12. Rodríguez-Padrón, D.; Zhao, D.; Carrillo-Carrion, C. Exploring the potential of biomass-templated Nb/ZnO nanocatalysts for the sustainable synthesis of N-heterocycles. *Catal. Today* **2021**, *368*, 243–249. [CrossRef]
13. Galarza, E.; Fermanelli, C.; Pierella, L.; Saux, C.; Renzini, M. Influence of the Sn incorporation method in ZSM-11 zeolites in the distribution of bio-oil products obtained from biomass pyrolysis. *J. Anal. Appl. Pyrolysis* **2021**, *156*, 105116. [CrossRef]
14. Liang, J.; Shan, G.; Sun, Y. Catalytic fast pyrolysis of lignocellulosic biomass: Critical role of zeolite catalysts. *Renew. Sustain. Energy Rev.* **2021**, *139*, 110707. [CrossRef]
15. Ribeiro, L.; de Melo Órfão, J.; Ribeiro Pereira, M. An overview of the hydrolytic hydrogenation of lignocellulosic biomass using carbon-supported metal catalysts. *Mater. Today Sustain.* **2020**, *11*, 100058. [CrossRef]
16. Safaei, M.; Foroughi, M.; Ebrahimpoor, N.; Jahani, S.; Omidi, A.; Khatami, M. A review on metal–organic frameworks: Synthesis and applications. *TrAC Trends Anal. Chem.* **2019**, *118*, 401–425. [CrossRef]
17. Zhang, L.; Li, S.; Xin, J. A non-enzymatic voltammetric xanthine sensor based on the use of platinum nanoparticles loaded with a metal–organic framework of type MIL-101(Cr). Application to simultaneous detection of dopamine, uric acid, xanthine and hypoxanthine. *Microchim. Acta* **2019**, *186*. [CrossRef]
18. Mirkovic, I.; Lei, L.; Ljubic, D.; Zhu, S. Crystal Growth of Mmetal-Organic Framework-5 around Cellulose-Based Fibers Having a Necklace Morphology. *ACS Omega* **2019**, *4*, 169–175. [CrossRef]
19. Zhang, L.; Jiang, K.; Zhang, J. Low-Cost and High-Performance Microporous Mmetal-Organic Framework for Separation of Acetylene from Carbon Dioxide. *ACS Sustain. Chem. Eng.* **2018**, *7*, 1667–1672. [CrossRef]
20. Huan, W.; Xing, M.; Cheng, C.; Li, J. Facile Fabrication of Magnetic Mmetal-Organic Framework Nanofibers for Specific Capture of Phosphorylated Peptides. *ACS Sustain. Chem. Eng.* **2018**, *7*, 2245–2254. [CrossRef]
21. Wei, X.; Wang, Y.; Chen, J. Poly(deep eutectic solvent)-functionalized magnetic metal organic framework composites coupled with solid-phase extraction for the selective separation of cationic dyes. *Anal. Chim. Acta* **2019**, *1056*, 47–61. [CrossRef] [PubMed]

22. Yi, X.; He, X.; Yin, F.; Chen, B.; Li, G.; Yin, H. Co-CoO-Co$_3$O$_4$/N-doped carbon derived from metal organic framework: The addition of carbon black for boosting oxygen electrocatalysis and Zn-Air battery. *Electrochim. Acta* **2019**, *295*, 966–977. [CrossRef]
23. Cao, A.; Zhang, L.; Wang, Y. 2D–2D Heterostructured UNiMOF/g-C$_3$N$_4$ for Enhanced Photocatalytic H2 Production under Visible-Light Irradiation. *ACS Sustain. Chem. Eng.* **2018**, *7*, 2492–2499. [CrossRef]
24. Baek, J.; Rungtaweevoranit, B.; Pei, X. Bioinspired Mmetal-Organic Framework Catalysts for Selective Methane Oxidation to Methanol. *J. Am. Chem. Soc.* **2018**, *140*, 18208–18216. [CrossRef] [PubMed]
25. Andrew, R.M. Global CO$_2$ emissions from cement production. *Earth Syst. Sci. Data Discuss.* **2018**, *10*, 195–217. [CrossRef]
26. Aresta, M. Carbon dioxide reduction to C$_1$ or C$_n$ molecules. In *Carbon Dioxide Recovery and Utilization*; Aresta, M., Ed.; Springer: Dordrecht, The Netherlands, 2003; pp. 293–312.
27. Whang, H.S.; Lim, J.; Choi, M.S.; Lee, J.; Lee, H. Heterogeneous catalysts for catalytic CO$_2$ conversion into value-added chemicals. *BMC Chem. Eng.* **2019**, *1*, 9. [CrossRef]
28. Wei, J.; Ge, Q.; Yao, R.; Wen, Z.; Fang, C.; Guo, L.; Xu, H.; Sun, J. Directly converting CO$_2$ into a gasoline fuel. *Nat. Commun.* **2017**, *8*, 15174. [CrossRef]
29. Porosoff, M.D.; Yan, B.; Chen, J.G. Catalytic reduction of CO$_2$ by H$_2$ for synthesis of CO, methanol and hydrocarbons: Challenges and opportunities. *Energy Environ. Sci.* **2016**, *9*, 62–73. [CrossRef]
30. Li, W.; Wang, H.; Jiang, X.; Zhu, J.; Liu, Z.; Guo, X.; Song, C. A short review of recent advances in CO$_2$ hydrogenation to hydrocarbons over heterogeneous catalysts. *RSC Adv.* **2018**, *8*, 7651–7669. [CrossRef]
31. Gao, P.; Dang, S.; Li, S.; Bu, X.; Liu, Z.; Qiu, M.; Yang, C.; Wang, H.; Zhong, L.; Han, Y.; et al. Direct production of lower olefins from CO$_2$ conversion via bifunctional catalysis. *ACS Catal.* **2018**, *8*, 571–578. [CrossRef]
32. Fujiwara, M.; Ando, H.; Tanaka, M.; Souma, Y. Hydrogenation of carbon dioxide over Cu Zn-chro mate/zeolite composite catalyst: The effects of reaction behavior of alkenes on hydrocarbon synthesis. *Appl. Catal. A Gen.* **1995**, *130*, 105–116. [CrossRef]
33. Dang, S.; Gao, P.; Liu, Z.; Chen, X.; Yang, C.; Wang, H.; Zhong, L.; Li, S.; Sun, Y. Role of zirconium in direct CO$_2$ hydrogenation to lower olefins on oxide/zeolite bifunctional catalysts. *J. Catal.* **2018**, *364*, 382–393. [CrossRef]
34. Gao, J.; Jia, C.; Liu, B. Direct and selective hydrogenation of CO$_2$ to ethylene and propene by bifunctional catalysts. *Catal. Sci. Technol.* **2017**, *7*, 5602–5607. [CrossRef]
35. Fujiwara, M.; Kieffer, R.; Ando, H.; Xu, Q.; Souma, Y. Change of catalytic properties of FeZnO/zeolite composite catalyst in the hydrogenation of carbon dioxide. *Appl. Catal. A Gen.* **1997**, *154*, 87–101. [CrossRef]
36. Helal, A.; Usman, M.; Arafat, M.E.; Abdelnaby, M.M. Allyl functionalized UiO-66 metal–organic framework as a catalyst for the synthesis of cyclic carbonates by CO$_2$ cycloaddition. *J. Ind. Eng. Chem.* **2020**, *89*, 104–110. [CrossRef]
37. Helal, A.; Cordova, K.E.; Arafat, M.E.; Usman, M.; Yamani, Z.H. Defect-engineeringa metal–organic framework for CO$_2$ fixation in the synthesis of bioactive oxazolidinones. *Inorg. Chem. Front.* **2020**, *7*, 3571–3577. [CrossRef]
38. Santibañez, L.; Escalona, N.; Torres, J.; Kremer, C.; Cancino, P.; Spodine, E. Cu^{II}- and Co^{II}-Based MOFs: {[La$_2$Cu$_3$(μ-H$_2$O)(ODA)$_6$(H$_2$O)$_3$]·3H$_2$O}$_n$ and {[La$_2$Co$_3$(ODA)$_6$(H$_2$O)$_6$]·12H$_2$O}$_n$. The Relevance of Physicochemical Properties on the Catalytic Aerobic Oxidation of Cyclohexene. *Catalysts* **2020**, *10*, 589. [CrossRef]
39. Cancino, P.; Santibañez, L.; Stevens, C.; Fuentealba, P.; Audebrand, N.; Aravena, D.; Torres, J.; Martinez, S.; Kremer, C.; Spodine, E. Influence of the channel size of isostructural 3d–4f MOFs on the catalytic aerobic oxidation of cycloalkenes. *New J. Chem.* **2019**, *43*, 11057–11064. [CrossRef]
40. Han, Y.; Xu, H.; Su, Y.; Xu, Z.-l.; Wang, K.; Wang, W. Noble metal (Pt, Au@Pd) nanoparticles supported on metal organic framework (MOF-74) nanoshuttles as high-selectivity CO$_2$ conversion catalysts. *J. Catal.* **2019**, *370*, 70–78. [CrossRef]
41. Wang, T.; Shi, L.; Tang, J.; Malgras, V.; Asahina, S.; Liu, G.; Zhang, H.; Meng, X.; Chang, K.; He, J.; et al. A Co$_3$O$_4$-embedded porous ZnO rhombic dodecahedron prepared using zeolitic imidazolate frameworks as precursors for CO$_2$ photoreduction. *Nanoscale* **2016**, *8*, 6712–6720. [CrossRef]
42. Zhang, H.; Wang, T.; Wang, J.; Liu, H.D.; Dao, T.; Li, M.; Liu, G.; Meng, X.; Chang, K.; Shi, L.; et al. Surface-Plasmon-Enhanced Photodriven CO$_2$ Reduction Catalyzed by Mmetal-Organic-Framework-Derived Iron Nanoparticles Encapsulated by Ultrathin Carbon Layers. *Adv. Mater.* **2016**, *28*, 3703–3710. [CrossRef] [PubMed]
43. Wang, S.; Guan, B.Y.; Lu, Y.; Lou, X.W.D. Formation of Hierarchical In$_2$S$_3$–CdIn$_2$S$_4$ Heterostructured Nanotubes for Efficient and Stable Visible Light CO$_2$ Reduction. *J. Am. Chem. Soc.* **2017**, *139*, 17305–17308. [CrossRef] [PubMed]
44. Hu, S.; Liu, M.; Ding, F.; Song, C.; Zhang, G.; Guo, X. Hydrothermally stable MOFs for CO$_2$ hydrogenation over iron-based catalyst to light olefins. *J. CO$_2$ Util.* **2016**, *15*, 89–95. [CrossRef]
45. Tarasov, A.L.; Isaeva, V.I.; Tkachenko, O.P.; Chernyshev, V.V.; Kustov, L.M. Conversion of CO$_2$ into liquid hydrocarbons in the presence of a Cocontaining catalyst based on the microporous metal–organic framework MIL-53(Al)Fuel. *Proc. Tech.* **2018**, *176*, 101–106. [CrossRef]
46. Zhao, Z.-W.; Zhou, X.; Liu, Y.-N.; Shen, C.-C.; Yuan, C.-Z.; Jiang, Y.-F.; Zhao, S.-J.; Ma, L.-B.; Cheang, T.-Y.; Xu, A.-W. Ultrasmall Ni nanoparticles embedded in Zr-based MOFs provide high selectivity for CO$_2$ hydrogenation to methane at low temperatures. *Catal. Sci. Technol.* **2018**, *8*, 3160–3165. [CrossRef]
47. Zhen, W.; Gao, F.; Tian, B.; Ding, B.; Deng, Y.; Li, Z.; Gao, H.; Lu, G. Enhancing activity for carbon dioxide methanation by encapsulating (111) facet Ni particle in metal–organic frameworks at low temperature. *J. Catal.* **2017**, *348*, 200–211. [CrossRef]
48. Zhen, W.; Li, B.; Lu, G.; Ma, J. Enhancing catalytic activity and stability for CO$_2$ methanation on Ni@MOF-5 via control of active species dispersion. *Chem. Commun.* **2015**, *51*, 1728–1731. [CrossRef]

49. Rungtaweevoranit, B.; Baek, J.; Araujo, J.R.; Archanjo, B.S.; Yaghi, O.M.; Somorjai, G.A. Copper Nanocrystals Encapsulated in Zr-based Mmetal-Organic Frameworks for Highly Selective CO_2 Hydrogenation to Methanol. *Nano Lett.* **2016**, *16*, 7645–7649. [CrossRef]

50. An, B.; Zhang, J.; Cheng, K.; Ji, P.; Wang, C.; Lin, W. Confinement of Ultrasmall Cu/ZnOx Nanoparticles in Mmetal-Organic Frameworks for Selective Methanol Synthesis from Catalytic Hydrogenation of CO_2. *J. Am. Chem. Soc.* **2017**, *139*, 3834–3840. [CrossRef]

51. Liu, J.; Zhang, A.; Liu, M.; Hu, S.; Ding, F.; Song, C.; Guo, X. Fe-MOF-derived highly active catalysts for carbon dioxide hydrogenation to valuable hydrocarbons. *J. CO_2 Util.* **2017**, *21*, 100–107. [CrossRef]

52. Wang, Y.; Kazumi, S.; Gao, W.; Gao, X.; Li, H.; Guo, X.; Yoneyama, Y.; Yang, G.; Tsubaki, N. Direct conversion of CO_2 to aromatics with high yield via a modified Fischer-Tropsch synthesis pathway. *Appl. Catal. B Env.* **2020**, *269*, 118792–118801. [CrossRef]

53. Martín, N.; Portillo, A.; Ateka, A.; Cirujano, F.G.; Oar-Arteta, L.; Aguayo, A.T.; Dusselier, M. MOF-derived/zeolite hybrid catalyst for the production of light olefins from CO_2. *ChemCatChem* **2020**, *12*, 5750–5758. [CrossRef]

54. Ramirez, A.; Gevers, L.; Bavykina, A.; Ould-Chikh, S.; Gascon, J. Metal Organic Framework-Derived Iron Catalysts for the Direct Hydrogenation of CO_2 to Short Chain Olefins. *ACS Catal.* **2018**, *8*, 9174–9182. [CrossRef]

55. Chen, C.S.; Cheng, W.H.; Lin, S. Mechanism of CO formation in reverse water–gas shift reaction over Cu/Al_2O_3 catalyst. *Catal. Lett.* **2000**, *68*, 45–48. [CrossRef]

56. Porosoff, M.D.; Baldwin, J.; Peng, X.; Mpourmpakis, G.; Willauer, H.D. Potassium Promoted Molybdenum Carbide as a Highly Active and Selective Catalyst for CO_2 Conversion to CO. *ChemSusChem* **2017**, *10*, 2408–2415. [CrossRef] [PubMed]

57. Matsubu, J.C.; Yang, V.N.; Christopher, P. Isolated Metal Active Site Concentration and Stability Control Catalytic CO_2 Reduction Selectivity. *J. Am. Chem. Soc.* **2015**, *137*, 3076–3084. [CrossRef] [PubMed]

58. Kim, S.S.; Lee, H.H.; Hong, S.C. A study on the effect of support's reducibility on the reverse water-gas shift reaction over Pt catalysts. *Appl. Catal. A* **2012**, *423*, 100–107. [CrossRef]

59. Lu, B.; Kawamoto, K. Preparation of mesoporous CeO_2 and monodispersed NiO particles in CeO_2, and enhanced selectivity of NiO/CeO_2 for reverse water gas shift reaction. *Mater. Res. Bull.* **2014**, *53*, 70–78. [CrossRef]

60. Loiland, J.A.; Wulfers, M.J.; Marinkovic, N.S.; Lobo, R.F. $Fe/\gamma\text{-}Al_2O_3$ and $Fe{-}K/\gamma\text{-}Al_2O_3$ as reverse water-gas shift catalysts. *Catal. Sci. Technol.* **2016**, *6*, 5267. [CrossRef]

61. Pour, A.N.; Shahri, S.M.K.; Bozorgzadeh, H.R.; Zamani, Y.; Tavasoli, A.; Marvast, M.A. Effect of Mg, La and Ca promoters on the structure and catalytic behavior of iron-based catalysts in Fischer–Tropsch synthesis. *Appl. Catal. A* **2008**, *348*, 201–208. [CrossRef]

62. Li, J.; Cheng, X.; Zhang, C.; Wang, J.; Dong, W.; Yang, Y.; Li, Y. Alkalis in iron-based Fischer–Tropsch synthesis catalysts; distribution, migration and promotion. *J. Chem. Technol. Biotechnol.* **2017**, *92*, 1472–1480. [CrossRef]

63. Dorner, R.W.; Hardy, D.R.; Williams, F.W.; Willauer, H.D. K and Mn doped iron-based CO_2 hydrogenation catalysts; Detection of $KAlH_4$ as part of the catalyst's active phase. *Appl. Catal. A* **2010**, *373*, 112–121. [CrossRef]

64. Rodemerck, U.; Holeňá, M.; Wagner, E.; Smejkal, Q.; Barkschat, A.; Baerns, M. Catalyst Development for CO_2 Hydrogenation to Fuels. *ChemCatChem* **2013**, *5*, 1948–1955. [CrossRef]

65. Numpilai, T.; Witoon, T.; Chanlek, N.; Limphirat, W.; Bonura, G.; Chareonpanich, M.; Limtrakul, L. Structure–activity relationships of $Fe\text{-}Co/K\text{-}Al_2O_3$ catalysts calcined at different temperatures for CO_2 hydrogenation to light olefins. *Appl. Catal. A* **2017**, *547*, 219–229. [CrossRef]

66. Zhao, T.; Hui, Y.; Li, Z. Controllable preparation of ZIF-67 derived catalyst for CO_2 methanation. *Mol. Catal.* **2019**, *474*, 110421–110430. [CrossRef]

67. Lin, X.; Wang, S.; Tu, W.; Hu, Z.; Ding, Z.; Hou, Y.; Xu, R.; Dai, W. MOF-derived hierarchical hollow spheres composed of carbon-confined Ni nanoparticles for efficient CO_2 methanation. *Catal. Sci. Technol.* **2019**, *9*, 731–738. [CrossRef]

68. Fang, R.; Chen, L.; Shen, Z.; Li, Y. Efficient hydrogenation of furfural to fufuryl alcohol over hierarchical MOF immobilized metal catalysts. *Catal. Today* **2021**, *368*, 217–223. [CrossRef]

69. Song, Y.; Feng, X.; Chen, J.S.; Brzezinski, C.; Xu, Z.; Lin, W. Multistep Engineering of Synergistic Catalysts in a Metal–Organic Framework for Tandem C-O Bond Cleavage. *J. Am. Chem. Soc.* **2020**, *142*, 4872–4882. [CrossRef]

70. Phan, D.P.; Lee, E.Y. Phosphoric acid enhancement in a Pt-encapsulated Metal–Organic Framework (MOF) bifunctional catalyst for efficient hydro-deoxygenation of oleic acid from biomass. *J. Catal.* **2020**, *386*, 19–29. [CrossRef]

71. Yuan, Q.; Zhang, D.; Van Haandel, L.; Ye, F.; Xue, T.; Hensen, E.J.M.; Guan, Y. Selective liquid phase hydrogenation of furfural to furfuryl alcohol by Ru/Zr-MOFs. *J. Mol. Catal. A Chem.* **2015**, *406*, 58–64. [CrossRef]

72. Fulajtárova, K.; Soták, T.; Hronec, M.; Vávra, I.; Dobročka, E.; Omastová, M. Aqueous phase hydrogenation of furfural to furfuryl alcohol over Pd-Cu catalysts. *Appl. Catal. A Gen.* **2015**, *502*, 78–85. [CrossRef]

73. Feng, J.; Li, M.; Zhong, Y.; Xu, Y.; Meng, X.; Zhao, Z.; Feng, C. Hydrogenation of levulinic acid to γ-valerolactone over Pd@UiO-66-NH2 with high metal dispersion and excellent reusability. *Microporous Mesoporous Mater.* **2020**, *294*, 109858–109867. [CrossRef]

74. Aijaz, A.; Karkamkar, A.; Choi, Y.J.; Tsumori, N.; Rönnebro, E.; Autrey, T.H.; Shioyama, H.; Xu, Q. Immobilizing Highly Catalytically Active Pt Nanoparticles inside the Pores of Metal–Organic Framework. *J. Am. Chem. Soc.* **2012**, *134*, 13–16. [CrossRef] [PubMed]

75. Lin, S.; Kumar Reddy, D.H.; Bediako, J.K.; Song, M.H.; Wei, W.; Kim, J.A.; Yun, Y.S. Effective adsorption of Pd(II), Pt(IV) and Au(III) by Zr(IV)-based metal–organic frameworks from strongly acidic solutions. *J. Mater. Chem. A* **2017**, *5*, 13557–13564. [CrossRef]

76. Li, X.; Guo, Z.; Xiao, C.; Goh, T.W.; Tesfagaber, D.; Huang, W. Tandem catalysis by palladium nanoclusters encapsulated in metal–organic frameworks. *ACS Catal.* **2014**, *4*, 3490–3497. [CrossRef]

77. Klet, R.C.; Liu, Y.; Wang, T.C.; Hupp, J.T.; Farha, O.K. Evaluation of Brønsted acidity and proton topology in Zr- and Hf-based metal–organic frameworks using potentiometric acid-base titration. *J. Mater. Chem. A* **2016**, *4*, 1479–1485. [CrossRef]

78. Cirujano, F.G.; Corma, A.; Llabrés, I.; Xamena, F.X. Zirconium-containing metal organic frameworks as solid acid catalysts for the esterification of free fatty acids: Synthesis of biodiesel and other compounds of interest. *Catal. Today* **2015**, *257*, 213–220. [CrossRef]

79. Li, X.; Deng, Q.; Yu, L.; Gao, R.; Tong, Z.; Lu, C.; Wang, J.; Zeng, Z.; Zou, J.J.; Deng, S. Double-metal cyanide as an acid and hydrogenation catalyst for the highly selective ring-rearrangement of biomass-derived furfuryl alcohol to cyclopentenone compounds. *Green Chem.* **2020**, *22*, 2549–2557. [CrossRef]

80. Kim, D.W.; Kim, H.G.; Cho, D.H. Catalytic performance of MIL-100 (Fe, Cr) and MIL-101 (Fe, Cr) in the isomerization of endo- to exo-dicyclopentadiene. *Catal. Commun.* **2016**, *73*, 69–73. [CrossRef]

81. Rinaldi, R.; Schüth, F. Design of solid catalysts for the conversion of biomass. *Energy Environ. Sci.* **2009**, *2*, 610–626. [CrossRef]

82. Geilen, F.M.A.; Engendahl, B.; Harwardt, A.; Marquardt, W.; Klankermayer, J.; Leitner, W. Selective and flexible transformation of biomass-derived platform chemicals by a multifunctional catalytic system. *Angew. Chem. Int. Ed.* **2010**, *49*, 5510–5514. [CrossRef] [PubMed]

83. Zhang, D.; Ye, F.; Guan, Y.; Wang, Y.; Hensen, E.J.M. Hydrogenation of γ-valerolactone in ethanol over Pd nanoparticles supported on sulfonic acid functionalized MIL-101. *RSC Adv.* **2014**, *4*, 39558–39564. [CrossRef]

84. Valekar, A.H.; Lee, M.; Yoon, J.W.; Kwak, J.; Hong, D.Y.; Oh, K.R.; Cha, G.Y.; Kwon, Y.U.; Jung, J.; Chang, J.S.; et al. Catalytic Transfer Hydrogenation of Furfural to Furfuryl Alcohol under Mild Conditions over Zr-MOFs: Exploring the Role of Metal Node Coordination and Modification. *ACS Catal.* **2020**, *10*, 3720–3732. [CrossRef]

85. Bai, Y.; Dou, Y.; Xie, L.H.; Rutledge, W.; Li, J.R.; Zhou, H.C. Zr-based metal–organic frameworks: Design, synthesis, structure, and applications. *Chem. Soc. Rev.* **2016**, *45*, 2327–2367. [CrossRef] [PubMed]

86. Li, H.; He, J.; Riisager, A.; Saravanamurugan, S.; Song, B.; Yang, S. Acid-base bifunctional zirconium N-alkyltriphosphate nanohybrid for hydrogen transfer of biomass-derived carboxides. *ACS Catal.* **2016**, *6*, 7722–7727. [CrossRef]

87. Li, H.; Liu, X.; Yang, T.; Zhao, W.; Saravanamurugan, S.; Yang, S. Porous Zirconium–Furandicarboxylate Microspheres for Efficient Redox Conversion of Biofuranics. *ChemSusChem* **2017**, *10*, 1761–1770. [CrossRef]

88. Yang, D.; Bernales, V.; Islamoglu, T.; Farha, O.K.; Hupp, J.T.; Cramer, C.J.; Gagliardi, L.; Gates, B.C. Tuning the Surface Chemistry of Metal Organic Framework Nodes: Proton Topology of the Metal-Oxide-Like Zr6 Nodes of UiO-66 and NU-1000. *J. Am. Chem. Soc.* **2016**, *138*, 15189–15196. [CrossRef]

89. Yang, D.; Gaggioli, C.A.; Ray, D.; Babucci, M.; Gagliardi, L.; Gates, B.C. Tuning Catalytic Sites on Zr6O8 Metal–Organic Framework Nodes via Ligand and Defect Chemistry Probed with tert-Butyl Alcohol Dehydration to Isobutylene. *J. Am. Chem. Soc.* **2020**, *142*, 8044–8056. [CrossRef]

90. Jung, K.D.; Bell, A.T. Role of Hydrogen Spillover in Methanol Synthesis over Cu/ZrO$_2$. *J. Catal.* **2000**, *193*, 207–223. [CrossRef]

91. Li, D.; Xu, H.Q.; Jiao, L.; Jiang, H.L. Metal-organic frameworks for catalysis: State of the art, challenges, and opportunities. *EnergyChem* **2019**, *1*, 100005. [CrossRef]

92. Han, A.; Wang, B.; Kumar, A.; Qin, Y.; Jin, J.; Wang, X.; Yang, C.; Dong, B.; Jia, Y.; Liu, J.; et al. Recent Advances for MOF-Derived Carbon-Supported Single-Atom Catalysts. *Small Methods* **2019**, *3*, 1800471–1800492. [CrossRef]

93. Salunkhe, R.R.; Kaneti, Y.V.; Kim, J.; Kim, J.H.; Yamauchi, Y. Nanoarchitectures for Metal–Organic Framework-Derived Nanoporous Carbons toward Supercapacitor Applications. *Acc. Chem. Res.* **2016**, *49*, 2796–2806. [CrossRef] [PubMed]

94. Wang, Q.; Astruc, D. State of the Art and Prospects in Metal–Organic Framework (MOF)-Based and MOF-Derived Nanocatalysis. *Chem. Rev.* **2020**, *120*, 1438–1511. [CrossRef] [PubMed]

95. Yang, S.; Peng, L.; Bulut, S.; Queen, W.L. Recent Advances of MOFs and MOF-Derived Materials in Thermally Driven Organic Transformations. *Chem. Eur. J.* **2019**, *25*, 2161–2178. [CrossRef]

96. Chaikittisilp, W.; Hu, M.; Wang, H.; Huang, H.S.; Fujita, T.; Wu, K.C.W.; Chen, L.C.; Yamauchi, Y.; Ariga, K. Nanoporous carbons through direct carbonization of a zeolitic imidazolate framework for supercapacitor electrodes. *Chem. Commun.* **2012**, *48*, 7259–7261. [CrossRef]

97. Fang, R.; Dhakshinamoorthy, A.; Li, Y.; Garcia, H. Metal organic frameworks for biomass conversion. *Chem. Soc. Rev.* **2020**, *49*, 3638–3687. [CrossRef]

98. Liao, Y.; Van Chi, N.; Ishiguro, N.; Young, A.P.; Tsung, C.; Wu, K.C. Environmental Engineering a homogeneous alloy-oxide interface derived from metal- organic frameworks for selective oxidation of 5-hydroxymethylfurfural to 2, 5-furandicarboxylic acid. *Applied Catalysis B* **2020**, *270*, 118805–118817. [CrossRef]

99. Liao, Y.; Chen, J.E.; Isida, Y.; Yonezawa, T.; Chang, W. De Novo Synthesis of Gold-Nanoparticle-Embedded, Nitrogen-Doped Nanoporous Carbon Nanoparticles (Au@NC) with Enhanced Reduction Ability. *ChemCatChem* **2016**, *44*, 502–509. [CrossRef]

100. Liao, Y.; Huang, Y.; Chen, H.M.; Komaguchi, K.; Hou, H.; Henzie, J.; Yamauchi, Y.; Ide, Y.; Wu, K.C.; Liao, Y.; et al. Mesoporous TiO$_2$ Embedded with a Uniform Distribution of CuO Exhibit Enhanced Charge Separation and Photocatalytic Efficiency Mesoporous TiO$_2$ Embedded with a Uniform Distribution of CuO Exhibit Enhanced Charge Separation and Photocatalytic Efficiency. *ACS Appl. Mater. Interfaces* **2017**, *9*, 42425–42429. [CrossRef]

101. Abad, A.; Concepción, P.; Corma, A. A Collaborative Effect between Gold and a Support Induces the Selective Oxidation of Alcohols. *Angew. Chem. Int. Ed.* **2005**, *44*, 4066–4069. [CrossRef]

102. Kim, B.Y.; Shim, I.B.; Araci, Z.O.; Scott Saavedra, S.; Monti, O.L.A.; Armstrong, N.R.; Sahoo, R.; Srivastava, D.N.; Pyun, J. Synthesis and colloidal polymerization of ferromagnetic Au-Co nanoparticles into Au-Co$_3$O$_4$ nanowires. *J. Am. Chem. Soc.* **2010**, *132*, 3234–3235. [CrossRef] [PubMed]

103. Wu, Z.; Deng, J.; Liu, Y.; Xie, S.; Jiang, Y.; Zhao, X.; Yang, J.; Arandiyan, H.; Guo, G.; Dai, H. Three-dimensionally ordered mesoporous Co$_3$O$_4$-supported Au—Pd alloy nanoparticles: High-performance catalysts for methane combustion. *J. Catal.* **2015**, *332*, 13–24. [CrossRef]

104. Casanova, O.; Iborra, S.; Corma, A. Biomass into Chemicals: Aerobic Oxidation of 5-Hydroxy-methyl-2-furfural into 2, 5-Furandicarboxylic Acid with Gold Nanoparticle Catalysts. *ChemSusChem* **2009**, *2*, 1138–1144. [CrossRef] [PubMed]

105. Ardemani, L.; Cibin, G.; Dent, A.J.; Isaacs, M.A.; Kyriakou, G.; Lee, A.F.; Parlett, C.M.A.; Parry, S.A.; Wilson, K. Solid base catalysed 5-HMF oxidation to 2,5-FDCA over Au/hydrotalcites: Fact or fiction? *Chem. Sci.* **2015**, *6*, 4940–4945. [CrossRef]

106. Van Nguyen, C.; Liao, Y.T.; Kang, T.C.; Chen, J.E.; Yoshikawa, T.; Nakasaka, Y.; Masuda, T.; Wu, K.C.W. A Metal-Free, High Nitrogen-Doped Nanoporous Graphitic Carbon Catalyst for an Effective Aerobic HMF-to-FDCA Conversion. *Green Chem.* **2016**, *18*, 5957–5961. [CrossRef]

107. Zhou, C.; Deng, W.; Wan, X.; Zhang, Q.; Yang, Y. Functionalized Carbon Nanotubes for Biomass Conversion: The Base-Free Aerobic Oxidation of 5-Hydroxymethyl- furfural to 2,5-Furandicarboxylic Acid over Platinum Supported on a Carbon Nanotube Catalyst. *ChemCatChem* **2015**, *7*, 2853–2863. [CrossRef]

108. Feng, Y.; Jia, W.; Yan, G.; Zeng, X.; Sperry, J.; Xu, B.; Sun, Y.; Tang, X.; Lei, T.; Lin, L. Insights into the active sites and catalytic mechanism of oxidative esterification of 5-hydroxymethylfurfural by metal–organic frameworks-derived N-doped carbon. *J. Catal.* **2020**, *381*, 570–578. [CrossRef]

109. Zhang, M.; Townend, I.; Cai, H.; He, J.; Mei, X. The influence of seasonal climate on the morphology of the mouth-bar in the Yangtze Estuary, China. Cont. *Shelf Res.* **2018**, *153*, 30–49. [CrossRef]

110. Zhong, W.; Liu, H.; Bai, C.; Liao, S.; Li, Y. Base-free oxidation of alcohols to esters at room temperature and atmospheric conditions using nanoscale Co-based catalysts. *ACS Catal.* **2015**, *5*, 1850–1856. [CrossRef]

111. Wu, G.; More, K.L.; Johnston, C.M.; Zelenay, P. High-performance electrocatalysts for oxygen reduction derived from polyaniline, iron, and cobalt. *Science* **2011**, *332*, 443–447. [CrossRef] [PubMed]

112. Deng, D.; Yu, L.; Chen, X.; Wang, G.; Jin, L.; Pan, X.; Deng, J.; Sun, G.; Bao, X. Iron Encapsulated within Pod-like Carbon Nanotubes for Oxygen Reduction Reaction. *Angew. Chem. Int. Ed.* **2013**, *52*, 371–375. [CrossRef]

113. Bao, L.; Sun, F.; Zhang, G.; Hu, T.L. High-efficient Aerobic Oxidation of Biomass-derived 5-Hydroxymethylfurfural to 2,5-Furandicarboxylic Acid over Holey 2D Mn$_2$O$_3$ Nanoflakes from a Mn-based MOF. *ChemSusChem* **2019**, *13*, 548–555. [CrossRef]

114. Fang, R.; Tian, P.; Yang, X.; Luque, R.; Li, Y. Encapsulation of ultra fi ne metal-oxide nanoparticles within mesopores for biomass-derived catalytic applications. *Chem. Sci.* **2018**, *9*, 1854–1859. [CrossRef]

115. Meng, J.; Niu, C.; Xu, L.; Li, J.; Liu, X.; Wang, X.; Wu, Y.; Xu, X.; Chen, W.; Li, Q.; et al. General Oriented Formation of Carbon Nanotubes from Metal–Organic Frameworks. *J. Am. Chem. Soc.* **2017**, *139*, 8212–8221. [CrossRef]

116. Fang, R.; Luque, R.; Li, Y. Selective aerobic oxidation of biomass-derived HMF to 2,5-diformylfuran using a MOF-derived magnetic hollow Fe–Co nanocatalyst. *Green Chem.* **2016**, *18*, 3152–3157. [CrossRef]

 *catalysts*

Article

Practical Approaches towards NO$_x$ Emission Mitigation from Fluid Catalytic Cracking (FCC) Units

Aleksei Vjunov *, Karl C. Kharas, Vasileios Komvokis, Amy Dundee and Bilge Yilmaz *

BASF Corporation, 25 Middlesex/Essex Turnpike, Iselin, NJ 08830, USA; karl.kharas@basf.com (K.C.K.); vasileios.komvokis@basf.com (V.K.); amy.dundee@basf.com (A.D.)

* Correspondence: aleksei.vjunov@basf.com (A.V.); bilge.yilmaz@basf.com (B.Y.); Tel.: +1-732-205-7003 (A.V.); +1-732-205-5232 (B.Y.)

Abstract: There appears to be consensus among the general public that curtailing harmful emissions resulting from industrial, petrochemical and transportation sectors is a common good. However, there is also a need for balancing operating expenditures for applying the required technical solutions and implementing advanced emission mitigation technologies to meet desired sustainability goals. The emission of NO$_x$ from Fluid Catalytic Cracking (FCC) units in refineries for petroleum processing is a major concern, especially for those units located in densely populated urban settings. In this work we strive to review options towards cost-efficient and pragmatic emissions mitigation using optimal amounts of precious metal while evaluating the potential benefits of typical promoter dopant packages. We demonstrate that at present catalyst development level the refinery is no longer forced to make a promoter selection based on preconceived notions regarding precious metal activity but can rather make decisions based on the best "total cost" financial impact to the operation without measurable loss of the CO/NO$_x$ emission selectivity.

Keywords: Pd-based promoter; NO$_x$ emission; environmental catalysis; refinery compliance; FCC

Citation: Vjunov, A.; Kharas, K.C.; Komvokis, V.; Dundee, A.; Yilmaz, B. Practical Approaches towards NO$_x$ Emission Mitigation from Fluid Catalytic Cracking (FCC) Units. *Catalysts* **2021**, *11*, 1146. https://doi.org/10.3390/catal11101146

Academic Editors: José Ignacio Lombraña, Héctor Valdés, Cristian Ferreiro and Jean-François Lamonier

Received: 30 July 2021
Accepted: 22 September 2021
Published: 24 September 2021

Publisher's Note: MDPI stays neutral with regard to jurisdictional claims in published maps and institutional affiliations.

1. Instructions

As society develops awareness and concern regarding the extent that environmental pollution, in particular that of the atmosphere, has a negative impact on the quality of life, there is an ever-increasing demand for the (petro)chemical industry to take action and contribute to the overall goal of environmental protection [1]. In an earlier communication we presented our view of the pragmatic approach towards CO oxidation in a fluid catalytic cracking unit, both from a catalyst design as well as implementation cost perspective [2]. There are, however, further considerations refineries may need to account for when operating in highly regulated areas, where additional focus is dedicated to NO$_x$ emissions [3].

For readers' reference, the schematic depiction of the FCC unit regenerator setup as well as the relevant catalyst movement steps under operating conditions are shown in Figure 1. Aside from CO$_2$, a greenhouse gas that is, unfortunately, emitted as part of the FCC process due to spent catalyst regeneration, the regenerator is also potentially a major source of CO emissions. These emissions are typically addressed by adding a CO promoter component, often Pt- or Pd-based, to the FCC catalyst formulation [4,5]. The specific promoter choice is often based on the FCC unit certification and existing emissions levels, as well as the financial commitment of the refiner towards environmental protection. Commonly, some 2–5 pounds of CO promoter per ton of FCC catalyst are necessary to maintain a reasonable (1–3 ppm) platinum group metal (PGM) concentration in the circulating catalyst inventory [6].

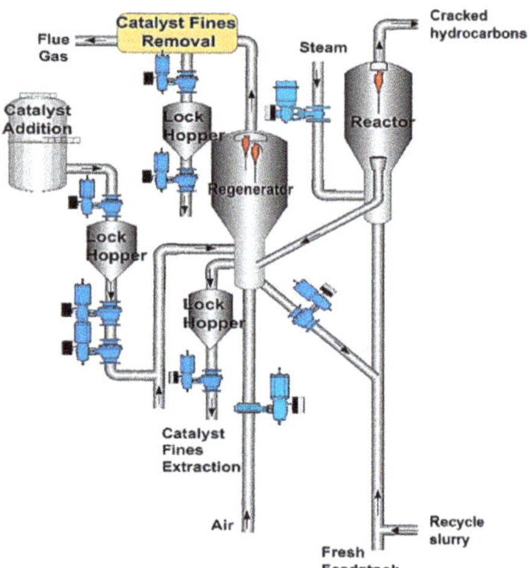

Figure 1. The FCC unit regenerator as well as the associated catalyst logistic streams are schematically shown. Image reprinted from a prior publication [2].

The critical downside of using promoter catalysts is that typically the PGM used to successfully convert CO to CO_2 is also prolific at oxidizing nitrogen-containing compounds to NO_x under the oxygen-rich conditions of the FCC regenerator [7,8]. The NO_x emissions from the regenerator may originate from a multitude of reaction pathways. As an example, nitrogen species may originate from crude oil and then be deposited as part of coke-species on the spent catalyst surface [9]. The combustion of coke in this case would lead to a release of this nitrogen in the form of NO_x once the catalyst is treated in the oxidizing environment of the regenerator. Alternatively, NO_x can also originate from nitrates and nitrites that can be present in the crude oil feedstock and are decomposed in the regenerator to form nitrous oxides. Finally, while not too common, residual ammonia species that were not removed in the riser can be oxidized to NO_x in the regenerator and thus contribute to the overall FCC unit NO_x emission footprint [10,11].

Regardless, the key concern is that the generated NO_x emissions are not only corrosive to the refinery equipment but are also highly harmful to the environment and human health [12]. Sometimes a refinery exhaust stack system will include a selective noncatalytic reduction (SNCR) process or selective catalytic reduction (SCR) emission system that allows for NO_x conversion to N_2 by means of treatment with NH_3 while invoking the help of a catalyst [13,14]. While the general notion is that the SCR units can successfully take care of NO_x, there is still significant concern around the consistency and concentration of the NO_x emissions over time as well as the ability of the SCR unit to accurately control the NH_3 dosing to match the ammonia demand to the NO_x emissions at any given time during the SCR unit operation [15]. The reason this aspect deserves significant attention is that the failure to comply with the emissions regulations may force a refinery FCC unit shutdown or may result in hefty financial penalties from the regulating authorities [16].

With the above considerations in mind, refineries located in areas with stringent emissions regulations are therefore required to carefully select the promoter package used in the FCC unit to achieve the delicate balance between CO oxidation and possible NO_x generation. Moreover, the refiner is also interested in a linear or at least as linear as possible, emissions profile, to avoid any NO_x emissions spikes that could potentially disrupt the SCR operation and lead to a NO_x breakthrough at the stack.

In the past, there have been several extensive and fundamental studies into the NO_x formation and mitigation chemistry [17,18]. Certainly, there is a significant degree of understanding and a number of strategies, especially in mobile emissions catalysis, as to ways for NO_x mitigation [19]. In this work, however, we will review and discuss the potential pathways towards improved NO_x generation control as well as the practical aspects of a catalyst design implementation in a refinery FCC unit. We will review approaches that allow suitable CO/NO_x selectivity, i.e., desired CO oxidation at acceptable NO_x generation levels. Furthermore, the impact of different types of dopants on the promoter activity and selectivity will be discussed. We will also share some of the considerations the typical refiner needs to account for to maintain both an economically viable as well as environmentally sustainable unit operation.

2. Results and Discussion

2.1. Promoter Pd-Loading

There is a general tendency for refiners to assume that when emissions related promoters containing PGM are concerned, a higher loading (wt. % or ppm) of the precious metal will yield higher activity, e.g., for CO oxidation. And, in fact, oftentimes this is the case; however, the obvious question that a refiner will ask next is regarding the cost. That is, is it worth the increased operating expenses?

To address this concern, we have prepared a series of catalysts with identical support but varying concentration of Pd as shown in Figure 2. All samples discussed in this work are summarized in Table 1 in the experimental section. The 250–1000 ppm Pd range is quite typical for industrial applications [9], with higher concentrations generally less viable in the refining industry due to the prohibitively high costs. As an example, 1000 ppm Pd in a promoter would result in an ~900 \$/kg Pd metal surcharge (at 2800 \$/oz.t. Pd price on 1 July 2021) [20]. In this work we have chosen a ceria/alumina support as basis for all testing since the extent to which one can achieve lower NO_x emissions has been shown to depend on the ceria-content of CO promoters [2]. Among other possible explanations, we want to mention here the potential of ceria to reduce a portion of the NO_x emissions by trapping some of the NO_2 in the regenerator and storing it in the form of cerium nitrate which can then be decomposed in the FCC unit riser. Examples of successful implementation of this strategy are well known from prior reports of mobile emissions catalyst systems [21–23].

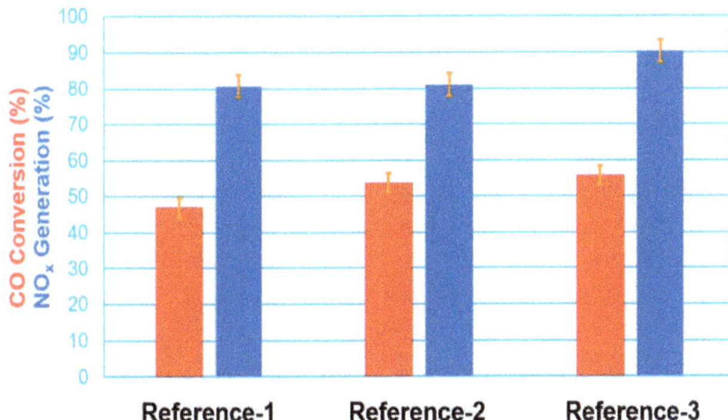

Figure 2. CO conversion (%) and NO_x generation emission (%) values for identical CO promoters varying only in Pd content are reported. The values are generated following the procedure described in Section 3.2.

Table 1. The sample composition as well as measured CO conversion, NO_x generation and CO/NO_x ratio values for the samples discussed in this contribution are reported.

Promoter	CeO_2 (wt. %)	CuO (wt. %)	SrO (wt. %)	Pd (ppm)	Pt (ppm)	Total Surface Area (m^2/g) [2]	CO Conversion (%) [1]	NO_x Generation (%) [1]	CO/NO_x Ratio
Reference-1	10	0	0	250	0	87	47.0	80.6	0.58
Reference-2	10	0	0	500	0	87	53.7	80.9	0.67
Reference-3	10	0	0	1000	0	87	55.7	90.2	0.62
Catalyst A	10	0.6	0	0	0	88	23.5	35.3	0.67
Catalyst B	10	0.3	0	500	0	86	46.8	65.1	0.72
Catalyst C	10	0.6	0	500	0	88	49.8	67.5	0.74
Catalyst D	10	0	0.6	0	0	90	4.8	6.8	0.71
Catalyst E	10	0	0.6	500	0	90	47.7	67.5	0.71
Catalyst F	10	0	1.2	500	0	86	47.6	79.1	0.60
Catalyst G	10	0	0	0	300	87	72.5	107.6	0.67

[1] Test: 99% spent FCC catalyst + 1% fresh promoter, plug flow reactor operating at 1 L/min flow rate with a feed of 2 vol. % O_2 in N_2 at 700 °C. Values determined by comparing to a promoter-free base case experiment. The method used to calculate CO conversion and NO_x generation is reported in the next chapter. [2] The same batch of alumina support was used for all samples allowing for comparable pore size distribution and structure for all catalysts. The typical total pore volume for this support is 0.2 cm^3/g.

From Figure 2, Reference-1 (250 ppm Pd) appears to be only slightly inferior to Reference-2 (500 ppm Pd) when the CO conversion efficiency is concerned. There is also minimal to no variability in NO_x emission, which one could consider surprising at first, yet there may be a plausible explanation to this observation. As reported previously [24,25], the effective PGM surface area available for the reaction does not necessarily scale proportionally with the total PGM content in the sample. That is, while in a 1 nm Pd particle ~50% of Pd atoms form the surface, the surface Pd atoms in at 20 nm particle account for only ~5% of the total Pd present in the particle. Therefore, it should not be surprising that the activity benefit observed from the higher Pd content in Reference-2 and Reference-3 (Figure 2) offers are minor benefit to the overall promoter performance. This is especially concerning from a practical utilization aspect since the Reference-2 and Reference-3, due to their Pd content, come at a two- and 4-fold premium in PGM cost, respectively, when compared to Reference-1.

Before we move on, we also need to emphasize that one of the key parameters which may be used to assess promoter efficiency is the CO/NO_x ratio, i.e., the NO_x generation penalty or cost for CO oxidation. Unsurprisingly, higher values would be more desirable, and even seemingly minor shifts in this ratio can significantly affect the refinery's decision regarding the type of promoter/catalyst to be used in the unit. The CO/NO_x ratios for the reference promoters are listed in Table 1 with the Reference-2 design exhibiting the best value at 0.67. With the above in mind, as well as considering the fact that in this contribution we aim to demonstrate options that the typical refinery faces, and also in order to stay within the reasonable PGM content limits that are typical for the industry, we will use Reference-2 in subsequent comparisons.

2.2. Impact of Dopant Package

Based on the findings in the previous section, further testing is performed using catalysts that contain 500 ppm Pd in all cases. The focus here is the impact of dopants which may or may not affect the CO and NO_x activity of the promoters. Figure 3 shows the results observed for several samples that were prepared and tested in this study.

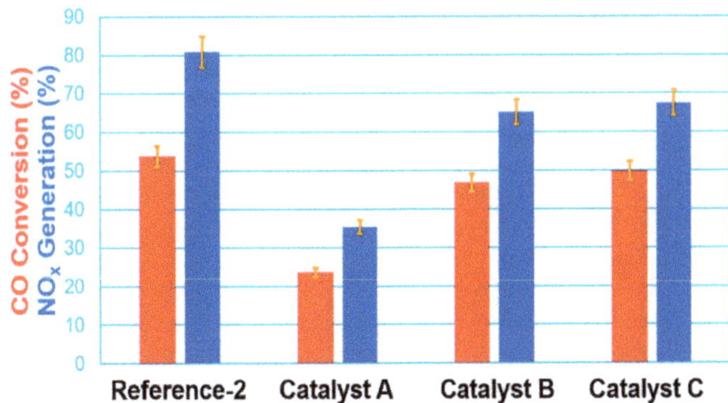

Figure 3. CO conversion (%) and NO$_x$ generation emission (%) values for 500 ppm Pd promoters with varying dopant packages are reported. The values are generated following the procedure described in Section 3.2.

The benchmark promoter in this series is the dopant-free Reference-2 catalyst that demonstrates a 53.7% CO conversion efficiency at an 80.9% NO$_x$ generation level, which yields value of 0.67 CO/NO$_x$. The addition of just 0.3% CuO, Catalyst B in Figure 3, leads to a decrease of CO conversion to 46.8% while generating only 65.1% NO$_x$ in the process, i.e., we observe a 0.72 CO/NO$_x$ yield. Interestingly, while the nominal yield of CO/NO$_x$ has gone up from 0.67 observed for the undoped Reference-2 sample, the overall activity with regards to absolute emission value has dropped significantly. This observation suggests that CuO acts as a poison for the Pd that was deposited on the support surface. We note, however, that this observation is not totally surprising as copper affinity to PGM is well known [26,27] and has been earlier shown to be generally detrimental to Pd performance in mobile emissions catalysts [21]. Cu can alloy with Pd, possibly forming Pd-core Cu-shell structures as have been reported previously [28]. The encapsulation of Pd, i.e., deactivation of the surface, would also explain the drop in NO$_x$ generation observed for Catalyst B.

Following the logic of the observation above, it would be reasonable to expect that further addition of CuO would lead to an even stronger degree of Pd encapsulation and deactivation. However, this is not the case. When doping 500 ppm Pd with 0.6% CuO (Catalyst C), the promoter appears to gain some CO activity (49.8%) and develops additional NO$_x$ (67.5%). As a result, a CO/NO$_x$ ratio of 0.74 is observed, which appears to be an improvement, in fact to be the best sample discussed in this set so far. The results suggest that CuO may be a CO promoter that just happens to partially deactivate Pd, at least to a certain degree.

To further explore the behavior of CuO as a catalyst under chosen reaction conditions, a Pd-free 0.6% CuO Catalyst A is tested (Figure 3). With a CO conversion level of 23.5% at a NO$_x$ generation of 35.3% the catalyst exhibits a CO/NO$_x$ of 0.67, which, coincidentally, is same as the Cu-free Reference-2 catalyst. This observation suggests that the PGM-free catalyst exhibits the same selectivity for CO as the Pd-design yet achieves only ~44% of the total activity of the PGM option (Reference-2). The importance of this observation is that beyond the simple "highest conversion catalyst" choice one now needs to also address concerns of a more practical nature to determine the "best" solution for the application in each specific case.

In most cases, the refinery will first need to consider whether the performance of a promoter, present typically at a 0.5–1 wt. % loading in the FCC catalyst blend, is sufficient to meet the set emissions targets. In this case, one would expect CuO-based, or similar in nature, catalysts to dominate the CO promoter field. Alternatively, the refinery could choose to accept a blend with a much higher promoter loading, yet that could come at the expense of the FCC activity [29]. Furthermore, the addition of higher promoter amounts tends to effectively serve as a cracking catalyst dilutant and, therefore, may result in reduction of the unit conversion as well as gasoline and LPG yields, a compromise most refiners would not be willing to accept. Furthermore, increasing the dosing of the CuO-based promoter, e.g., such as Catalyst A, would lead to significant CuO-levels in the equilibrium catalyst (ECat) that is periodically extracted from the FCC unit. Blending of Catalyst A and Reference-2 to achieve reduced Cu levels is described in detail in the Supporting Information section (Figure S1).

The equilibrium catalyst, also referred to as purchased equilibrium catalyst (ECat or PCat), is not necessarily a refinery waste stream. On the contrary, this material is a commodity that the refinery can often sell to another refinery that operates under different constraints or has a very limited operating budget, in which case the purchase of the relatively cheap PCat may be a financially viable option to maintain business. In fact, under the COVID-19 economic conditions, when the demand for transportation fuels dropped to historically low levels [30], a significant number of refineries were forced to switch to elevated levels of PCat purchasing from certain large and complex FCC units globally to be able to sustain operations. One of the main concerns around using CuO in FCC units is the strong tendency to dehydrogenate hydrocarbons, which in turn leads to significant amounts of H_2 and coke as well as has a detrimental effect on gasoline and LCO yields [31,32]. For those reasons, the general notion in the refining community is to avoid adding metals such as Cu, Ni or Co to the FCC units. Certainly, this makes the CuO-containing CO promoters less attractive and becomes a concern for potential buyers of PCat.

A further concern is that the increase of CuO concentration in PCat can trigger several environmental health and safety (EH&S) concerns around the handling and disposition of spent catalyst, with Cu known to be a potential hazard for aquatic species when exposed to marine environments or landfilled in significant concentrations [33,34].

With the limitations for CuO-based promoter use as described above, a more common approach for a refinery is to choose a PGM-based promoter, often using Pt or Pd as the key catalytic species. This in turn brings us back to the results described above where the Pd-based CuO-free system (Reference-2) offers a CO/NO_x selectivity of 0.67 at an overall CO conversion level of ~53.7%. That is, the catalyst has the necessary activity density such that when it is added to the FCC blend at a maximum level of 1 wt. %, the desired emissions targets can be met without diluting the cracking catalyst.

The cost of the catalyst support in the example studied here would be essentially the same in all cases as we have purposefully used the same ceria/alumina support throughout the study. Hence, the differentiation of the cost is predominantly driven by the cost of the PGM and/or dopants, if these are present. To keep the overall analysis transparent, we need to make an assumption that the fixed cost of PGM and/or CuO deposition on the support is identical and will thus not affect the overall financial estimates. That is, we focus here on the cost of raw materials while omitting the large-scale manufacturing concerns to simplify the discussion. We also use a Pd metal spot value of 2800 $/oz.t. (reasonable average value for the late Q2 2021) [20] and a Cu-nitrate solution (28%) value of 3 $/kg. With the constraints in mind, Reference-2 with the CO/NO_x selectivity of 0.67 costs ~450 $/kg in Pd for a promoter achieving a total CO conversion of ~53.7%. The refinery may choose to go for a higher CO/NO_x selectivity, e.g., 0.74 achieved by Catalyst C (500 ppm Pd + 0.6% CuO), which would still cost ~450 $/kg because of the negligible (<0.1 $/kg) cost impact from addition of the CuO in this case. Therefore, at first glance, it would seem the shift in selectivity is essentially free, however, as discussed earlier, CuO has some tendency to poison Pd, i.e. the shift in selectivity also leads to a decrease of the overall CO conversion efficiency from ~53.7 to ~49.8%. While seemingly minor, this ~7.2% relative decrease of activity must be considered. If the refinery is operating well below the CO emission limit while being close to the maximum allowed NO_x emission, the refinery may choose to pay the premium and accept the less-than-perfect Pd utilization. On the contrary, if the refinery is somewhat concerned about NO_x, but aims to comply with the CO emissions regulation, the unit would be forced to use a more selective catalyst but increase the dosing. For example, to match the CO activity of the CuO-free catalyst, one would need to dose 7–8% more of the Pd/CuO promoter, which in turn further increases cost to ~480–490 \$/kg, a roughly 30 – 40 \$/kg premium for the improved CO/NO_x selectivity. Considering the promoter consumption across a network of refineries on an annual basis, especially in light of recent margin pressure, a decision to purchase premium promoters with better selectivity as opposed to a "good-enough" promoter, becomes quite costly.

In addition to addressing the NO_x emissions in the regenerator through a careful choice of a promoter catalyst, one can also choose to trap and store NO_x using a NO_x-adsorber, which can then be regenerated in the riser, where NO_x would be reduced to N_2 and NH_3, both of which can be removed at the top of the riser column. The concept of trapping NO_x is well established in the field of mobile emissions control, where elements such as Sr and Ba are used to selectively trap NO_2, e.g. in diesel motor emissions control systems [35–37]. The adsorber functions in a stoichiometric way and, ideally, should not introduce any adverse performance effects to the overall promoter system. To explore the potential of the trapping concept, we have decided to invoke Sr (Catalyst D, E & F) and compare it's impacts on the promoter activity with the Pd-only (Reference-2) system. Sr was chosen over Ba due to its molecular weight (87.62 g/mol for Sr compared to 137.33 g/mol for Ba), which means that at the same nominal wt. % in the promoter formulation we can achieve a significantly higher number of Sr-sites, and consequently an ~32% higher potential nitrate capacity (SrO vs. BaO) as both elements can bind two NO_2 molecules [38]. The CO conversion % as well as NO_x generation % values for the compared samples are reported in the Figure 4.

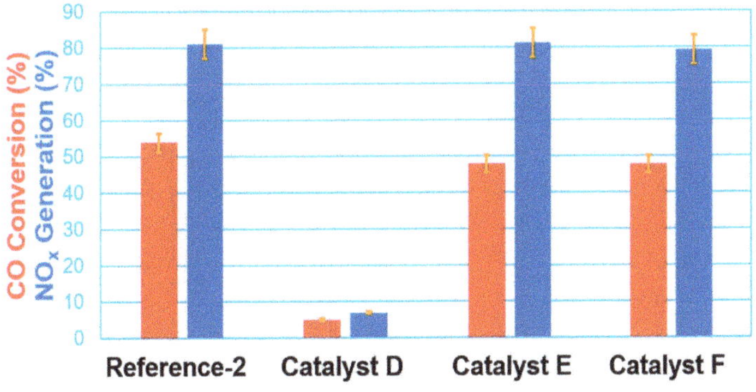

Figure 4. CO conversion (%) and NO_x generation emission (%) values for select promoters with varying dopant packages are reported. The values are generated following the procedure described in Section 3.2.

Since Sr is known to have no redox activity [38,39], it is a reasonable assumption that the addition of Sr does not lead to an increase in the CO oxidation activity. It is also not likely that the addition of a relatively small (0.6 wt. %) amount of SrO dopant could markedly block CeO_2 (or rather the mixture of Ce_2O_3 and CeO_2) that is part of the catalyst support, from delivering some "background" CO oxidation activity. This is exemplified in Figure 4 where Catalyst D (PGM-free SrO/CeO_2-based design) shows a 4.79% CO conversion efficiency and a 6.82% NO_x generation. CO/NO_x ratio in this case is 0.71, however, we do not deem it to be significant since the overall activity is extremely low. The unanswered question, nonetheless, is whether the addition of Sr has any benefit. That is, whether Sr can capture NO_x, but the amount generated on the ceria support is overwhelming the trap or whether the trap cannot effectively function due to the high operating conditions of the regenerator. This concern gains importance since the traditional NO_x trap operating window in mobile emissions applications, e.g. for Fuel Cut NO_x Trap

(FCNT) applications, is between ~300 and 500 °C with temperature above ~650–700 °C used for the trap regeneration [40].

To explore this concern further we have prepared the promoter Catalyst E with 500 ppm Pd and 0.6 wt. % SrO on the same particle, see Figure 4. Interestingly, there is no benefit in NO_x emission reduction, but there appears to be a reduction in the CO oxidation activity. Whether this observation suggests that Sr affect the Pd/support interaction and limits the PGM ability to make CO_2 remains unclear at this time. To further probe the concept, we have also prepared Catalyst F (500 ppm Pd with 1.2% SrO), which seems to maintain the CO activity same as Catalyst E, but now allows a slight decrease in NO_x generation. While conceptually possible, we suggest that further increase of SrO content becomes impractical for refineries for reasons like those discussed for CuO earlier. SrO is certainly less of an EH&S concern when it comes to catalyst handling, however, there is still the potential for Sr-mobility. Specifically, there is the concern of Sr forming carbonate deposits around valves and fittings in an FCC unit, which while not highly likely, is something refiners would consider, especially when the Sr levels are increasing above ~1–2 wt. % in the promoter catalyst.

2.3. Pt vs. Pd Comparison

Let us now turn to the obvious question the reader may have after the Pd-based catalyst analysis: how does a Pd-based promoter compare to a Pt-based promoter? In the past it was common to argue that NO_x-sensitive FCC units ought to use a Pd promoter and all other units can use Pt-based catalysts. While there may have been a performance-related rationale in the past, when emissions standards were less stringent and the promoter catalyst designs were very simple, e.g. PGM/Al_2O_3. Furthermore, Pd used to be more affordable in the past compared to Pt [19,20]. This trend reversed in 2015 with the decrease in consumer preference for diesel-powered light-duty vehicles, which led to a decline in the global Pt prices [41]. Today, in 2021, with Pd price is more than double that of Pt, the number of refineries willing to utilize Pd-based promoters is declining. To address the question at hand, we have compared Reference-2 with a Pt-based promoter (Catalyst G) using the design and sample synthesis methods reported in a previous publication [2]. Because the support material is ceria-alumina in both cases, we can specifically probe the impact of the PGM on the observed performance. The comparison is reported in Figure 5.

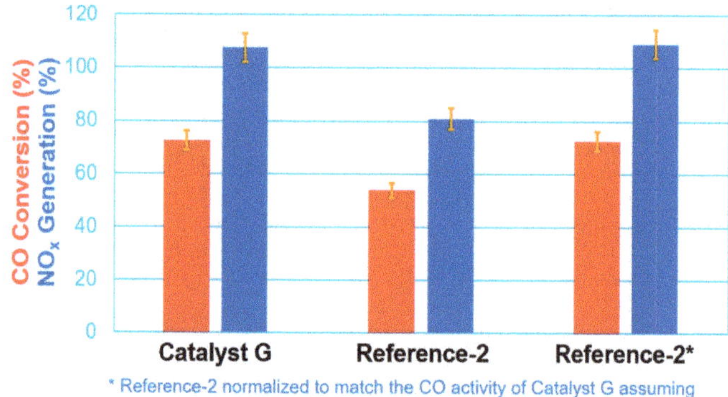

Figure 5. CO conversion (%) and NO_x generation emission (%) values for select Pt- and Pd-based promoters are reported. The values are generated following the procedure described in Section 3.2.

Catalyst G exhibits strong CO conversion at 72.5% while generating 107.6% NO_x emission, which yields a CO/NO_x ratio of 0.67. Incidentally, and quite surprisingly, this is exactly the CO/NO_x ratio observed for the Pd-based Reference-2. For readers convenience, we have also included a linear extrapolation of the performance of Reference-2 (reported as Reference-2* is Figure 5) performance at an elevated usage in the unit assuming the goal of the experiment were to match CO activity of the Pd-based design to that of Catalyst G. Not surprisingly, the NO_x is almost identical to what is observed for the Pt-based catalyst G since the CO/NO_x ratio in this assumption remains unchanged. Considering the PGM cost of Reference-2 is ~450 \$/kg, increasing the dosage by ~26% to match the CO activity to Catalyst G (300 ppm Pt design), the PGM cost of Reference-2* is estimated at ~567 \$/kg. Note that Catalyst G, a Pt-based CO promoter, has a PGM cost of only ~107 \$/kg at current Pt (1135 \$/oz t. in July 2021) value [20], making Catalyst G very attractive from a cost to operate perspective.

What the observation also suggests is that because the CO/NO_x ratios of the two promoters, Catalyst G and Reference-2, are identical, one can now choose the suitable PGM based on refinery environmental regulatory (e.g., EPA in the US) consent decrees or PGM price, rather than NO_x performance. In other words, the former notion of having to use a Pd-based promoter for FCC units that are NO_x constrained is not necessarily true at present. This in turn is excellent for the refineries, who can now maximize the cost savings by choosing the product based on current market PGM-price. That is, in Q2 2021 Pt-based promoters are strongly preferred, however, in the future, should Pd once again be lower priced than Pt, the Pd-based promoters can once again become financially attractive. Moreover, it appears there is flexibility in the choice of PGM, after all, the two catalysts compared here are supported on the exact same kind of ceria-alumina support.

3. Experimental

3.1. Material Synthesis

A series of reference and experimental catalysts was prepared-the full list is provided in Table 1. In all cases, γ-Al_2O_3 microspheres with a typical D_{50} of 80 μm and a total surface area (TSA) of 90 m^2/g were used as the alumina support. Commercially available cerium nitrate, copper nitrate and strontium nitrate, industrial grade in all cases, were used as precursors to form the doping package for the alumina support. An aqueous solution of Pd-nitrate, with a typical Pd-content of 20 wt. %, and Pt-nitrate, with a typical Pt content of 10 wt. % was used as PGM source. Both dopants as well as PGM were deposited via incipient wetness impregnation to afford even metal distribution on the support.

Reference catalysts 1–3 were prepared by impregnating the alumina support with the aqueous palladium nitrate solution such that the desired final metal concentration is achieved. The reference catalysts included in this study (reported in Table 1) are Pd/Al_2O_3 with 250–1000 ppm precious metal loading, as earlier reported as industry standard and reference point in the past [9]. Experimental Catalysts A–G were prepared by first impregnating the alumina support with cerium-nitrate such that the desired final CeO_2 concentration is achieved. The support is then calcined at 550 °C for 2 h. For Catalyst A, the ceria-alumina support was further impregnated with Cu-nitrate at incipient wetness and the material was then calcined at 550 °C for 2 h. Catalyst B is prepared like catalyst A but using half the amount of Cu-nitrate to achieve the desired CuO-concentration. The material is then calcined at 550 °C for 2 h. Subsequently, the desired amount of the Pd nitrate solution is impregnated onto the $CeO_2/CuO/Al_2O_3$ support. Catalyst C is prepared by impregnating Catalyst A with the desired amount of the Pd nitrate solution such that the final Pd concentration target is met. Catalyst D is prepared by impregnating the ceria/alumina support with strontium nitrate followed by a calcination at 550 °C for 2 h. Catalyst E is prepared by impregnating Catalyst D with the desired amount of the Pd nitrate solution such that the final Pd concentration target is met. Catalyst F is prepared by impregnating the ceria/alumina support with strontium nitrate followed by a calcination at 550 °C for 2 h. Subsequently, the material is impregnated with the desired amount of the Pd nitrate solution such that the final Pd concentration target is met. Catalyst G is prepared by impregnating the ceria/alumina support with the desired amount of the Pt nitrate solution. For all catalysts, upon PGM deposition, samples were calcined at 550 °C for 2 h.

The chemical composition of the catalysts was verified using a combination of X-ray Fluorescence (XRF) (PANalytical, Westborough, MA, USA), used for base and first transition row metal quantification, as well as Inductively Coupled Plasma (ICP) (Agilent Technologies, Santa Clara, CA, USA), used for Pd and Pt quantification. The measured values are essentially equivalent to those reported in Table 1, with a typical uncertainty of ~0.1 wt. % for base and first transition row metals and ~15 ppm for Pd and Pt. As there is no indication that the reported levels of experimental uncertainty in the chemical composition analysis could significantly affect the observations or conclusions presented in this work, these values were rounded up when reported in Table 1.

3.2. Catalyst Ageing and Testing

All catalyst testing work was performed as the Chemical Process & Energy Resources Institute (CPERI) located in Thessaloniki, Greece. The testing was in accord with the industry standard protocol of CPERI; detailed description of this test protocol was previously reported [42].

The protocol for the evaluation includes mixing spent FCC catalyst with the additive (1 wt. %) and loading this mixture in a fluidized bed reactor. The above mixture is then regenerated at 700 °C by 2 vol. % O_2 diluted in N_2. The NO_x and CO emissions were measured during the regeneration procedure. The reduction in CO emissions resulted in increased NO_x emissions.

The effect of catalyst mixtures on NO_x and CO emissions during regeneration of FCC catalyst was tested using a fluidized-bed reactor. A three-zone radiant heater furnace is used to heat up the reactor to the desired temperature, achieved via standard temperature controllers. Reaction temperature is measured by a thermocouple placed inside the catalytic bed. The bench-scale unit is equipped with a gas feed system, capable to supply the following gas components: O_2 and N_2. The volume flow rates of individual components, at the laboratory temperature, are monitored by mass flow controllers. The analysis section of the unit involves a FT-IR gas analyzer from MKS Instruments (MKS-MG2030). For the purposes of this study the signals from the NO_x and CO were recorded and stored in the PC every 5 s. Integration of the gas concentration vs. time curves provided the cumulative amounts of gases produced or consumed.

All of the additives were evaluated in respect to their ability to affect NO_x and curtail CO emissions during the regeneration of the spent FCC catalyst. The pure spent catalyst was used for the base case tests, while for the evaluation studies mixtures of 1 wt. % of the additive and 99 wt. % of the spent catalyst were loaded on the reactor. All catalytic materials were sieved at 63–106 μm. The reaction conditions for this protocol are summarized in Table 2. Pure spent catalyst was tested as a base case and each experiment was carried out more than once for repeatability reasons. All results showed great repeatability, for the chosen reactor setup and under the specific experimental conditions reported in this study, an average variation of ±5% was determined. For readers' reference, this variation is also reported as error bars in the Figures presented in this report. The NO_x generation and CO conversion, defined as follows from integrated amounts (gmol/gr of spent catalyst), is also calculated in relation to the base case. All catalysts were tested after a simulated steam-ageing, which is used to simulate the regenerator portion of the FCC operation cycle and has been demonstrated sufficient to discern trends and draw reasonable conclusions for promoter catalyst previously [2]. The chosen procedure is a 12 h ageing with a T_{max} = 787 °C using a closed, fluidized steam reactor that using 90% H_2O-steam/10% N_2 mixture.

Table 2. The CO promoter testing parameters are reported.

Reactor Type	Fluid Bed
Reactor loading	10 gr
Catalyst mixture	99 wt. % spent FCC catalyst + 1 wt. % CO promoter
Inlet gas flow rate	1 L/min
Inlet gas composition	2 vol. % O_2 in N_2
Reactor bed temperature	700 °C

The CO conversion (% decrease) values were calculated using the formula reported in Equation (1) below, where the "Base Case" refers to the CO emissions as measured for the non-promoted base system, i.e., the spent FCC catalyst without any additional environmental additives. In this "Base Case" the spent FCC catalyst is diluted with 1 wt. % inert microsphere phase to allow for correct catalyst amount for subsequent comparisons (inert replaced with active ingredients). The "Promoter" refers to the CO emissions observed when the spent FCC catalyst is promoted with the respective additive at a 1 wt. % doping level. In this way, the measured results can directly be related to the additive performance.

$$CO\ (\%\ decrease) = 100 \times \frac{CO_{Base\ Case} - CO_{Promoter}}{CO_{Base\ Case}} \tag{1}$$

The NO_x generation (% increase) values were calculated using the formula reported in Equation (2) below, where the "Base Case" refers to the NO_x emissions as measured for the non-promoted base system, i.e., the spent FCC catalyst without any additional environmental additives. Similar to the approach described above, in the "Base Case" the spent FCC catalyst is diluted with 1 wt. % inert microsphere phase to allow for correct catalyst amount for subsequent comparisons (inert replaced with active ingredients). The "Promoter" refers to the NO_x emissions observed when the spent FCC catalyst is promoted with the respective additive at a 1 wt. % doping level. In this way, the measured results can directly be related to the additive performance. Note that the general trend for CO promoters to oxidize CO to CO_2 and N-species to NO_x, in the context of Equation (2), results in positive values for NO_x generation, as it describes the additional NO_x forming as a result of promoter addition to the FCC catalyst.

$$NO_x\ (\%\ increase) = 100 \times \frac{NO_{x\,Promoter} - NO_{x\,Base\ Case}}{NO_{x\,Base\ Case}} \tag{2}$$

4. Conclusions

In this work we have considered the conceptual approaches toward NO_x mitigation from a refinery FCC unit and have found that the key consideration when choosing a catalyst is not only based on the absolute activity values, but also on the CO/NO_x selectivity as well as the overall promoter cost to the refinery. There are clear benefits of using dopants, e.g. Cu and Sr, however their practical application is limited due to a number of operational, economic as well as EH&S concerns. With the current generation of promoter catalysts as well as the existing emissions regulations in mind, it is now viable to use Pt-based promoters instead of Pd-based ones, that were preferred in the past when the Pt/Pd cost spread was significantly in favor of Pd.

Supplementary Materials: The following are available online at https://www.mdpi.com/article/10.3390/catal11101146/s1, Figure S1: The CO conversion and NO_x generation emission values observed for selected catalysts are reported. The values are generated following the procedure described in Section 2.2. Table S1: The estimated sample cost as well as measured CO conversion, NO_x generation and CO/NO_x ratio values for the samples discussed in this contribution are reported.

Author Contributions: Conceptualization, A.V. and B.Y.; methodology, A.V., K.C.K. and V.K.; formal analysis, A.V.; investigation, V.K.; resources, B.Y.; data curation, A.D.; writing—original draft preparation, A.V. and B.Y.; writing—review and editing, A.V., K.C.K. and B.Y.; project administration, B.Y. All authors have read and agreed to the published version of the manuscript.

Funding: This research received no external funding.

Data Availability Statement: The data presented in this study are available on request from the corresponding author.

Acknowledgments: The authors would like to recognize the support and high quality of performed CO promoter testing by Angelos Lappas and Eleni Pachatouridou at the Chemical Process & Energy Resources Institute (CPERI) located in Thessaloniki, Greece.

Conflicts of Interest: The authors declare no conflict of interest.

Abbreviations

Platinum group metal (PGM), oxygen storage component (OSC), fluid catalytic cracking (FCC).

References

1. Petroleum Sector (NAICS 324), US EPA. Available online: https://www.epa.gov/regulatory-information-sector/petroleum-sector-naics-324 (accessed on 21 September 2021).
2. Vjunov, A.; Kharas, K.C.; Komvokis, V.; Dundee, A.; Zhang, C.C.; Yilmaz, B. Pragmatic Approach toward Catalytic CO Emission Mitigation in Fluid Catalytic Cracking (FCC) Units. *Catalysts* **2021**, *11*, 707. [CrossRef]
3. Air Emissions—Protecting Air Quality, Chevron Sustainability Roadmap. Available online: https://www.chevron.com/sustainability/environment/air-emissions (accessed on 21 September 2021).
4. Lin, J.; Wang, X.; Zhang, T. Recent progress in CO oxidation over Pt-group-metal catalysts at low temperatures. *Chin. J. Catal.* **2016**, *37*, 1805–1813. [CrossRef]
5. van Spronsen, M.A.; Frenken, J.W.M.; Groot, I.M.N. Surface science under reaction conditions: CO oxidation on Pt and Pd model catalysts. *Chem. Soc. Rev.* **2017**, *46*, 4347–4374. [CrossRef] [PubMed]
6. BASF USP™ CO Promoter, Ultra Stable Promoter (USP) | BASF Catalysts. Available online: https://catalysts.basf.com/products/ultra-stable-promoter-ups (accessed on 21 September 2021).
7. Iliopoulou, E.F.; Efthimiadis, E.A.; Vasalos, I.A.; Barth, J.-O.; Lercher, J.A. Effect of Rh-based additives on NO and CO formed during regeneration of spent FCC catalyst. *Appl. Catal. B* **2004**, *47*, 165–175. [CrossRef]
8. Dishman, K.L.; Doolin, P.K.; Tullock, L.D. NO_x Emissions in Fluid Catalytic Cracking Catalyst Regeneration. *Ind. Eng. Chem. Res.* **1998**, *37*, 4631–4636. [CrossRef]
9. Chester, A.W. Chapter 6—CO combustion promoters: Past and present. *Stud. Surf. Sci. Cat.* **2007**, *166*, 67–77. [CrossRef]
10. Stockwell, D.M. Chapter 6—CO combustion promoters: Past and present. *Stud. Surf. Sci. Cat.* **2007**, *166*, 79–102.
11. Cheng, W.-C.; Kim, G.; Peters, A.W.; Zhao, X.; Rajagolapan, K.; Ziebarth, M.S.; Pereira, C.J. Environmental Fluid Catalytic Cracking Technology. *Catal. Rev.* **1998**, *40*, 39–79. [CrossRef]
12. Gary, J.H.; Handwerk, G.E. *Petroleum Refining: Technology and Economics*, 4th ed.; CRC Press: Boca Raton, FL, USA, 2001; ISBN 0824704827.
13. Sadeghbeigi, R. *Fluid Catalytic Cracking Handbook*, 3rd ed.; Chapter 14; Butterworth-Heinemann: Oxford, UK, 2012; ISBN 9780123869654.
14. Babich, I.V.; Seshan, K.; Lefferts, L. Nature of nitrogen specie in coke and their role in NO_x formation during FCC catalyst regeneration. *Appl. Catal. B Environ.* **2005**, *59*, 205–211. [CrossRef]
15. Ye, X.; Schmidt, J.E.; Wang, R.-P.; van Ravenhorst, I.K.; Oord, R.; Chen, T.; de Groot, F.; Meirer, F.; Weckhuysen, B.M. Deactivation of Cu-Exchanged Automotive-Emission NH_3-SCR Catalysts Elucidated with Nanoscale Resolution Using Scanning Transmission X-ray Microscopy. *Angew. Chem. Int. Ed.* **2020**, *59*, 15610–15617. [CrossRef]
16. US EPA, Civil Cases and Settlements by Statute. Available online: https://cfpub.epa.gov/enforcement/cases/index.cfm?templatePage=12&ID=1&sortby=TYPE_OF_ORDER (accessed on 21 September 2021).
17. Zhao, X.; Peters, A.W.; Weatherbee, G.W. Nitrogen Chemistry and NO_x Control in a Fluid Catalytic Cracking Regenerator. *Ind. Eng. Chem. Res.* **1997**, *36*, 4535–4542. [CrossRef]
18. Barth, J.-O.; Jentys, A.; Lercher, J.A. Elementary Reactions and Intermediate Species Formed during the Oxidative Regeneration of Spent Fluid Catalytic Cracking Catalysts. *Ind. Eng. Chem. Res.* **2004**, *43*, 3097–3104. [CrossRef]
19. Heck, R.M.; Farrauto, R.J.; Gulati, S.T. *Catalytic Air Pollution Control*; John Wiley & Sons: Hoboken, NJ, USA, 2009; ISBN 9781118397749. [CrossRef]

20. APMEX Precious Metals Trading. Available online: https://www.apmex.com/platinum-price (accessed on 21 September 2021).
21. Campbell, L.E.; Danzinger, R.; Guth, E.D.; Padron, S. Process for the Reaction and Absorption of Gaseous Air Pollutants, Apparatus Therefor and Method of Making the Same. U.S. Patent 5,451,558, 19 September 1995.
22. Theis, J.; Lambert, C. The Effects of CO, C_2H_4, and H_2O on the NO_x Storage Performance of Low Temperature NO_x Adsorbers for Diesel Applications. *SAE Int. J. Engines* **2017**, *10*, 1627–1637. [CrossRef]
23. Ferré, G.; Aouine, M.; Bosselet, F.; Burel, L.; Cadete Santos Aires, F.J.; Geantet, C.; Ntais, S.; Maurer, F.; Casapu, M.; Grunwaldt, J.-D.; et al. Exploiting the dynamic properties of Pt on ceria for low-temperature CO oxidation. *Catal. Sci. Technol.* **2020**, *10*, 3904–3917. [CrossRef]
24. Soliman, N.K. Factors affecting CO oxidation reaction over nanosized materials: A review. *J. Mat. Res. Tech.* **2019**, *8*, 2395–2407. [CrossRef]
25. Fox, E.B.; Velu, S.; Engelhard, M.H.; Chin, Y.-H.; Miller, J.T.; Kropf, J.; Song, C. Characterization of CeO_2-supported Cu–Pd bimetallic catalyst for the oxygen-assisted water–gas shift reaction. *J. Catal.* **2008**, *260*, 358–370. [CrossRef]
26. Koryabkina, N.A.; Phatak, A.A.; Ruettinger, W.F.; Farrauto, R.J.; Ribeiro, F.H. Determination of kinetic parameters for the water–gas shift reaction on copper catalysts under realistic conditions for fuel cell applications. *J. Catal.* **2003**, *217*, 233–239. [CrossRef]
27. Rej, S.; Wang, H.-J.; Huang, M.-X.; Hsu, S.-C.; Tan, C.-S.; Lin, F.-C.; Huang, J.-S.; Huang, M.H. Facet-dependent optical properties of Pd–Cu_2O core–shell nanocubes and octahedra. *Nanoscale* **2015**, *7*, 11135–11141. [CrossRef]
28. Alabdullah, M.A.; Rodriguez Gomez, A.; Vittenet, J.; Bendjeriou-Sedjerari, A.; Xu, W.; Abba, I.A.; Gascon, J. A Viewpoint on the Refinery of the Future: Catalyst and Process Challenges. *ACS Catal.* **2020**, *10*, 8131–8140. [CrossRef]
29. US Bureau of Labor Statistics. From the barrel to the pump: The impact of the COVID-19 pandemic on prices for petroleum products, From the Barrel to the Pump: The Impact of the COVID-19 Pandemic on Prices for Petroleum Products. Available online: https://www.bls.gov/opub/mlr/2020/article/from-the-barrel-to-the-pump.htm (accessed on 21 September 2021).
30. US Energy Information Administration. COVID-19 Mitigation Efforts Result in the Lowest U.S. Petroleum Consumption in Decades, COVID-19 Mitigation Efforts Result in the Lowest U.S. Petroleum Consumption in Decades. Available online: https://www.eia.gov/todayinenergy/detail.php?id=43455 (accessed on 21 September 2021).
31. Mills, G.A. Aging of Cracking Catalysts. *Ind. Eng. Chem.* **1950**, *42*, 182–187. [CrossRef]
32. Venuto, P.B.; Habib, T. Catalyst-Feedstock-Engineering Interactions in Fluid Catalytic Cracking. Catal. *Rev. Sci. Eng.* **1978**, *18*, 1–150. [CrossRef]
33. Kleldsen, P.; Barlaz, M.A.; Rooker, A.P.; Baun, A.; Ledin, A.; Christensen, T.H. Present and Long-Term Composition of MSW Landfill Leachate: A Review. *Critic. Rev. Environ. Sci. Technol.* **2002**, *32*, 297–336. [CrossRef]
34. Kiaune, L.; Singhasemanon, N. Pesticidal Copper(I) Oxide: Environmental Fate and Aquatic Toxicity. *Rev. Environ. Contam. Toxicol.* **2011**, *213*, 1–26. [CrossRef]
35. Wittka, T.; Holderbaum, B.; Dittmann, P.; Pischinger, S. Experimental Investigation of Combined LNT + SCR Diesel Exhaust Aftertreatment. *Emiss. Contr. Sci. Technol.* **2015**, *1*, 167–182. [CrossRef]
36. Onrubia-Calvo, J.A.; Pereda-Ayo, B.; Caravca, A.; De-La-Torre, U.; Vernoux, P.; Gonzalez-Velasco, J.R. Tailoring perovskite surface composition to design efficient lean NO_x trap Pd–$La_{1-x}A_xCoO_3/Al_2O_3$-type catalysts (with A = Sr or Ba). *Appl. Catal. B Environ.* **2020**, *266*, 118628. [CrossRef]
37. Zhang, Y.; Liu, D.; Meng, M.; Jiang, Z.; Zhang, S. A Highly Active and Stable Non-Platinic Lean NO_x Trap Catalyst MnO_x–$K_2CO_3/K_2Ti_8O_{17}$ with Ultra-Low NO_x to N_2O Selectivity. *Ind. Eng. Chem. Res.* **2014**, *53*, 8416–8425. [CrossRef]
38. Cotton, F.A.; Wilkinson, G.; Gaus, P.L. *Basic Inorganic Chemistry*, 3rd ed.; John Wiley & Sons: Hoboken, NJ, USA, 1995; ISBN 9780471505327. [CrossRef]
39. Luther, G.W., III. *Inorganic Chemistry for Geochemistry and Environmental Sciences: Fundamentals and Applications*; John Wiley & Sons: Hoboken, NJ, USA, 2016; ISBN 978-1-118-85137-1. [CrossRef]
40. Choi, M.; Song, J.; Lee, E.; Ma, S.; Lee, S.; Seo, J.; Yoo, S.; Lee, J. *The Development of a NO_x Reduction System during the Fuel Cut Period for Gasoline Vehicles*; SAE Technical Paper 2019-01-1292; SAE International: Washington, DC, USA, 2019. [CrossRef]
41. Hachenberg, B.; Kiesel, F.; Schiereck, D. Dieselgate and its expected consequences on the European auto ABS market. *Econom. Lett.* **2018**, *171*, 180–182. [CrossRef]
42. Efthimiadis, E.A.; Iliopoulou, E.F.; Lappas, A.A.; Iatridis, D.K.; Vasalos, I.A. NO Reduction Studies in the FCC Process. Evaluaiton of NO Reduciton Additives for FCCU in Bench- and Pilot Plant-Scale Reactors. *Ind. Eng. Chem. Res.* **2002**, *41*, 5401–5409. [CrossRef]

*catalysts*

MDPI

Article

Application of a Combined Adsorption—Ozonation Process for Phenolic Wastewater Treatment in a Continuous Fixed-Bed Reactor

Cristian Ferreiro [1,*], Ana de Luis [2], Natalia Villota [3], Jose María Lomas [3], José Ignacio Lombraña [1] and Luis Miguel Camarero [3]

[1] Department of Chemical Engineering, Faculty of Science and Technology, University of the Basque Country UPV/EHU, Barrio Sarriena s/n, 48940 Leioa, Spain; ji.lombrana@ehu.eus

[2] Department of Chemical and Environmental Engineering, Faculty of Engineering in Bilbao, University of the Basque Country UPV/EHU, Plaza Ingeniero Torres Quevedo, 1, 48013 Bilbao, Spain; ana.deluis@ehu.eus

[3] Department of Environmental and Chemical Engineering, Faculty of Engineering of Vitoria-Gasteiz, University of the Basque Country UPV/EHU, 01006 Vitoria-Gasteiz, Spain; natalia.villota@ehu.eus (N.V.); josemaria.lomas@ehu.eus (J.M.L.); luismiguel.camarero@ehu.eus (L.M.C.)

* Correspondence: cristian.ferreiro@ehu.eus; Tel.: +34-946-012-512

Abstract: This work studied the removal of phenol from industrial effluents through catalytic ozonation in the presence of granular activated carbon in a continuous fixed-bed reactor. Phenol was chosen as model pollutant because of its environmental impact and high toxicity. Based on the evolution of total organic carbon (TOC) and phenol concentration, a kinetic model was proposed to study the effect of the operational variables on the combined adsorption–oxidation (Ad/Ox) process. The proposed three-phase model expressed the oxidation phenomena in the liquid and the adsorption and oxidation on the surface of the granular activated carbon in the form of two kinetic constants, k_1 and k_2 respectively. The interpretation of the constants allow to study the benefits and behaviour of the use of activated carbon during the ozonisation process under different conditions affecting adsorption, oxidation, and mass transfer. Additionally, the calculated kinetic parameters helped to explain the observed changes in treatment efficiency. The results showed that phenol would be completely removed at an effective contact time of 3.71 min, operating at an alkaline pH of 11.0 and an ozone gas concentration of 19.0 mg L^{-1}. Under these conditions, a 97.0% decrease in the initial total organic carbon was observed.

Keywords: catalytic ozonation; three-phase modelling; fixed-bed reactor; wastewater treatment; phenol; granular activated carbon; Ad/Ox

Citation: Ferreiro, C.; de Luis, A.; Villota, N.; Lomas, J.M.; Lombraña, J.I.; Camarero, L.M. Application of a Combined Adsorption—Ozonation Process for Phenolic Wastewater Treatment in a Continuous Fixed-Bed Reactor. *Catalysts* **2021**, *11*, 1014. https://doi.org/10.3390/catal11081014

Academic Editor: Pedro B. Tavares

Received: 26 July 2021
Accepted: 21 August 2021
Published: 22 August 2021

1. Introduction

Increases in the world population and the industrial revolution have brought many advantages to humanity. However, the intensive use and pollution of natural resources is leading many developed countries on the European and American continents towards an ecological deficit by the first third of the 21st century [1]. For this reason, national governments are encouraging the improvement of manufacturing production processes to increase the efficiency of water resources, as well as raw materials, in order to minimise the environmental impact of goods produced. In the context of water scarcity, the responsible use and management of wastewater is a desirable objective [2].

To safeguard the environment and public health, it is necessary the development and implementation of effective wastewater treatment that allow us to exceed the quality standards regulated by the U.S. Environmental Protection Agency's (EPA) Water Quality Standards Regulation (WQSR) [3] or the Water Framework Directive (Directive 2000/60/EC) of the European Union [4]. This is one of the highest-priority challenges that society must solve in the next decade. In fact, in 2015, the United Nations (UN) included SDG 6 on Clean

Water and Sanitation in the Sustainable Development Goals (SDGs) for the safe treatment of wastewater, to ensure the availability of water and its sustainable management for future generations [5].

A common example of this type of pollution is phenolic wastewater from the cellulose, oil refining, metallurgical, and plastics industries [6]. Phenol is an EPA priority compound listed on the Contaminant Candidate List 4 (CCL4) [7] of substances that are of high toxicity to humans and aquatic life. In humans, phenol has been reported as a potential source of skin irritation and kidney problems, and it has been associated with leukaemia and other mutagenic diseases [8]. Toxicity levels harmful to humans and aquatic life are between 9 and 25 mg L^{-1} [9]. In many countries, such as Brazil, its discharge into the environment is limited to a concentration of 0.5 mg L^{-1} and in the USA, to a concentration of less than 0.001 mg L^{-1}.

Different treatment methods have been proposed for the removal of phenols from wastewater, the most traditional being chemical oxidation by Fenton reaction (100%) [10], adsorption (26%) [6], extraction by liquid membrane emulsion (72%) [11], coagulation and precipitation (36.8%) [12], or activated sludge (87%) [9]. Other biological treatments [13], especially enzymatic treatments, using peroxidases (98%) [14] or tyrosinase (25%) [15] reduced the phenol concentration after 30 and 2 h respectively. Activated sludge is one of the most widely used treatments due to its low cost and ease of handling, but it is limited in applicability because the microorganisms, despite prior acclimatisation, are incapable of treating phenol concentrations of more than 100 mg L^{-1}. This is due to the low biodegradability of these effluents and the inhibitory effects that these concentrations of phenol have on the microorganisms [16]. Unfortunately, many wastewaters from chemical and petrochemical industries far exceed these concentrations. For example, in the coke industry and petrochemical plants concentrations in the range of 28–3900 mg L^{-1} are common [17]. Other treatments based on adsorption on activated carbons could be a feasible alternative. Adsorption is considered one of the most efficient and effective methods to separate emerging pollutants such as diclofenac [18] and petrochemical effluents containing benzene and toluene [19], due to its simplicity and flexibility of use, its high porosity, large specific surface area and, the high degree of surface interactions [20]. However, the presence of organic content in the wastewater could be a potential limitation because it could interfere in pollutants removal efficiency by competing for adsorption active sites on activated carbon [21]. Consequently, given the complexity of typical aqueous effluents in the industry, adsorption alone would not be the most suitable method.

In order to provide efficient solution for the treatment of these effluents, advanced oxidation processes (AOPs) have attracted the interest of many researchers due to their advantages [22,23]. Among the many existing AOPs, ozonation is a process with a high oxidation potential, which can lead to efficient removal of organic compounds, such as pharmaceuticals, personal care products, pesticides, solvents and surfactants, even at low ozone concentrations [24–26]. However, certain groups of organic compounds are particularly refractory to oxidation by ozone, such as carboxylic, oxalic, and pyruvic acids [27].

A combination of ozonisation and adsorption processes with activated carbon (AC) could be more efficient and sustainable treatment for wastewaters containing refractory organic pollutants. Some of the previous studies in which catalytic ozonisation processes with activated carbons were used for the removal of different organic compounds are listed in Table S1. According with Table S1, catalytic ozonation with granular activated carbons (GAC) could overcome the limitations of ozonation due to the adsorption capacity, high surface area, and the catalytic activity.

The catalytic mechanism of ozonisation in the presence of GAC is still unclear, but recent results suggest that the carbon essentially promotes the decomposition of ozone with a consequent increase in the production of radicals. These hydroxyl radicals would not be bound to the surface, remaining free to react in the aqueous phase. Therefore, the activated carbon would behave as an initiator of the radical-like chain reaction that transforms ozone into hydroxyl radicals, which in turn react with the organic compounds in the bulk [28].

On the other hand, other authors, such as Nawrocki and Kasprzyk-Hordern [29] and Guo et al. [30], have postulated that the activated carbon initiates the decomposition of ozone into hydroxyl radicals, and then the ozone reacts with the superficial oxygenated groups to generate H_2O_2, which in turn reacts with the ozone in the bulk to produce hydroxyl radicals. In other words, in this model, the catalyst plays a dual role during the catalytic ozonisation process. First, it adsorbs and decomposes the ozone, leading to the formation of active species, and then the active species reacts with the non-chemisorbed organic compounds. In addition, the activated carbon adsorbs organic compounds, and then reacts with the oxidising species generated on the catalyst's surface via the Criegee mechanism [31].

These mechanistic reactions of ozone decomposition and radical reactions, depending on the pH of the medium, can be summarised, according to Beltrán et al. [26]:

- Homogeneous reaction (at the liquid level):

$$O_3 + OH^- \rightarrow HO_2^- + O_2 \tag{1}$$

$$O_3 + HO_2^- \rightarrow HO_2^\bullet + O_3^{\bullet-} \tag{2}$$

$$O_3 + \text{Initiator} \rightarrow O_3^{\bullet-} + \text{Initiator}^+ \tag{3}$$

$$HO_2^\bullet \rightleftarrows O_2^{\bullet-} + H^+ \tag{4}$$

- Heterogeneous reactions (at the level of the solid):

For an acid pH:

$$O_3 + GAC \rightleftarrows O_3 - GAC \tag{5}$$

$$O_3 - GAC \rightleftarrows O - GAC + O_2 \tag{6}$$

$$O_3 + O - GAC \rightleftarrows 2O_2 + GAC \tag{7}$$

For an alkaline pH:

$$OH^- + GAC \rightleftarrows OH - GAC \tag{8}$$

$$O_3 + OH - GAC \rightleftarrows O_3^\bullet - GAC + HO^\bullet \tag{9}$$

$$O_3 + O^\bullet - GAC \rightleftarrows O_2^{\bullet-} + GAC + O_2 \tag{10}$$

- Homogeneous propagation and termination reactions:

$$O_3 + O_2^- \rightarrow O_3^{\bullet-} + O_2 \tag{11}$$

$$O_3^{\bullet-} + H^+ \rightarrow HO_3^\bullet \tag{12}$$

$$HO_3^\bullet \rightarrow HO^\bullet + O_2 \tag{13}$$

$$HO_3^\bullet + \text{Scavenger} \rightarrow \text{Oxidation products} \tag{14}$$

According with this, the integration of a GAC catalyst into a continuous adsorption–ozonisation (Ad/Ox) process would lead to a complete removal and mineralisation of phenol containing waters. Ad/Ox process is a complex system involving different aspects, such as mass transfer and radical generation or adsorption equilibria, among others. Few studies deepened the kinetics of the process with the prospect of implementation of this catalytic technology on an industrial scale. Thus, a study of the behaviour of the system operating in a fixed-bed under different operating conditions through a kinetic model could provide the necessary information to achieve the desired final scale-up.

Given the excessive number of variables involved in Ad/Ox process, some of which are unknown, it is impossible to develop a rigorous model, let alone propose a detailed reaction mechanism. Consequently, from a practical point of view, a model employing experimental data derived under different operational conditions can serve as a basis for real applications, as long as the model adequately simulates the experimental data.

Therefore, this study aimed to develop a three-phase kinetic model of a continuous process in a fixed-bed catalytic ozonation (Ad/Ox) treatment with granular activated carbon (GAC). Phenol was chosen as a model organic pollutant because of its environmental impact, high toxicity, and occurrence in the industry. It has been proposed a kinetic model that includes mass transfer parameters, adsorption equilibria, and reaction rate constants at the solid and liquid surface, to analyze the effect of operating conditions and to identify the operational strategies that will lead to increased degradation and mineralisation rates.

2. Results and Discussion

2.1. Removal of Phenol and Mineralisation

In order to study the influence of GAC on phenol removal, a preliminary experiment was performed with ozone alone in a fixed-bed reactor, with an inert material (glass spheres) and adsorption or ozonation only in the presence of GAC, in order to evaluate the improvement achieved by activated carbon. The transitory profiles of the primary degradation and mineralisation of the three systems are compared in Figure 1.

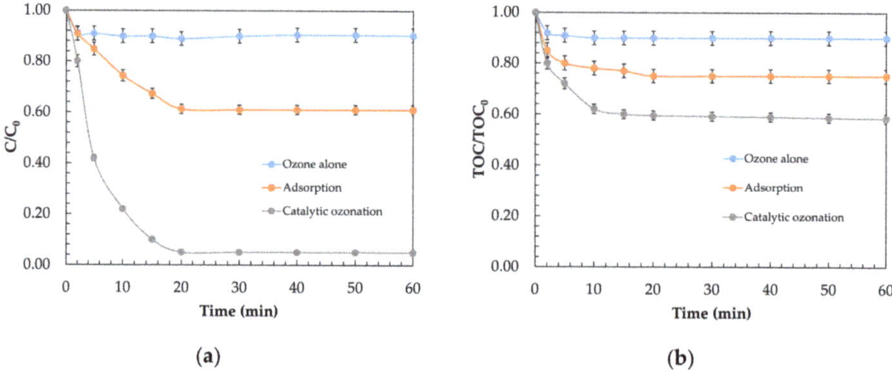

Figure 1. Comparison of ozonation alone, adsorption, and catalytic ozonation processes with GAC in a fixed-bed reactor for phenol removal. Evolution of (**a**) primary degradation and (**b**) mineralisation in terms of TOC. Experimental conditions: $C_0 = 250$ mg L^{-1}; $Q_G = 0.05$ L h^{-1}; $C_{O_3,G} = 12.0$ mg L^{-1}; $Q_L = 12$ mL min^{-1}; pH = 6.5; $M_{CAT} = 70.8$ g L^{-1}; $P = 1.0$ atm; $T = 20$ °C; $V = 0.14$ L.

Ozonation alone achieved a primary degradation of only 10% (see Figure 1). This low degradation could be due to the oxidation potential of ozone being lower (E° = 2.07 V) than that of the hydroxyl radicals (E° = 2.80 V) generated by the indirect reactions associated with the decomposition of ozone in the presence of GAC [22]. The mineralisation of phenol in the catalytic system was more efficient compared with ozonation alone or adsorption, with a 41.7% mineralisation achieved 20 min after reaching the steady state.

As can be seen in Figure 1b, the mineralisation obtained by catalytic ozonisation with GAC does not match the sum (35.0%) of the efficiencies obtained by ozonisation or adsorption with GAC. According to Lin et al. [32], this could be because during the catalytic ozonisation adsorption, reaction, and desorption processes of oxidised pollutants were involved, in contrast to the ozonisation. In overall, the results indicate that GAC had some catalytic activity to increase the generation of oxidative species responsible for the degradation of phenol. This same effect was observed by Xiong et al. [33], who obtained 26.1% additional mineralisation using the catalytic system with GAC in a basket reactor, compared with adsorption or ozonisation alone. According with their research, the lower molecular weight of the oxidation by-products adsorbed onto the GAC could explain the higher removal obtained in comparison with ozone alone.

2.2. Kinetic Model of the Ad/Ox Process Operated in a Continuous Fixed-Bed Reactor

The implementation of this system on an industrial scale requires predictions of the system's behaviour. Therefore, the development of the three-phase model was based on the considerations made by Ferreiro et al. [17], in which the G–L–S ozone mass transfer, the adsorption process of both ozone and phenol, and the parallel chemical reaction occurring at both the liquid and solid levels were taken into account. The application of the Ad/Ox model in a continuous system was based on the following considerations:

- The overall oxidation rate of the process is represented by the ozone consumption in the parallel reaction process (both at the liquid and solid level);
- The oxidation rate of the parallel stages, in the liquid phase and on the GAC, were represented by pseudo-first-order kinetics with respect to phenol;
- The GAC was considered a sufficiently porous material, where the diffusion of ozone and phenol into the catalyst particles took place;
- The adsorption kinetics during the Ad/Ox process are represented by a pseudo-second-order kinetic equation (Equation (20));
- The kinetic constant of phenol removal in the solid incorporates the degradation and desorption of organic compounds;
- The degradation kinetics in both liquids and solids, as well as the adsorption, are influenced by the operational conditions of the ozonisation process.

Taking into account the above, the combined process of the physical adsorption and oxidation of phenol must be described through its correlation with ozone consumption. Therefore, the following expression was defined, which relates the rate of phenol oxidation, $-r_P$, to the ozone consumed:

$$- r_P = -\frac{dF_P}{dV} = z \times N_{O_3},\tag{15}$$

where z is the stoichiometric coefficient of the reaction between transferred ozone and oxidised phenol, V is the bed volume, F_P is the phenol mass flow and N_{O_3} is the total ozone consumption. From the ozone consumption in the liquid ($N^I_{O_3}$) and the production of ozone at the GAC's surface ($N^{II}_{O_3}$), the following expressions (Equations (16)–(18)) were obtained:

$$N_{O_3} = N^I_{O_3} + N^{II}_{O_3} = K_L a \times \left(C^*_{O_3,L} - C_{O_3,L} \right)\tag{16}$$

$$N^I_{O_3} = k_{c,L} \times C_{O_3,L} \times C_P = k_{c,L} \times C_P \times \left(C_{O_3,L} - 0 \right) = k_{c,L} \times C_P \times \left(\frac{P^*_{O_3}}{He} - 0 \right)\tag{17}$$

$$N^{II}_{O_3} = k_{c,S} \times \frac{C_{O_3,L}}{m} \times Z_P \times M_{CAT},\tag{18}$$

where $K_L a$ is the volumetric ozone mass transfer coefficient, $C^*_{O_3,L}$ is the concentration of dissolved ozone in the liquid phase at saturation conditions, $k_{c,L}$ is the kinetic constant of the reaction between phenol and ozone at the liquid level, $C_{O_3,L}$ is the ozone concentration in the liquid phase, C_P is the phenol concentration, $P^*_{O_3}$ is the partial pressure of ozone in equilibrium with the ozone concentration in the liquid phase, He is Henry's constant, Z_P is the concentration of phenol adsorbed on the GAC, and M_{CAT} is the GAC concentration. Equation (18) assumes that the ozone adsorbed on the solid phase catalyst is in equilibrium with the ozone concentration in the liquid, which can be expressed as $C_{O_3,L} = m \times C^*_{O_3,S}$, where m is the slope of the equilibrium line between the liquid and solid phases.

After the description of the phenol degradation process through ozone consumption, the adsorption equilibrium must have been taken into account in the kinetic model of the Ad/Ox process, because the catalytically functional adsorbent used in this system was GAC, a material with high porosity and large available specific surface area. The equilibrium was assumed to be a Freundlich isotherm, in accordance with the adsorption experiments carried out in previous studies of the same GAC [17]. Freundlich equation was used because the adsorption step in these systems is usually a quick process [34]. Freundlich's equation is given by the following expression [24]:

$$Z_{P,\infty} = K_F \times C_P^{1/n_F} \tag{19}$$

where $Z_{P,\infty}$ is the concentration of phenol in equilibrium with the concentration of the liquid phase, K_F is the Freundlich constant and n_F is a factor describing the adsorption intensity. Consequently, the kinetics corresponding to the adsorption process could be described through pseudo-second-order kinetics, according to the following equation [35]:

$$\left(-\frac{dZ_P}{dt}\right)_{ads} = k_{ads} \times (Z_{P,\infty} - Z_P)^2 \times M_{CAT} \tag{20}$$

To describe the evolution of the phenol concentration over time at each longitudinal position of the fixed-bed tubular system, It have been used the axial dispersion model of Alhemed et al. [36]. The measurement of axial dispersion considers the possible deviation from ideal flow due to turbulence, as well as changes in bed characteristics and gas presence. For a tubular system, with one-dimensional flow and first-order kinetics, the axial dispersion is given by the following Equation (21):

$$v \times \frac{dC_P}{dL} - D_L \times \frac{d^2C_P}{dL^2} = 0 \tag{21}$$

where D_L is the axial dispersion coefficient, L is the length of tubular reactor (GAC bed height), and v is the linear flow velocity of the fluid. Through Equations (19)–(21), the overall velocity, incorporating the kinetic terms of adsorption, chemical reaction and dispersion, can be inferred from the following general expression:

$$\left(-\frac{dC_P}{dt}\right) = \left(-\frac{dC_P}{dt}\right)_{disp} + \left(-\frac{dC_P}{dt}\right)_{ads} + \left(-\frac{dF_P}{dV}\right)_{ox}, \tag{22}$$

To determine the evolution of phenol oxidation with reaction time and reactor position during a continuous catalytic ozonation process (Ad/Ox) on a GAC bed, for a dL volume it was obtained the following Equation (23):

$$v \times \frac{\partial C_P}{\partial L} - D_L \times \frac{\partial^2 C_P}{\partial L^2} + \frac{\partial C_P}{\partial t} + \left(\frac{1-\varepsilon}{\varepsilon}\right) \times \frac{\partial Z_P}{\partial t} + k_P \times C_P^n = 0, \tag{23}$$

where ε is the bed porosity, k_P is the kinetic constant that relates the reaction of ozone to the phenol in both the liquid and the solid, and n is the kinetic order reaction. Considering that the chemical reaction takes place both in the liquid and at the surface of the GAC, it have been obtained the following Equation (24) from Equations (17) and (18):

$$-r_{O_3} = k_{c,L} \times C_{O_3,L} \times C_P + k_{c,S} \times \frac{C_{O_3,L}}{m} \times Z_P = k_{c,L} \times C_{O_3,L} \times C_P + k_{c,S} \times \frac{C_{O_3,L}}{m} \times M_{CAT}^2 \times K_F \times C_P^{1/n_F} \tag{24}$$

Assuming that the ozone concentration is constant after an initial transitory period, and that the ozone distribution in the liquid and solid is proportional to its consumption [17,37], the following global kinetic constants, k_1 and k_2, were defined, which incorporated the chemical reaction and mass transfer, leading to the following expression:

$$v \times \frac{\partial C_{O_3,L}}{\partial L} - D_L \times \frac{\partial^2 C_P}{\partial L^2} + \frac{\partial C_{O_3,L}}{\partial t} + k_1 \times C_P + k_2 \times C_P^{1/n_F} = 0, \tag{25}$$

where k_1 and k_2 are the kinetic constants of phenol removal, referring to the liquid and solid phases, respectively. The combination of Equations (19)–(21) and (25) offers a description of the evolution of the phenol concentration in the system through the determination of the kinetic, fluodynamic, and equilibrium parameters.

2.3. Determination of the Characteristic Parameters of the Continuous Ad/Ox System

Due to the high complexity of the three-phase Ad/Ox model, before solving the equations describing the process, it was necessary to determine the characteristic parameters of the system. The characteristic parameters it have been considered the interstitial velocity, dispersion coefficient, kinetic constants of adsorption and equilibrium parameters.

The interstitial velocity was determined from the velocity at which the liquid flows through the voids in the bed, according to Equation (26):

$$v = \frac{Q_L}{A \times \varepsilon} \tag{26}$$

where Q_L is the liquid flow rate through the bed voids and A is the cross-sectional area of the reactor. With the studied flow rate (Q_L = 12 mL min^{-1}) and void fraction (ε = 0.32), an interstitial velocity of 5.8 cm min^{-1} was obtained. Another characteristic parameter was the axial dispersion coefficient, which characterises the degree of back-mixing of the flow. For a flow rate Q_G = 0.2 mL min^{-1}, a dispersion coefficient D_L = 12.8 cm^2 min^{-1} was obtained.

Regarding the residence time, the ratio of liquid and gas volumes was considered to be proportional to their respective flow rates. Consequently, both liquid and gas take the same time to circulate through the GAC fixed-bed column. For the studied flow rate, an empty bed contact time (EBCT) of 11.6 min was estimated. This value coincided with the time commonly used (10–30 min) in industrial water treatment processes in real plants [38].

With respect to the determination of the mass transfer coefficient of the reaction system ($K_L a$) was estimated from ozone concentration in the gas (see Figure S1) for each pressure and ozone flow rate with a determination coefficient of $R^2 \cong 0.99$ (see Figures S2 and S3). It was observed that with increasing gas flow rate, the mass transfer coefficient increased slightly from 0.130 to 0.183 min^{-1} at a flow rate of 0.05 and 0.4 L h^{-1} respectively. Regarding the pressure it was observed an increased from 0.110 to 0.125 min^{-1} at a pressure of 1.0 and 2.5 atm respectively. Obtained $K_L a$ values were according with other ozonation systems of literature [24,27].

Finally, it have been discussed the equilibrium and kinetic adsorption parameters necessary for the resolution of Equations (19)–(21) and (25). The adsorptive characteristics of the GAC at different pH conditions between 3.0 and 11.0 are shown in Figure 2.

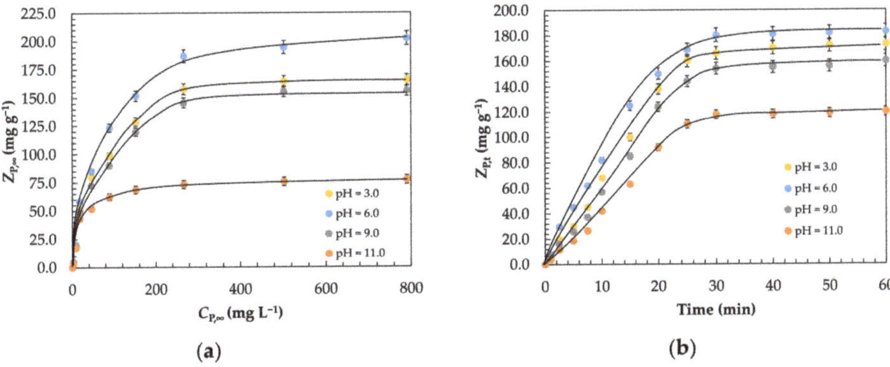

Figure 2. Determination of the adsorption parameters for the different pH values studied and at a temperature of 20 °C. (a) Adsorption isotherms of phenol on GAC; (b) evolution of adsorption kinetics fitted to a pseudo-second order model for different hydraulic retention times.

All phenol adsorption isotherms in Figure 2a were fitted by the empirical Freundlich multilayer adsorption model, which assumes the existence of interactions between the adsorbed molecules on the adsorbent [39]. Figure 2b shows the adsorption kinetic profiles fitted to a pseudo-second order model, which assumes that the chemical reaction is significant. The adsorption parameters necessary to the Ad/Ox kinetic model were listed in Table 1. A regression analysis of the experimental data showed a coefficient of determination $R^2 > 0.98$, indicating that both the Freundlich isotherm and the pseudo-second-order kinetic model adequately described the adsorption phenomena. The amount of phenol adsorbed on the GAC surface was greater ($K_F = 2.01$ (mg g^{-1}) (L mg^{-1})$^{1/n_F}$) at neutral pH than that at an alkaline pH ($K_F = 1.49$ (mg g^{-1}) (L mg^{-1})$^{1/n_F}$). This could be because when the pH of the solution is higher, the electrostatic interactions of attraction between the phenol and the GAC are lower [40].

Table 1. Summary of the kinetic parameters obtained from the pseudo-second-order and equilibrium model determined via the Freundlich isotherm model for the adsorption of phenol on the GAC's surface at a temperature of 20 °C and different pH values.

	pH = 3.0	pH = 6.0	pH = 9.0	pH = 11.0
Equilibrium				
K_F, (mg g^{-1}) (L mg^{-1})$^{1/n_F}$	1.89	2.01	1.82	1.49
n_F	1.12	1.19	1.10	1.06
R^2	0.981	0.995	0.984	0.991
Kinetic				
k_{ads}, g mg^{-1} min^{-1}	2.77×10^{-4}	3.10×10^{-4}	2.52×10^{-4}	4.85×10^{-5}
R^2	0.992	0.987	0.996	0.993

In order to explain the effect of pH on the adsorption phenomenon in more detail, the zeta potential was determined over a wide range of pH (2.5–11.5) (Figure 3). Based on obtained zeta potential values, at pH = 7.0, it was observed that the attraction due to electrostatic charge was greater than at other acidic or alkaline pH values. This is because their potential value (32.2 mV) was higher than that observed at pH = 3.0 (30.53 mV) or pH = 9.0 (10.47 mV). In contrast, at pH = 11.0, a lower adsorption capacity was achieved because of the repulsive forces between the negative charges of the phenol and the GAC surface difficult the adsorption of phenol. For this reason, the zeta potential value was −2.2 mV. Although higher adsorption of phenol would be achieved at neutral or acidic pH, oxidation by catalytic ozonation will be more effective at alkaline pH values because the generation of more hydroxyl radicals is promoted. Consequently, it will be necessary to work at higher pH values even if the adsorptive properties of the GAC are lost.

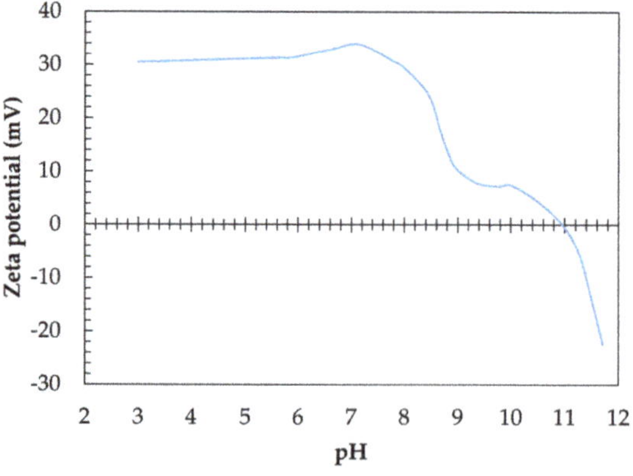

Figure 3. Evolution of the GAC's zeta potential at different pH values.

2.4. Effect of Operational Conditions and Kinetic Model Validation

In order to establish the most favourable operational conditions leading to higher mineralisation and primary degradation, the effects of system pressure, ozone dose, ozone flow rate, initial phenol concentration, and pH on the fixed-bed reaction system with GAC were analysed. Figure 4 shows the transient profiles of the obtained primary degradation and mineralization, fitted to the proposed Ad/Ox model.

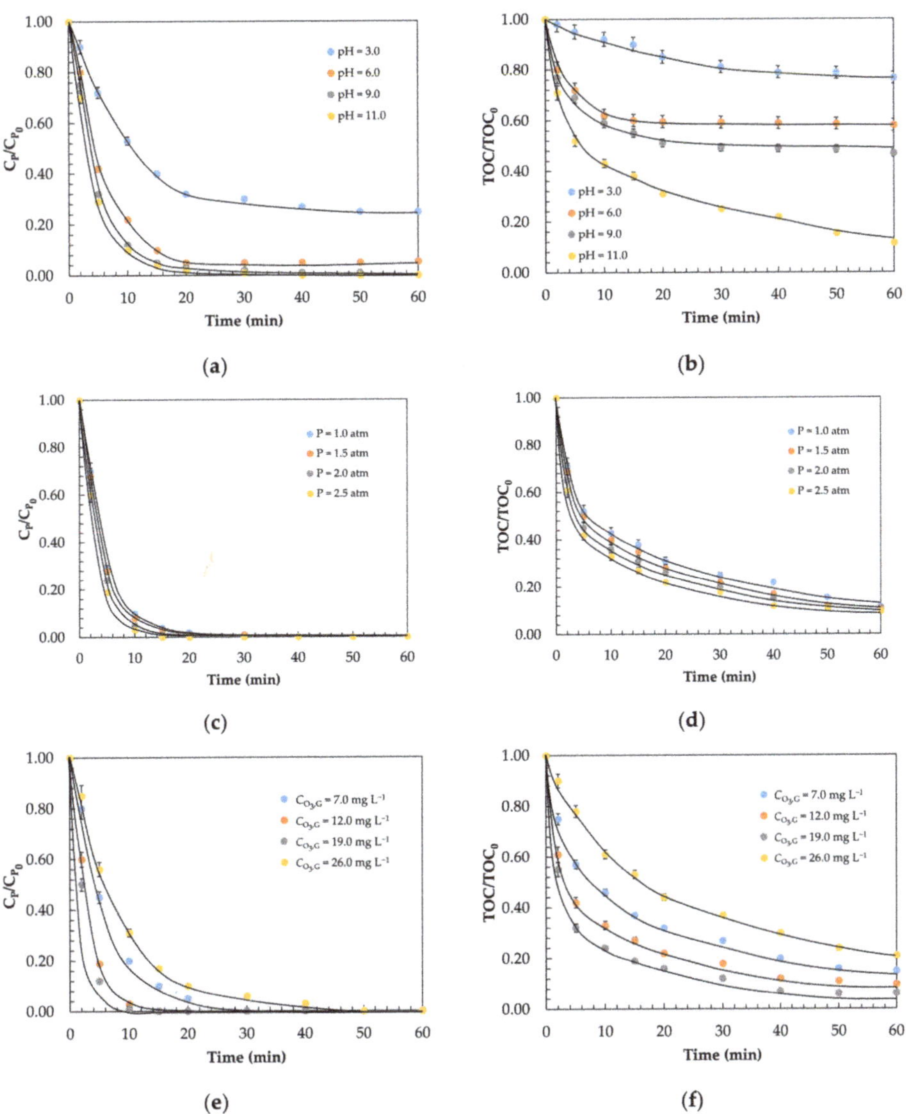

Figure 4. Effect of operational conditions on the primary degradation and mineralisation of phenol-containing waters during a catalytic ozonation process with GAC: (**a,b**) pH [1], (**c,d**) pressure [2], and (**e,f**) ozone concentration at inlet [3], showing the experimental profiles and fitted to the Ad/Ox kinetic model. Experimental conditions: [1] C_{P_0} = 250.0 mg L^{-1}, P = 1.0 atm, $C_{O_3,G}$ = 12.0 mg L^{-1}, Q_L = 12 mL min^{-1}, Q_G = 0.05 L h^{-1}, V = 0.14 L; [2] C_{P_0} = 250.0 mg L^{-1}, pH = 11.0, $C_{O_3,G}$ = 12.0 mg L^{-1}, Q_L = 12 mL min^{-1}, Q_G = 0.05 L h^{-1}, V = 0.14 L; [3] C_{P_0} = 250.0 mg L^{-1}, P = 2.5 atm, pH = 11.0, Q_L = 12 mL min^{-1}, Q_G = 0.05 L h^{-1}, V = 0.14 L.

According to Figure 4a,b, the pH of the solution plays an important role in the catalytic ozonation of phenol, as well as in the mass transfer and its subsequent decomposition [24]. As the pH of the solution increased, the rate of degradation and mineralisation increased, becoming higher than at more acidic pH. This is because at more alkaline pH, the oxidation mechanism involved is predominantly radicalary and generates more $HO^\bullet$ due to the higher concentration of hydroxide anions, and, consequently, a higher oxidation potential ($E^\circ = 2.80$ V) [41] as shown in Figure S4a. In contrast, at weakly acidic pH levels, the predominant mechanism is the direct reaction of ozone, as this pathway is more selective [27]. This statement could lead to an error, given that at pH = 11.0, the participation of the direct reaction of ozone with phenol is insignificant. This is because depending on the pH, phenol can dissociate, according to the equilibrium shown in Equation (27) [17]:

$$C_6H_5OH(aq) + H_2O(l) \rightleftharpoons C_6H_5O^-(aq) + H_3O^+(aq) \quad pK_a = 9.95 \tag{27}$$

According to Equation (27), the nature of these species could affect the reactivity with ozone. At acidic pH, electrophilic substitution reactions with ozone (direct reaction) would be promoted due to the character of the substituent groups. However, since the –O– group of phenolate ion is more reactive group ($k = 1.4 \times 10^9$ M^{-1} s^{-1}) than –OH ($k = 1300$ M^{-1} s^{-1}) of phenol [22], the reactivity of ozone increases with pH to attack the pollutant. Consequently, at pH = 11.0, complete primary degradation and mineralisation was achieved, with a total organic carbon reduction of 88.5%.

This same behaviour was observed by Xiong et al. [33]—in a slurry-type reactor. A mineralisation of 70.2% was achieved operating at an alkaline pH. Oyher authors such as, Chand et al. [42] observed an increase of 7.2% in phenol mineralisation operating at a slightly alkaline pH (pH = 9.0). The difference in efficiency observed could be due to the pH plays a determining role in the charge of the surface hydroxyl groups on the catalyst. In this respect, the adsorptive properties of the GAC could contribute to the greater adsorption of the degradation by-products, and, consequently, to a greater reduction in TOC. In any case, the contribution of adsorption during the catalytic ozonisation of phenol-containing waters did not comprise more than 5–23% of the TOC removal [43].

In order to validate the Ad/Ox kinetic model for the proposed continuous catalytic ozonisation process, the dynamics were adjusted for the different operational conditions. In Table 2, the values of the kinetic constants obtained after adjusting the profiles for the analyses of both primary degradation and mineralisation are shown. According to the relative standard error values, in general terms, the fit was good, showing an *RSE* of approximately 5% for the primary degradation and mineralisation profiles. According to the kinetic constant k_2, referring to the solid, slight increases were observed at neutral (4.6×10^{-3} (mg L^{-1}) (mg g^{-1} GAC min)$^{-1}$) and acidic (3.8×10^{-3} (mg L^{-1}) (mg g^{-1} GAC min)$^{-1}$) pH values, agreeing with the adsorption phenomenon described above. Respect the kinetic constant of the liquid (k_1), at alkaline pH ($k_1 = 0.20$ min^{-1}), it was observed in Figure S4a, an increased production of radical species capable of degrading a phenol-containing effluent. The kinetic constants obtained from the evolution of the mineralisation were lower due to the more refractory nature of the degradation products such as p-benzoquinone ($k = 2.5 \times 10^3$ M^{-1} s^{-1}) or catechol ($k = 5.2 \times 10^5$ M^{-1} s^{-1}) according with Figure S5 [22].

Table 2. Summary of the adsorption and oxidation kinetic constants of the Ad/Ox kinetic model related to the removal and mineralisation of phenol during catalytic ozonation with GAC.

Parameter/Evolution	Effect of pH [1]			
	pH = 3.0	pH = 6.0	pH = 9.0	pH = 11.0
Phenol removal				
k_1 (min^{-1})	0.05	0.09	0.12	0.20
$k_2 \times 10^3$ (mg L^{-1})(mg g^{-1} GAC min)$^{-1}$	3.8	4.6	3.7	2.1
RSE (%)	4.5	5.3	5.0	4.8
Mineralisation				
$k_1 \times 10^1$ (min^{-1})	0.009	0.017	0.022	0.038
$k_2 \times 10^4$ (mg L^{-1})(mg g^{-1} GAC min)$^{-1}$	0.71	0.86	0.69	0.39
RSE (%)	3.7	4.1	3.9	4.4
	Effect of pressure [2]			
	P = 1.0 atm	P = 1.5 atm	P = 2.0 atm	P = 2.5 atm
Phenol removal				
k_1 (min^{-1})	0.20	0.21	0.22	0.23
$k_2 \times 10^3$ (mg L^{-1})(mg g^{-1} GAC min)$^{-1}$	2.1	2.1	2.1	2.2
RSE (%)	4.1	4.6	4.0	5.2
Mineralisation				
$k_1 \times 10^1$ (min^{-1})	0.038	0.039	0.041	0.044
$k_2 \times 10^4$ (mg L^{-1})(mg g^{-1} GAC min)$^{-1}$	0.39	0.39	0.39	0.42
RSE (%)	4.5	4.2	4.7	4.3
	Effect of ozone gas concentration [3]			
	$C_{O_3,G}$ = 7.0 mg L^{-1}	$C_{O_3,G}$ = 12.0 mg L^{-1}	$C_{O_3,G}$ = 19.0 mg L^{-1}	$C_{O_3,G}$ = 26.0 mg L^{-1}
Phenol removal				
k_1 (min^{-1})	0.19	0.23	0.24	0.15
$k_2 \times 10^3$ (mg L^{-1})(mg g^{-1} GAC min)$^{-1}$	2.0	2.2	2.4	1.8
RSE (%)	5.3	4.9	4.9	5.1
Mineralisation				
$k_1 \times 10^1$ (min^{-1})	0.036	0.044	0.045	0.028
$k_2 \times 10^4$ (mg L^{-1})(mg g^{-1} GAC min)$^{-1}$	0.37	0.42	0.45	0.34
RSE (%)	4.8	4.2	4.6	4.7
	Effect of ozone flow rate [4]			
	Q_G = 0.05 L h^{-1}	Q_G = 0.1 L h^{-1}	Q_G = 0.2 L h^{-1}	Q_G = 0.4 L h^{-1}
Phenol removal				
k_1 (min^{-1})	0.24	0.25	0.26	0.27
$k_2 \times 10^3$ (mg L^{-1})(mg g^{-1} GAC min)$^{-1}$	2.4	2.4	2.5	2.5
RSE (%)	4.7	4.9	4.4	4.2
Mineralisation				
$k_1 \times 10^1$ (min^{-1})	0.045	0.047	0.049	0.051
$k_2 \times 10^4$ (mg L^{-1})(mg g^{-1} GAC min)$^{-1}$	0.45	0.45	0.47	0.47
RSE (%)	5.0	4.4	4.8	4.3
	Effect of initial phenol concentration [5]			
	C_{P_0} = 250.0 mg L^{-1}	C_{P_0} = 500.0 mg L^{-1}	C_{P_0} = 750.0 mg L^{-1}	C_{P_0} = 1000.0 mg L^{-1}
Phenol removal				
k_1 (min^{-1})	0.27	0.19	0.17	0.16
$k_2 \times 10^3$ (mg L^{-1})(mg g^{-1} GAC min)$^{-1}$	2.5	2.7	3.0	3.2
RSE (%)	4.0	4.3	4.8	4.5
Mineralisation				
$k_1 \times 10^1$ (min^{-1})	0.051	0.036	0.031	0.030
$k_2 \times 10^4$ (mg L^{-1})(mg g^{-1} GAC min)$^{-1}$	0.47	0.51	0.56	0.60
RSE (%)	4.2	4.6	3.9	4.4

Experimental conditions: [1] C_{P_0} = 250.0 mg L^{-1}, P = 1.0 atm, $C_{O_3,G}$ = 12.0 mg L^{-1}, Q_L = 12 mL min^{-1}, Q_G = 0.05 L h^{-1}, V = 0.14 L. [2] C_{P_0} = 250.0 mg L^{-1}, pH = 11.0, $C_{O_3,G}$ = 12.0 mg L^{-1}, Q_L = 12 mL min^{-1}, Q_G = 0.05 L h^{-1}, V = 0.14 L. [3] C_{P_0} = 250.0 mg L^{-1}, P = 2.5 atm, pH = 11.0, Q_L = 12 mL min^{-1}, Q_G = 0.05 L h^{-1}, V = 0.14 L. [4] C_{P0} = 250.0 mg L^{-1}, P = 2.5 atm, pH = 11.0; $C_{O_3,G}$ = 19.0 mg L^{-1}, Q_L = 12 mL min^{-1}, V = 0.14 L. [5] pH = 11.0, P = 2.5 atm; $C_{O_3,G}$ = 19.0 mg L^{-1}, Q_L = 12 mL min^{-1}, Q_G = 0.4 L h^{-1}, V = 0.14 L.

The effect of system pressure on the enhancement of degradation and mineralisation was studied in the range of 1 to 2.5 atm. Pressure is a operational variable that could affect mass transfer and, consequently, improve the contact amongst the ozone molecules in the gaseous and liquid phases [44,45] (see Figure S2). As shown in Figure 4c,d, the increase in pressure to 2.5 atm had a slightly positive effect, but not as evident as the effect of pH, because it only improved mineralisation by 1.8%, which was within the experimental error. The observed improvement could be due to the generation of slightly smaller bubbles, which could lead to an increase in the specific contact surface area between the liquid and gas phases. The same effect was observed for the kinetic constants of the Ad/Ox model. This could be due to the internal pressure in the gas microbubbles being much higher than the external pressure applied to them.

Figure 4e,f show the effects of ozone dose. Increasing the ozone dose from 7.0 to 19.0 mg L^{-1} increased the force gradient, thus improving the mass transfer and ultimately significantly increasing the rate of phenol removal until complete degradation; subsequently, mineralisation improved from 85.0% to 94.0% after reaching a steady state at 60 min. However, higher ozone doses led to a decrease in mineralisation from 94.0% to 79.0% at 26.0 mg L^{-1}. This can be explained via the mechanism proposed by Buhler et al. [46], as shown in Equations (28) and (29):

$$O_3 + HO^{\bullet} \rightarrow HO_4^{\bullet} \quad k = 2 \times 10^9 \text{ M}^{-1}\text{s}^{-1} \tag{28}$$

$$HO_4^{\bullet} \rightarrow HO_2^{\bullet} + O_2 \quad k = 2.8 \times 10^4 \text{ s}^{-1} \tag{29}$$

According to Equations (27) and (28), when there is an ozone dose above the critical value of 19.0 mg L^{-1}, ozone reacts with the generated hydroxyl radicals, producing radicals with a lower oxidative capacity (HO$_2$$^{\bullet}$ and O$_2$) than hydroxyl radical, according with Figure S4b. On the other hand, according to Rekhate and Srivastava [41], another feasible explanation could be that, when applying an excessive ozone dose, its utilisation is limited by the number of active sites available on the GAC catalyst's surface. Consequently, the resulting excess of ozone would react only through the direct pathway.

This same effect was observed by Nawaz et al. [47] who reported TOC removal rates of 50.6, 85.2, and 79.5% at ozone doses of 10, 20 and 50 mg L^{-1}, respectively. However, increasing the ozone dose did not lead to total phenol mineralisation.

After analysing the kinetic constants of the degradation and mineralisation profiles, the use of dose of 19.0 mg L^{-1} enhanced the kinetics at the solid level (k_2 = 2.4 × 10^{-3} (mg L^{-1}) (mg g^{-1} GAC min)$^{-1}$) compared with lower doses ($C_{O_3,G}$ = 7.0 mg L^{-1}), where a constant of 2.0 × 10^{-3} (mg L^{-1}) (mg g^{-1} GAC min)$^{-1}$ was obtained. This could have been because the pore structure of the activated carbon may have been affected by the ozonisation treatment, as described by Guelli Ulson de Souza et al. [6] and confirmed in previous studies with pristine and TiO$_2$-doped GACs [17,24].

Another operational variable that could have an effect on phenol removal is the ozone flow rate. Different flow rates between 0.05 and 0.4 L h^{-1} were evaluated (Figure 5a,b). Increasing the ozone flow rate led to a increase in phenol degradation rate and mineralisation efficiency (97.0%) at a flow rate of 0.4 L h^{-1}. This increase could be due to the influence of the flow rate on the ozone mass transfer from the gas to the liquid phase (see Figure S3), and therefore on the effective ozone utilisation. This effect was negligible; a better explanation might be that the quantity of ozone transferred to the liquid phase was higher than that used for oxidation and subsequent decomposition. As was the case with pressure, the improvement in the kinetic constants was not clearly evident here, except the one for the oxidation in the liquid (k_1), which increased from 0.045 to 0.051 min^{-1} in the mineralisation.

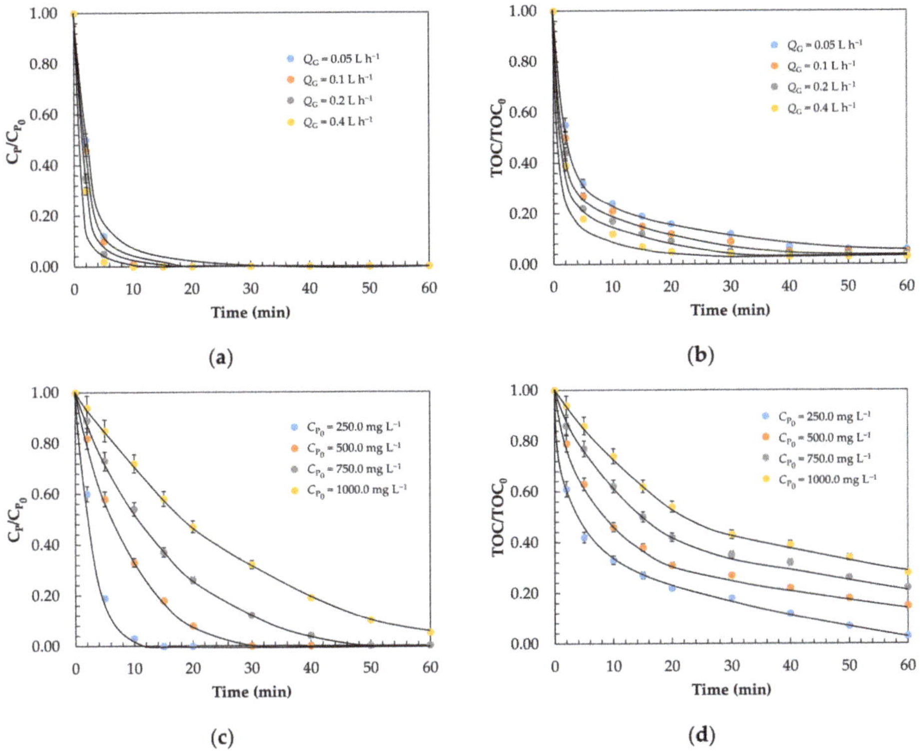

Figure 5. Analysis of C_P and TOC concentration profiles during the initial transitory period (t < 70 min) and adjustment to the Ad/Ox kinetic model. Effect of (**a**,**b**) ozone flow rate [1] and (**c**,**d**) concentration of the pollutant load [2]. Experimental conditions: [1] $C_{P_0} = 250.0$ mg L^{-1}, $P = 2.5$ atm, pH = 11.0; $C_{O_3,G} = 19.0$ mg L^{-1}, $Q_L = 12$ mL min^{-1}, $V = 0.14$ L; [2] pH = 11.0, $P = 2.5$ atm; $C_{O_3,G} = 19.0$ mg L^{-1}, $Q_L = 12$ mL min^{-1}, $Q_G = 0.05$ L h^{-1}, $V = 0.14$ L.

This same effect was observed by Yang et al. [48] during the treatment of an industrial effluent. A significant reduction in COD was initially observed up to a critical value of 2.0 L min^{-1}. After that, a negative effect was seen.

In order to study the feasibility of the catalytic ozonation treatment in this continuous system, a wider range of initial phenol concentrations that are common in industrial effluents (up to 1000.0 mg L^{-1}) was evaluated [17]. As shown in Figure 5c,d, the time required for complete degradation and TOC removal was greater at higher initial phenol concentrations [49]. Evaluating the transitory profiles, it can be seen that after 70 min of reaction, the same levels of complete degradation and 97.0% mineralisation were reached, and kinetic constants for the initial concentration of 1000.0 mg L^{-1} of 0.27 min^{-1} and 2.5×10^{-3} (mg L^{-1}) (mg g^{-1} GAC min)$^{-1}$ were obtained.

As a comparison of the obtained efficiencies, Lin and Wang [50] studied the removal of phenol through a catalytic ozonation process in a basket reactor with AC as the catalyst. At a temperature of 30 °C and after 2 h of reaction, they achieved complete phenol removal and a COD reduction rate of 85%. Beltrán et al. [43] evaluated the use of AC catalysts with other oxides, such as Fe, Co, and alumina, to treat a phenol-containing effluent via catalytic ozonation in a slurry reactor. However, no significant improvement was observed in the catalytic activities of the AC and the other three composites in terms of TOC reduction, which was about 90% after 5 h of reaction. Chaichanawong et al. [39] degraded an aqueous solution containing phenol in a slurry-type reactor, via a catalytic ozonisation process with three ACs with different physico-chemical properties. With a mesoporous carbon, the complete degradation of phenol and a mineralisation of 85.7% were obtained during the ozonisation and simultaneous adsorption process over 2 h. Guelli Ulson de Souza et al. [6] studied the simultaneous application of ozonation and adsorption processes in a packed bed of AC for the removal of phenol from an industrial stream. With a catalyst loading of 592.22 g L^{-1}, it achieved complete degradation and a COD reduction of 60.67%.

Finally, the toxicity of the effluent was determined in order to evaluate the feasibility of the discharge of the treated effluent into the river. According to De Luis et al. [16], a toxicity value less than 1 TU indicates that an effluent is non-toxic or exhibits low toxicity. Thus, the effluent before treatment showed a value of 18.02 TU, while after treatment carried out under favourable operational conditions its toxicity decreased to 0.09 TU.

In overall terms, taking into account that complete degradation, a mineralisation of 97.0% and a toxicity-free effluent were obtained with an empty bed contact time of 11.6 min and a catalyst load of 432.14 g L^{-1}. The adequately combine adsorption and ozonisation processes shows that the proposed continuous Ad/Ox system would be viable for scaling up into a real process in accordance with the technical criteria.

3. Materials and Methods

3.1. Materials

The granular activated carbon Kemisorb$^{\circledR}$ 530 GR 12 × 40 (Kemira Ibérica, Barcelona, Spain) was used as the catalytic material, with an average particle size of 1.0 mm. The activated carbon used throughout the experiments was characterised in a previous work [17]. Table 3 summarises the main physical properties of the GAC Kemisorb$^{\circledR}$ 530.

Table 3. Textural and chemical surface properties of GAC Kemisorb$^{\circledR}$ 530.

Property	Kemisorb$^{\circledR}$ 530
S_{BET} (m^2 g^{-1})	961.5
S_{ext} (m^2 g^{-1})	410.4
V_T (cm^3 g^{-1})	0.38
V_{micro} (cm^3 g^{-1})	0.24
V_{meso} (cm^3 g^{-1})	0.14
V_{meso}/V_T (%)	36.8
V_{micro}/V_T (%)	63.2
D_P (Å)	27.9
pH$_{pzc}$	10.95
Ash (%)	11.99
Apparent density (kg m^{-3})	432.1

S_{BET}—BET surface area; S_{ext}—external surface area; V_T—total pore volume; V_{micro}—micropore volume; V_{meso}—mesopore volume; V_{meso}/V_T × 100—mesopore percentage; V_{micro}/V_T × 100—micropore percentage; D_P—average pore diameter.

The specific surface area (S_{BET}), total (V_T), mesopore (V_{meso}) and micropore (V_{micro}) volumes, and the average pore diameter (D_P) were obtained using the BJH model by observing N$_2$ adsorption–desorption isotherms at 77 K [51]. The point of zero charge (pH$_{pzc}$) was determined via electrokinetic zeta potential (ς) measurements [52]. The composition of GAC Kemisorb$^{\circledR}$ 530 was measured using X-ray fluorescence (XRF) [24]. The XRF results indicate that GAC is mainly composed of SiO$_2$ (7.72%) and Al$_2$O$_3$ (2.64%), followed by Fe$_2$O$_3$ (0.40%), CaO (0.33%), SO$_3$ (0.19%), MgO (0.07%), Na$_2$O (0.06%), K$_2$O (0.04%), P$_2$O$_5$ (0.09%), TiO$_2$ (0.11%), and MnO (0.001%).

3.2. Analytical Methods

Phenol concentration and primary intermediates was measured using a Waters Alliance 2695 high-performance liquid chromatograph system (Waters, Milford, CT, USA) equipped with a Teknokroma Mediterranea SEA C18 threaded column (150 mm × 4.6 mm, 1.8 µm, Teknokroma Analitica, Sant Cugat del Vallès, Barcelona, Spain) and a guard column working at 20 °C under isocratic elution of a water/methanol mixture (60:40 v/v) containing acetic acid (1%), and a flow rate of 1 mL min^{-1} was used. A Waters 2487 UV/Vis detector was used at a wavelength of 220 nm for phenol, hydroquinone, catechol, oxalic acid, formic acid and at 254 nm for *p*-benzoquinone [53].

The degree of mineralisation was quantified by total organic carbon (TOC) analysis on a Shimadzu TOC-VSCH analyser (Izasa Scientific, Alcobendas, Spain). Toxicity was evaluated in duplicate using the Microtox® bioassay in a Microtox® toxicity analyser, Azur 500 model (Microbics Corp., New Castle, DE, USA). The measurements were carried out according to ISO 11348-3 (1998), "Water Quality—Determination of the inhibitory effect of water samples on the light emission of *Aliivibrio fischeri* (Luminescent bacteria test)—Part 3: Method using freeze-dried bacteria" [54]. The concentration of I_3^-, proportional to the concentration of oxidising agents such as hydroxyl radicals was measured following the KI dosimetry method described by Alfonso-Muniozguren et al. [55].

Zeta potential measurements were carried out with a ZetaSizer Ultra (Malvern Panalytical, Malvern, UK) for suspensions of 1g L^{-1} GAC in distilled water at different pHs.

The composition of the GAC was determined by X-ray fluorescence spectroscopy (XRF). From the pulverised sample, a borate glass bead was prepared by melting in an induction micro furnace, and mixing the flux Spectromelt A12 (Merck KGaA, Darmstadt, Germany) and the sample to a ratio of 20:1. An oxidising agent was added to promote the removal (in the oxidation phase of the process) of all the organic parts of the carbon and the fixation of the inorganic oxides. The chemical analysis of the beads was carried out in a vacuum atmosphere, using an AXIOS model sequential wavelength dispersive X-ray fluorescence spectrometer (WDXRF—Panalytical). The fluorometer was equipped with an Rh and three detectors (gaseous flux, scintillation, and Xe) (Malvern Panalytical) [24].

3.3. Experimental Setup in the Continuous Fixed-Bed Catalytic System

Phenol removal via a continuous Ad/Ox process was carried out in a polyvinyl chloride (PVC) column filled with GAC with a 25 mm outer diameter, 21.2 mm inner diameter and 40 cm length (Figure 6). The single ozonation and Ad/Ox experiments were performed at a constant ozone flow rate (Q_G = 0.05 L h^{-1}) and a temperature of 20 °C, with a fresh GAC bed of 60.5 g for each experiment and a dissolution pumping rate of Q_L = 12 mL min^{-1}. The pH was measured at different initial values (between 3.0 and 11.0), initial phenol concentrations (250, 500, 750, and 1000 mg L^{-1}), and a pressure between 1.0 and 2.5 atm (depending on the experiment).

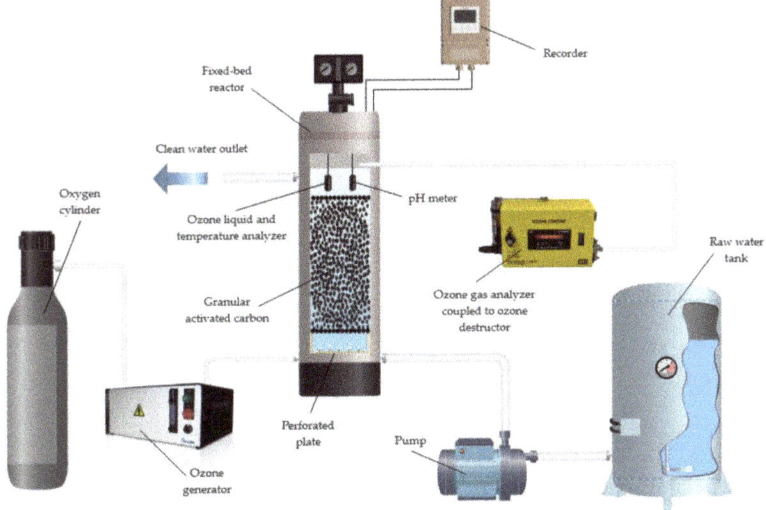

Figure 6. Experimental setup used to carry out ozonation and Ad/Ox tests.

A typical Ad/Ox or simple ozonisation experiment consisted of a continuous introduction of ozone generated in situ from ultrapure oxygen using a TRIOGEN LAB2B ozone generator (BIO UV, Lunel, France) through a porous plate placed at the bottom of the bed column. Here, the liquid flow of the phenol-containing solution was introduced through a PRECIFLOW peristaltic pump (LAMBDA Laboratory Instruments, Baar, Switzerland), through the lower inlet port on the side of the column. After ascending though the column, the gas and liquid flows passed through the granular activated carbon bed. Afterwards, the liquid flow was collected through one of the outlets downstream of the GAC packed bed.

The ozone concentration in the gas phase was measured with a BMT 964C ozone analyser (BMT MESSTECHNIK GMBH, Stahnsdorf, Germany) located on the side of the column. Dissolved ozone concentration and temperature were measured with a Rosemount 499AOZ-54 dissolved sensor (Emerson, Alcobendas, Spain). The pH value was recorded with a Rosemount Analytical model 399 sensor integrated into a Rosemount Analytical Solu Comp II recorder (Emerson). Residual ozone gas was removed using a Zonosistem thermocatalytic ozone destructor (Ingeniería del Ozono S.L.U, Cádiz, Spain).

Mass transfer characterization of the reactor was performed using deionized water in the presence of GAC, following a procedure previously described by Rodríguez et al. [37]. Operating conditions were kept similar to those used in the presence of phenol.

3.4. Statistical Analysis

The data were analysed with the SPSS software (IBM SPSS Statistics 27, SPSS Inc., Armonk, NY, USA). In each set of experiments, each experiment was performed in triplicate. Then, the relative standard deviation (RSE,%) of the data corresponding to conversions between 5 and 95% was calculated for the operational conditions established by Equation (30) [56]. Each measurement was replicated at least six times, and further replicates were carried out when the variation between each measurement exceeded 5%:

$$RSE = \sqrt{\frac{\sum\limits_{i=1}^{N}\left(\frac{C_{exp}-C_{mod}}{C_{exp}}\right)^2}{N-1}} \times 100 \tag{30}$$

4. Conclusions

A three-phase reaction kinetic model (Ad/Ox) for the description of G–L transfer within the liquid and on the catalyst's surface during the adsorption and ozonisation steps has been proposed. The model allows us to analyse catalytic ozonation, in the presence of GAC, for the removal of phenolic waters in a continuous process.

The combination of the simultaneous adsorption and ozonation processes with GAC resulted in an improvement of both phenol degradation kinetics and mineralisation efficiency, compared with an ozonation or adsorption process alone. The Ad/Ox kinetic model was verified via the experimental results of the catalytic ozonation process under a wide variety of operating conditions affecting the adsorption phenomena, the mass transfer, and the chemical reaction itself, providing a good fit with the experimental data, with a residual standard error of no more than 5% in most cases.

The estimation of the oxidation constants allowed us to study the role of GAC in the ozonisation process and its interaction with phenol. Depending on the degree of phenol dissociation, as a function of pH, the reactivity of ozone was different. At an alkaline pH (values over 11.0), greater degradation and mineralisation were obtained, with kinetic constants of 0.20 min^{-1} and 2.1 $\times$ 10^{-3} (mg L^{-1})(mg g^{-1} GAC min)$^{-1}$ for the liquid and solid, respectively. The use of an ozone dose above a critical value of 19.0 mg L^{-1} limited the kinetics and adsorption capacity of phenol, and its oxidation products by extension, leading to a decrease in mineralisation efficiency. On the other hand, at moderated ozone doses, a stronger influence of GAC adsorption mechanisms was observed, as the kinetic constant of the solid increased slightly to $k_2 = 2.4 \times 10^{-3}$ (mg L^{-1})(mg g^{-1} GAC min)$^{-1}$.

Increases in the pressure and gas flow rate did not lead to significant improvements, due to the insufficiency of the excess ozone transferred.

The most favourable operating conditions for the enhancement of the catalytic and adsorptive action of GAC were pH = 11.0, ozone dose = 19.0 mg L^{-1}, gas flow rate = 0.4 L h^{-1} and pressure = 2.5 atm. The GAC adsorbed the pollutant, subsequently exposing the phenol to attacks by ozone through the hydroxyl radicals generated on its surface. Consequently, the most favourable phenol removal conditions may involve a balance between the radical-generating and adsorptive functions of GAC.

The proposed model could be applied for the prediction of the operating behaviour of a continuous fixed-bed system under different working conditions, making it easily scalable to the industrial level.

Supplementary Materials: The following are available online at https://www.mdpi.com/article/10.3390/catal11081014/s1. Table S1: Previous studies on the treatment of wastewater through catalytic ozonation processes in the presence of activated carbon. Figure S1: Evolution of ozone concentration at the reactor outlet gas stream. Figure S2: Determination of mass transfer coefficient of the experimental system for various pressures. Figure S3: Determination of mass transfer coefficient of the experimental system for various ozone flow rates. Figure S4: $I_3{}^-$ concentration as a function of time for 0.1 M KI. Figure S5: Analysis of the main degradation by-products during catalytic ozonation of phenol.

Author Contributions: J.I.L., A.d.L. and C.F. performed the conceptualisation; A.d.L., N.V. and J.I.L. carried out the design of the methodology and analyses; C.F. and N.V. contributed to the model validation; C.F., N.V. and L.M.C. carried out the formal analysis; C.F. and A.d.L. performed the investigation; C.F., J.I.L. and L.M.C. prepared the original draft; J.M.L., C.F. and N.V. reviewed and edited the manuscript; J.I.L. and A.d.L. supervised the experimentation; J.I.L., N.V. and J.M.L. acquired the funding. All authors have read and agreed to the published version of the manuscript.

Funding: The authors are grateful to the University of the Basque Country for their financial support of this study through the GIU20/56 project and C. Ferreiro's predoctoral PIF grant (PIF16/367).

Acknowledgments: The authors are thankful for the technical and human support provided by General X-ray Service and Macrobehaviour–Mesostructure–Nanotechnology Service of the General Research Services (SGIker) of the UPV/EHU.

Nomenclature

ε	Bed porosity, m^3 m^{-3}
v	Linear velocity of fluid flow, cm min^{-1}
A	Reactor cross-sectional area, cm^2
$C^*{}_{O_3,L}$	Concentration of ozone in the equilibrium with the ozone adsorbed on the activated carbon, mg L^{-1}
$C^*{}_{O_3,s}$	Concentration of ozone on the catalyst in equilibrium with the liquid ozone concentration, mg L^{-1}
$C^*{}_p$	Calculated pollutant concentration in the liquid in terms of total organic carbon, mg L^{-1}
C_{exp}	Experimental pollutant concentration in the liquid, mg L^{-1}
C_{mod}	Modelled pollutant concentration in the liquid, mg L^{-1}
$C_{O_3,G}$	Concentrations of ozone in the gas phase, mg L^{-1}
$C_{O_3,L}$	Ozone concentration in liquid, mg L^{-1}
C_p	Pollutant concentration in the liquid, mg L^{-1}
D_L	Axial dispersion coefficient, cm^2 min^{-1}
F_G	Ozone mass flow, g O$_3$ h^{-1}
F_P	Phenol mass flow, g h^{-1}
He	Henry's constant, bar L mg^{-1}
k_1	Kinetic oxidation constant of phenol in the liquid, min^{-1}
k_2	Kinetic oxidation constant of phenol in the solid, (mg L^{-1}) (mg g^{-1} GAC min)$^{-1}$

k_{ads}	Kinetic constant of phenol adsorption, g mg^{-1} min^{-1}
$k_{c,L}$	Elemental kinetic constant for the ozonation in the liquid, L mg^{-1} min^{-1}
$k_{c,S}$	Elemental kinetic constant for the ozonation in the solid, L mg^{-1} min^{-1}
K_F	Freundlich constant, (mg g^{-1}) (L mg^{-1})$^{1/n}$
$K_L a$	Volumetric ozone mass transfer coefficient, min^{-1}
k_P	Overall kinetic reaction constant referring to the removal of phenol in both the liquid and the solid, min^{-1}
L	Length of tubular reactor, cm
m	Slope of the equilibrium line between the liquid and solid phase
m_{CAT}	Catalyst's mass, g
M_{CAT}	Concentration of catalyst, g L^{-1}
n	Kinetic reaction order
N	Number of experimental values
n_F	Heterogeneity factor, dimensionless
N^{II}_{O3}	Ozone consumption in the solid, mg L^{-1} min^{-1}
N^{I}_{O3}	Ozone consumption in the liquid, mg L^{-1} min^{-1}
N_{O3}	Whole ozone consumption, mg L^{-1} min^{-1}
P	Pressure, atm
$P^*_{O_3}$	Partial pressure of the ozone in equilibrium with the adsorbed ozone on the solid, bar
P_{O_3}	Partial pressure of ozone in the gas phase, bar
Q_G	Ozone gas flow, L min^{-1}
Q_L	Flow rate of liquid through the bed, mL min^{-1}
r_{O_3}	Chemical reaction rate of phenol removal reaction by catalytic ozonation, mg L^{-1} min^{-1}
r_p	Degradation of the pollutant in the liquid, mg L^{-1} min^{-1}
RSE	Relative standard deviation, %
t	Time, min
V	Reactor volume, L
z	Stoichiometric coefficient of the reaction between phenol and ozone
Z_p	Concentration of pollutant in the solid, mg g^{-1}
$Z_{p,\infty}$	Amount of pollutant adsorbed in the solid in equilibrium, mg g^{-1}

References

1. Akadiri, S.S.; Alola, A.A.; Alola, U.V.; Nwambe, C.S. The role of ecological footprint and the changes in degree days on environmental sustainability in the USA. *Environ. Sci. Pollut. Res.* **2020**, *27*, 24929–24938. [CrossRef]
2. Aitken, D.; Rivera, D.; Godoy-Faúndez, A.; Holzapfel, E. Water Scarcity and the Impact of the Mining and Agricultural Sectors in Chile. *Sustainability* **2016**, *8*, 128. [CrossRef]
3. Electronic Code of Federal Regulations (eCFR). Available online: https://www.ecfr.gov/ (accessed on 27 April 2021).
4. European Comission. Directive 2000/60/EC of the European Parliament and of the Council of 23 October 2000 Establishing a Framework for Community Action in the Field of Water Policy. *Official Journal* **2000**, *327*, 00001–0073. Available online: data.europa.eu/eli/dir/2000/60/oj (accessed on 22 December 2000).
5. United Nations About the Sustainable Development Goals. Available online: https://www.un.org/sustainabledevelopment/sustainable-development-goals/ (accessed on 7 February 2020).
6. de Arruda Guelli Ulson de Souza, S.M.; de Souza, F.B.; Ulson de Souza, A.A. Application of Individual and Simultaneous Ozonation and Adsorption Processes in Batch and Fixed-Bed Reactors for Phenol Removal. *Ozone Sci. Eng.* **2012**, *34*, 259–268. [CrossRef]
7. US EPA. Contaminant Candidate List 4-CCL 4. Available online: https://www.epa.gov/ccl/contaminant-candidate-list-4-ccl-4-0 (accessed on 9 October 2018).
8. Babich, H.; Davis, D.L. Phenol: A review of environmental and health risks. *Regul. Toxicol. Pharmacol.* **1981**, *1*, 90–109. [CrossRef]
9. Ribeiro, H.B.; Bampi, J.; da Silva, T.C.; Dervanoski, A.; Milanesi, P.M.; Fuzinatto, C.F.; de Mello, J.M.M.; da Luz, C.; Vargas, G.D.L.P. Study of phenol biodegradation in different agitation systems and fixed bed column: Experimental, mathematical modeling, and numerical simulation. *Environ. Sci. Pollut. Res.* **2020**, *27*, 45250–45269. [CrossRef]
10. Esplugas, S.; Giménez, J.; Contreras, S.; Pascual, E.; Rodríguez, M. Comparison of different advanced oxidation processes for phenol degradation. *Water Res.* **2002**, *36*, 1034–1042. [CrossRef]
11. Sumalatha, B.; Narayana, A.V.; Kumar, K.K.; Babu, D.J.; Venkateswarulu, T.C. Phenol Removal from Industrial Effluent Using Emulsion Liquid Memebranes. *J. Pharm. Sci. Res.* **2016**, *8*, 307–312.
12. Chen, L.; Xu, Y.; Sun, Y. Combination of Coagulation and Ozone Catalytic Oxidation for Pretreating Coking Wastewater. *Int. J. Environ. Res. Public Health* **2019**, *16*, 1705. [CrossRef]
13. Pradeep, N.V.; Anupama, S.; Navya, K.; Shalini, H.N.; Idris, M.; Hampannavar, U.S. Biological removal of phenol from wastewaters: A mini review. *Appl. Water Sci.* **2015**, *5*, 105–112. [CrossRef]

14. Klibanov, A.M.; Tu, T.M.; Scott, K.P. Peroxidase-catalyzed removal of phenols from coal-conversion waste waters. *Science* **1983**, *221*, 259–261. [CrossRef]
15. Bevilaqua, J.V.; Cammarota, M.C.; Freire, D.M.G.; Sant'Anna Jr., G.L. Phenol removal through combined biological and enzymatic treatments. *Braz. J. Chem. Eng.* **2002**, *19*, 151–158. [CrossRef]
16. De Luis, A.M.; Lombraña, J.I.; Menéndez, A.; Sanz, J. Analysis of the Toxicity of Phenol Solutions Treated with H_2O_2/UV and H_2O_2/Fe Oxidative Systems. *Ind. Eng. Chem. Res.* **2011**, *50*, 1928–1937. [CrossRef]
17. Ferreiro, C.; Villota, N.; de Luis, A.; Lombrana, J.I. Analysis of the effect of the operational variants in a combined adsorption-ozonation process with granular activated carbon for the treatment of phenol wastewater. *React. Chem. Eng.* **2020**, *5*, 760–778. [CrossRef]
18. Salvestrini, S.; Fenti, A.; Chianese, S.; Iovino, P.; Musmarra, D. Diclofenac sorption from synthetic water: Kinetic and thermodynamic analysis. *J. Environ. Chem. Eng.* **2020**, *8*, 104105. [CrossRef]
19. Erto, A.; Chianese, S.; Lancia, A.; Musmarra, D. On the mechanism of benzene and toluene adsorption in single-compound and binary systems: Energetic interactions and competitive effects. *Desalin. Water Treat.* **2017**, *86*, 259–265. [CrossRef]
20. Rout, P.R.; Zhang, T.C.; Bhunia, P.; Surampalli, R.Y. Treatment technologies for emerging contaminants in wastewater treatment plants: A review. *Sci. Total Environ.* **2021**, *753*, 141990. [CrossRef] [PubMed]
21. Rathi, B.S.; Kumar, P.S.; Show, P.-L. A review on effective removal of emerging contaminants from aquatic systems: Current trends and scope for further research. *J. Hazard. Mater.* **2021**, *409*, 124413. [CrossRef]
22. Von Sonntag, C.; Von Gunten, U. *Chemistry of Ozone in Water and Wastewater Treatment: From Basic Principles to Applications*; Iwa Publishing: London, UK, 2012; ISBN 978-1-84339-313-9.
23. Jamshidi, N.; Torabian, A.; Azimi, A.; Ghadimkhani, A. Degradation of Phenol in Aqueous Solution by Advanced Oxidation Process. *Asian J. Chem.* **2009**, *21*, 673–681.
24. Ferreiro, C.; Villota, N.; Lombraña, J.I.; Rivero, M.J. Heterogeneous Catalytic Ozonation of Aniline-Contaminated Waters: A Three-Phase Modelling Approach Using TiO_2/GAC. *Water* **2020**, *12*, 3448. [CrossRef]
25. Ameta, S. *Advanced Oxidation Processes for Wastewater Treatment: Emerging Green Chemical Technology*; Elsevier Science: San Diego, CA, USA, 2018; ISBN 978-0-12-810499-6.
26. Beltran, F.J. *Ozone Reaction Kinetics for Water and Wastewater Systems*; CRC Press: Boca Raton, FL, USA, 2003; ISBN 978-0-203-50917-3.
27. Rodríguez, A.; Rosal, R.; Perdigón-Melón, J.A.; Mezcua, M.; Agüera, A.; Hernando, M.D.; Letón, P.; Fernández-Alba, A.R.; García-Calvo, E. Ozone-Based Technologies in Water and Wastewater Treatment. In *Emerging Contaminants from Industrial and Municipal Waste*; Barceló, D., Petrovic, M., Eds.; Springer: Berlin/Heidelberg, Germany, 2008; Volume 5S/2, pp. 127–175, ISBN 978-3-540-79209-3.
28. Nawrocki, J. Catalytic ozonation in water: Controversies and questions. Discussion paper. *Appl. Catal. B Environ.* **2013**, *142–143*, 465–471. [CrossRef]
29. Nawrocki, J.; Kasprzyk-Hordern, B. The efficiency and mechanisms of catalytic ozonation. *Appl. Catal. B Environ.* **2010**, *99*, 27–42. [CrossRef]
30. Guo, Y.; Yang, L.; Wang, X. The Application and Reaction Mechanism of Catalytic Ozonation in Water Treatment. *J. Environ. Anal. Toxicol.* **2012**, *2*, 2161–0525. [CrossRef]
31. Criegee, R. Mechanism of Ozonolysis. *Angew. Chem. Int. Ed. Engl.* **1975**, *14*, 745–752. [CrossRef]
32. Lin, S.H.; Wang, C.H. Ozonation of phenolic wastewater in a gas-induced reactor with a fixed granular activated carbon bed. *Ind. Eng. Chem. Res.* **2003**, *42*, 1648–1653. [CrossRef]
33. Xiong, W.; Chen, N.; Feng, C.; Liu, Y.; Ma, N.; Deng, J.; Xing, L.; Gao, Y. Ozonation catalyzed by iron—and/or manganese—supported granular activated carbons for the treatment of phenol. *Environ. Sci. Pollut. Res. Int.* **2019**, *26*, 21022–21033. [CrossRef]
34. Shukla, S.; Kisku, G. Linear and Non-Linear Kinetic Modeling for Adsorption of Disperse Dye in Batch Process. *Res. J. Environ. Toxicol.* **2015**, *9*, 320–331. [CrossRef]
35. Ferreiro, C.; Gómez-Motos, I.; Lombraña, J.I.; de Luis, A.; Villota, N.; Ros, O.; Etxebarria, N. Contaminants of Emerging Concern Removal in an Effluent of Wastewater Treatment Plant under Biological and Continuous Mode Ultrafiltration Treatment. *Sustainability* **2020**, *12*, 725. [CrossRef]
36. Alhamed, Y.A. Adsorption kinetics and performance of packed bed adsorber for phenol removal using activated carbon from dates' stones. *J. Hazard. Mater.* **2009**, *170*, 763–770. [CrossRef]
37. Rodriguez, C.; Lombrana, J.I.; de Luis, A.; Sanz, J. Oxidizing efficiency analysis of an ozonation process to degrade the dye rhodamine 6G. *J. Chem. Technol. Biotechnol.* **2017**, *92*, 656–665. [CrossRef]
38. Gottschalk, C.; Saupe, A.; Libra, J.A. *Ozonation of Water and Waste Water: A practical Guide to Understanding Ozone and Its Application*; Wiley-VCH: Weinheim, Germany, 2010. [CrossRef]
39. Chaichanawong, J.; Yamamoto, T.; Ohmori, T. Enhancement effect of carbon adsorbent on ozonation of aqueous phenol. *J. Hazard. Mater.* **2010**, *175*, 673–679. [CrossRef] [PubMed]
40. Ferreiro, C.; Villota, N.; Lombraña, J.I.; Rivero, M.J.; Zúñiga, V.; Rituerto, J.M. Analysis of a Hybrid Suspended-Supported Photocatalytic Reactor for the Treatment of Wastewater Containing Benzothiazole and Aniline. *Water* **2019**, *11*, 337. [CrossRef]

41. Rekhate, C.V.; Srivastava, J.K. Recent advances in ozone-based advanced oxidation processes for treatment of wastewater—A review. *Chem. Eng. J. Adv.* **2020**, *3*, 100031. [CrossRef]
42. Chand, R.; Molina, R.; Johnson, I.; Hans, A.; Bremner, D.H. Activated carbon cloth: A potential adsorbing/oxidizing catalyst for phenolic wastewater. *Water Sci. Technol. J. Int. Assoc. Water Pollut. Res.* **2010**, *61*, 2817–2823. [CrossRef]
43. Beltrán, F.J.; Rivas, F.J.; Montero-de-Espinosa, R. Mineralization improvement of phenol aqueous solutions through heteroeneous catalytic ozonation. *J. Chem. Technol. Biotechnol.* **2003**, *78*, 1225–1233. [CrossRef]
44. Byun, S.; Cho, S.H.; Yoon, J.; Geissen, S.U.; Vogelpohl, A.; Kim, S.M. Influence of mass transfer on the ozonation of wastewater from the glass fiber industry. *Water Sci. Technol. J. Int. Assoc. Water Pollut. Res.* **2004**, *49*, 31–36. [CrossRef]
45. Rosen, H.M. Use of Ozone and Oxygen in Advanced Wastewater Treatment. *J. Water Pollut. Control Fed.* **1973**, *45*, 2521–2536.
46. Buhler, R.; Staehelin, J.; Hoigne, J. Ozone Decomposition in Water Studied by Pulse-Radiolysis 1. HO_2/O_2^- and HO_3/O_3^- as Intermediates. *J. Phys. Chem.* **1984**, *88*, 2560–2564. [CrossRef]
47. Nawaz, F.; Cao, H.; Xie, Y.; Xiao, J.; Chen, Y.; Ghazi, Z.A. Selection of active phase of MnO_2 for catalytic ozonation of 4-nitrophenol. *Chemosphere* **2017**, *168*, 1457–1466. [CrossRef]
48. Yang, L.; Sheng, M.; Li, Y.; Xue, W.; Li, K.; Cao, G. A hybrid process of Fe-based catalytic ozonation and biodegradation for the treatment of industrial wastewater reverse osmosis concentrate. *Chemosphere* **2020**, *238*, 124639. [CrossRef] [PubMed]
49. Hu, E.; Wu, X.; Shang, S.; Tao, X.; Jiang, S.; Gan, L. Catalytic ozonation of simulated textile dyeing wastewater using mesoporous carbon aerogel supported copper oxide catalyst. *J. Clean. Prod.* **2016**, *112*, 4710–4718. [CrossRef]
50. Lin, S.H.; Wang, C.H. Adsorption and catalytic oxidation of phenol in a new ozone reactor. *Environ. Technol.* **2003**, *24*, 1031–1039. [CrossRef] [PubMed]
51. Ribao, P.; Rivero, M.J.; Ortiz, I. TiO_2 structures doped with noble metals and/or graphene oxide to improve the photocatalytic degradation of dichloroacetic acid. *Environ. Sci. Pollut. Res. Int.* **2017**, *24*, 12628–12637. [CrossRef]
52. Kaledin, L.A.; Tepper, F.; Kaledin, T.G. Pristine point of zero charge (p.p.z.c.) and zeta potentials of boehmite's nanolayer and nanofiber surfaces. *Int. J. Smart Nano Mater.* **2016**, *7*, 1–21. [CrossRef]
53. Wu, C.; Liu, X.; Wei, D.; Fan, J.; Wang, L. Photosonochemical degradation of Phenol in water. *Water Res.* **2001**, *35*, 3927–3933. [CrossRef]
54. Association, A.P.H. *Standard Methods for the Examination of Water and Wastewater*; American Public Health Association: Washington, DC, USA, 2005; ISBN 978-0-87553-047-5.
55. Alfonso-Muniozguren, P.; Ferreiro, C.; Richard, E.; Bussemaker, M.; Lombraña, J.I.; Lee, J. Analysis of Ultrasonic Pre-treatment for the Ozonation of Humic Acids. *Ultrason. Sonochem.* **2020**, *71*, 105359. [CrossRef]
56. Sanchez, M.; Rivero, M.J.; Ortiz, I. Kinetics of dodecylbenzenesulphonate mineralisation by TiO_2 photocatalysis. *Appl. Catal. B Environ.* **2011**, *101*, 515–521. [CrossRef]

Article

Turbidity Changes during Carbamazepine Oxidation by Photo-Fenton

Natalia Villota [1],*, Cristian Ferreiro [2], Hussein A. Qulatein [3], Jose M. Lomas [1] and Jose Ignacio Lombraña [2]

[1] Department of Environmental and Chemical Engineering, Faculty of Engineering of Vitoria-Gasteiz, University of the Basque Country, UPV/EHU, 01006 Vitoria-Gasteiz, Spain; josemaria.lomas@ehu.eus

[2] Department of Chemical Engineering, Faculty of Science and Technology, University of the Basque Country, UPV/EHU, 48940 Leioa, Spain; cristian.ferreiro@ehu.eus (C.F.); ji.lombrana@ehu.eus (J.I.L.)

[3] Department of Chemical Engineering, Faculty of Engineering, Anadolu University, 26470 Eskişehir, Turkey; husseinqullatein@gmail.com

* Correspondence: natalia.villota@ehu.eus; Tel.: +34-945013248

Abstract: The objective of this study is to evaluate the turbidity generated during the Fenton photo-reaction applied to the oxidation of waters containing carbamazepine as a function of factors such as pH, H_2O_2 concentration and catalyst dosage. The results let establish the degradation pathways and the main decomposition byproducts. It is found that the pH affects the turbidity of the water. Working between pH = 2.0 and 2.5, the turbidity is under 1 NTU due to the fact that iron, added as a catalyst, is in the form of a ferrous ion. Operating at pH values above 3.0, the iron species in their oxidized state (mainly ferric hydroxide in suspension) would cause turbidity. The contribution of these ferric species is a function of the concentration of iron added to the process, verifying that the turbidity increases linearly according to a ratio of 0.616 NTU L/mg Fe. Performing with oxidant concentrations at (H_2O_2) = 2.0 mM, the turbidity undergoes a strong increase until reaching values around 98 NTU in the steady state. High turbidity levels can be originated by the formation of coordination complexes, consisting of the union of three molecules containing substituted carboxylic groups (BaQD), which act as ligands towards an iron atom with Fe^{3+} oxidation state.

Keywords: BaQD; carbamazepine; ferric coordination complex; photo-Fenton; turbidity

Citation: Villota, N.; Ferreiro, C.; Qulatein, H.A.; Lomas, J.M.; Lombraña, J.I. Turbidity Changes during Carbamazepine Oxidation by Photo-Fenton. *Catalysts* **2021**, *11*, 894. https://doi.org/10.3390/catal11080894

Academic Editor: Enric Brillas

Received: 25 June 2021
Accepted: 21 July 2021
Published: 24 July 2021

Publisher's Note: MDPI stays neutral with regard to jurisdictional claims in published maps and institutional affiliations.

1. Introduction

Over the last decade, special attention has been paid to the presence in waters (in relation to both their distribution and concentration) of certain organic compounds that, until now, had not been considered significant dangerous species. This is related to the improvement of analytical technique, as formerly undetected organic components are being more widely observed, considering that they have the potential to cause adverse effects both environmentally and in living beings [1].

Specifically, preventive measures are being adopted to control the emissions of pharmaceutically active products (PhACs) on environmental systems due to the harmful impacts that they can cause both on aquatic life and on human health [2] because they are recalcitrant compounds that generate toxicity [3,4]. The frequent presence of PhACs in freshwater and wastewater has promoted the establishment of water quality standards for periodic monitoring [5]. Thus, carbamazepine (CBZ) is proposed as an anthropogenic marker of water contamination, caused by its persistence in conventional water treatment plants, also being perceptible in some freshwater systems [6–8].

CBZ is an anticonvulsive and mood-stabilizing drug, which is used primarily in the treatment of epilepsy and bipolar disorder [9]. After consumption, around 10% of CBZ is excreted from the human body [10]. Besides, CBZ is the main cause of Stevens–Johnson syndrome that can cause toxic epidermal necrolysis [11]. This skin condition is potentially fatal, with a mortality rate of 30%, in which cell death causes the epidermis to separate from the dermis [12]. On the other hand, intrauterine exposure to CBZ is associated with a congenital defect of the spine and spinal cord, spina bifida [13] and problems with the neurodevelopmental embryo [14]. Moreover, higher fetal losses, as well as congenital malformation rates, have been reported among women consuming carbamazepine during pregnancy [15]. For these reasons, the presence of CBZ in drinking water and some groundwater is a cause for concern since it constitutes a risk factor as a possible route of access to the embryo and the infant through intrauterine exposure or breastfeeding.

A large part of the PhACs reach the wastewater through human body excretion and, if they are not effectively eliminated in the water treatment plants (WWTPs), both the effluent and the sludge lead to an important source of spreading PhACs in the environment [16,17]. In particular, conventional wastewater treatment plants remove less than 10% of the CBZ contained in the input influents [18–20]. Thus, WWTP effluents are an important gateway for CBZ accessing surface and groundwater. In general, the concentration of CBZ is higher in WWTPs than in exterior waters because the dilution phenomena and natural attenuation significantly reduce the concentration of these pollutants [21].

The need to effectively eliminate PhACs has promoted Advanced Oxidation Processes, known as AOPs [22]. Among the AOPs with the greatest applications stand out the technologies based on oxidation with ozone [23], electrochemical oxidation, photocatalysis based on the use of UV and Fenton processes [24] and photo-Fenton [25,26]. However, it should be noted that the oxidation with ozone, although highly reactive with organic compounds that have olefins or amines in their internal structure, is less effective when applied in the degradation of the CBZ and ibuprofen [27,28]. In electrochemical oxidation, the materials making up the electrodes are a limiting factor for industrial application. Besides, Fenton-like processes produce hydroxyl radicals, which are strong oxidizing agents capable of degrading a wide range of polluting organic compounds. However, the traditional Fenton reagent requires a continuous supply of Fe^{2+}, which produces an excess of iron in the generated sludge [29]. To alleviate this drawback, this work applies photo-Fenton technology since UV light increases the efficiency of the process. Therefore, the concentrations of Fe (II) utilized can be much smaller than in the conventional Fenton reaction.

This study evaluates the Fenton photo-reaction applied to the degradation of CBZ as a function of several factors, such as pH, hydrogen peroxide concentration and catalyst dosage. Experimental assays allow checking that during the oxidation treatment, the treated waters acquire high levels of turbidity depending on the operating conditions used in the tests. In this way, the aim of this work has been to establish the causes of turbidity in the treated water and the factors that affect it, relating the formation of turbidity with the degradation pathways and the main decomposition byproducts causing turbidity. Several references reported in the bibliography have studied the photo-Fenton treatment applied to carbamazepine degradation in domestic wastewater [30]. In general, the main objectives of these works are based on the intensification of ultraviolet technology combined with other AOPs as iron complexes or ultrasound waves [27,31] and the use of solar light improving the operational cost [32,33]. However, the novelty of this work is to analyze and establish the selectivity of the degradative routes of CBZ to water-turbidity generation as a function of the operating conditions.

2. Results

2.1. Turbidity Changes during Carbamazepine Oxidation

During the degradation of aqueous solutions containing different drugs, using a photo-Fenton process, it is found that turbidity appears in the treated water (see Figure 1a). The turbidity control of the water is closely related to the effectiveness of the disinfection processes, both chemical (chlorine or other biocides) and physical (UV radiation). This is due to the particles causing turbidity, which reduce the efficiency of the processes of chlorination in the elimination of pathogenic organisms, since they physically protect microorganisms from direct contact with the disinfectant. Although the direct effects of turbidity on health are not known yet, it affects the organoleptic properties of the water, which is why it often causes the rejection of consumers.

As shown in Figure 1a, the turbidity generated is a function of the type of pollutant that the water contains, as well as the operating conditions used in the oxidation treatment. Comparing these results with those shown in Figure 1b for the case of carbamazepine oxidation, when carrying out photo-oxidation using oxidant concentrations (H_2O_2) = 15.0 mM, the oxidized water presents turbidity levels of 4.6 NTU. Meanwhile, when using (H_2O_2) = 2.0 mM, the turbidity of the water increases to levels of 19.0 NTU after 120 min of reaction. For this reason, it is necessary to perform specific studies for each kind of effluent, since the turbidity will be determined by the presence of PhACs contained in the water, as a consequence of the human activities in the emission sources.

Globally, the World Health Organization (WHO) Quality Guidelines for Water for Human Consumption recommends a maximum of 5 NTU as a reference value, although the WHO indicates that, to achieve efficient disinfection, the water must have average turbidity lower than or equal to 1 NTU. Considering Spain, turbidity is a parameter included in current regulations, where its maximum permitted limits are regulated in Royal Decree 140/2003 [34] on hygienic–sanitary criteria of water for human consumption and Royal Decree 1620/2007 [35] on reuse of purified waters.

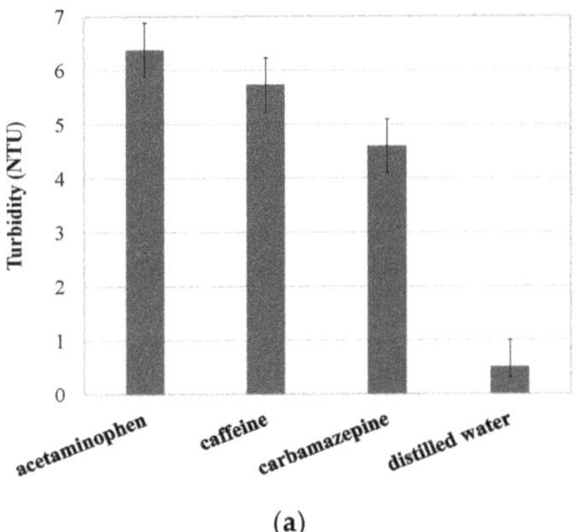

(a)

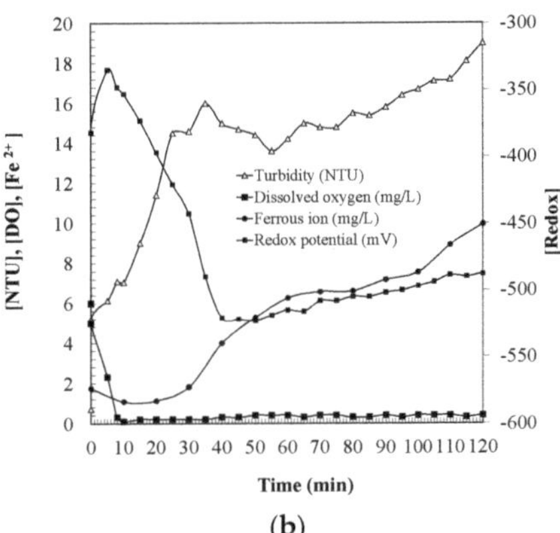

(b)

Figure 1. (a) Turbidity analyzed on aqueous solutions containing different PhACs oxidized by photo-Fenton. Experimental conditions: (C) = 50.0 mg/L; pH = 3.0; (H_2O_2) = 15.0 mM; (Fe) = 10.0 mg/L; (UV) = 150 W; T = 25 °C. (b) Water quality indicators analyzed during the carbamazepine oxidation by photo-Fenton. Experimental conditions: (CBZ) = 50.0 mg/L; pH = 3.0; (H_2O_2) = 2.0 mM; (Fe) = 10.0 mg/L; (UV) = 150 W; T = 25 °C.

On the other hand, Figure 1b represents turbidity as a function of other signs of water quality, such as the redox potential and the concentrations of ferrous ion and dissolved oxygen. As displayed, the results do not indicate a direct relationship with the formation of turbidity. Thus, a more in-depth analysis is necessary to estimate the effect of the main operating parameters of the photo-Fenton treatment on the formation of turbidity. In this work, pH, oxidant and catalyst dosage are considered.

2.2. pH Effect

Figure 2a shows the changes in turbidity of the aqueous solutions containing CBZ during their degradation, using the photo-Fenton process, where the pH of each test varied between pH = 2.0 and 5.0. It should be noted that the pH value has remained constant throughout the reaction. These results let verify that acidity affects the formation of turbidity in the water. However, its formation does not follow a linear relationship with the pH, but rather, three general ranges of operation are observed.

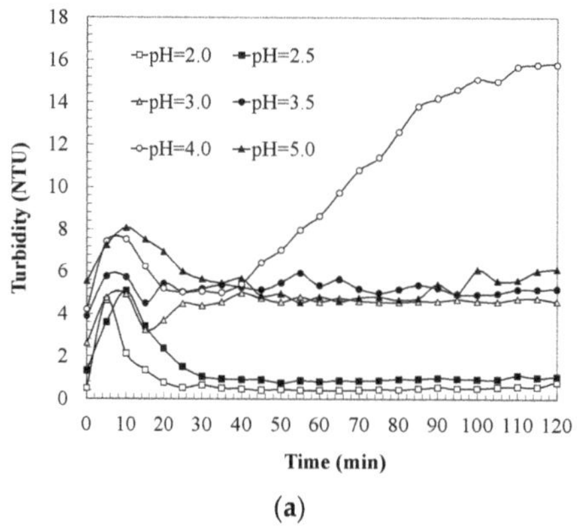

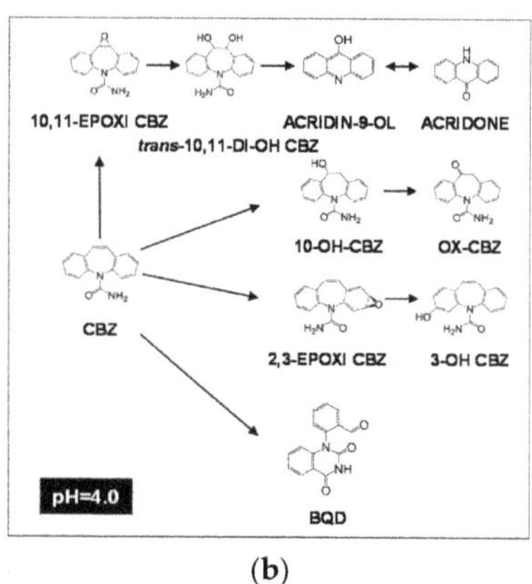

(a)   (b)

Figure 2. (a) pH effect on turbidity changes in a photo-Fenton system during the carbamazepine oxidation. (b) Equilibrium-species in carbamazepine aqueous solutions oxidized at pH = 4.0 by photo-Fenton treatment. Experimental conditions: (CBZ) = 50.0 mg/L; (H_2O_2) = 15.0 mM; (Fe) = 10.0 mg/L; (UV) = 150 W; T = 40 °C.

When applying acidity between pH = 2.0 and 2.5, the turbidity of the treated water is less than 1 NTU. This indicates that they are accepted by the legislation, which establishes the water quality criteria for both consumption and reuse. Oxidized water samples were analyzed to test carbamazepine degradation intermediates that coexist in the solution once a steady state is reached (see Appendix A, Table A1). The reason that these intermediate species cause low levels of turbidity is due to the fact that, performing at a controlled pH = 2.0–2.5, the iron species added in the form of a catalyst are present as ferrous ions. Iron species in their reduced state have a low capacity to react with organic matter, forming metallic complexes or inorganic hydroxides that cause turbidity.

The intermediates detected operating at pH = 2.0 allow proposing the degradation mechanism shown in Figure 3a, where CBZ would be oxidized through four main degradation routes. The dihydroxylation of the central benzene ring in the cis position of the hydroxyl groups, which would lead to obtaining acridones through the formation of hydroxylated acridines (Acridin-9-ol). On the other hand, the two benzene rings located at the extremes of the CBZ molecule would be hydroxylated, giving rise to the simultaneous formation of 3-hydroxy-carbamazepine (3-OH CBZ) and 2-hydroxy-carbamazepine (2-OH CBZ). Moreover, the attack of the aromatic ring of CBZ, according to the Criegee mechanism, would lead to the formation of 1-(2-benzaldehyde)-4-hydro-(1H,3H)-quinazoline-2-one(BQM) after intramolecular reactions and rearrangements. The reaction of BQM with hydroxyl radicals would lead to the formation of 1-(2-benzaldehyde)-(1H, 3H)-quinazoline-2,4-dione (BQD) [20,36]. Finally, the aldehyde group of the BQD could react with the hydroxyl radicals giving rise to the formation of the carboxyl group, generating the molecular structure 1-(2-benzoic acid)-(1H, 3H) -quinazoline-2,4-dione (BaQD), [37,38].

Conducting at pH values 3.0 and 3.5, turbidity around 5 NTU was observed, which would be the maximum limit value accepted by the water legislation. In tests performed at pH = 5.0, kinetic results were obtained that lead to similar turbidity. Besides, the intermediates that contain the oxidized CBZ samples were analyzed, operating at pH = 3.0 (see Table A2) and pH = 5.0 (see Table A4), in such a way that they allow to propose the potential degradation mechanisms of CBZ. It is found that, when carrying out the oxidation of CBZ both at pH = 3.0 and at pH = 5.0 controlled throughout the process, the four degradation pathways observed when operating at pH between 2.0 and 2.5 are kept, although with some nuances.

Figure 3b displays the CBZ degradation mechanism proposed for the assay conducted at a controlled pH = 5.0. In this case, the formation of the epoxide group in the central benzene ring, 10,11 epoxycarbamazepine (10,11-Epoxy CBZ), is detected, which leads to the formation of the two dihydroxylated isomers in cis positions 10,11-dihydroxycarbamazepine (cis 10,11-DiOH-CBZ) and trans 10,11-dihydroxycarbamazepine (trans 10,11-DiOH-CBZ). In turn, both are degraded, generating acridin-9-ol and acridone. On the other hand, the hydroxylation of the central benzene ring occurs, giving rise to the formation of 10-hydroxy-carbamazepine (10-OH CBZ), as well as the hydroxylation of the lateral ring generating 2-hydroxy-carbamazepine (2-OH CBZ). Moreover, the presence of (BQD) was tested. Given that the nature of the CBZ degradation intermediates analyzed does not present relevant structural differences with respect to the species detected in the previous case, operating at pH = 2.0, it should be considered that iron species could be the species directly affected by the change in the applied pH. In the case of conducting at pH = 3.0 and 5.0, iron would be found mainly in the form of ferric ions. However, when degrading intermediates of CBZ, the formation of metallic complexes between organic matter and ferric ions does not seem important. It would be more accurate to consider that the direct cause of turbidity formation would be the presence of ferric hydroxide in the solution, which would remain in suspension, and that would be a function of the concentration of iron added to the system. In this case, the tests were conducted at (Fe) = 10.0 mg/L, and the turbidity of the water was similar.

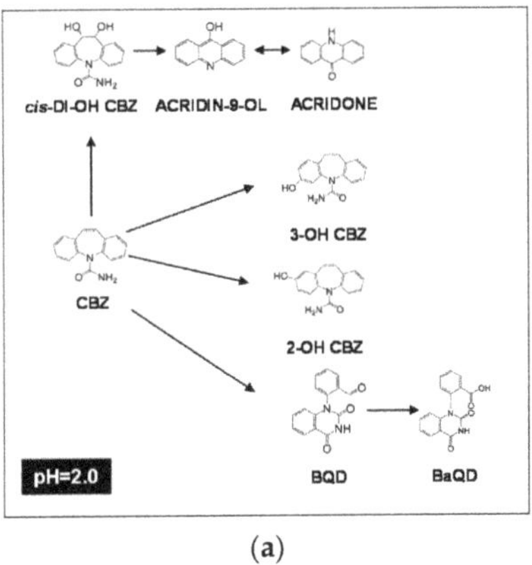

(a)

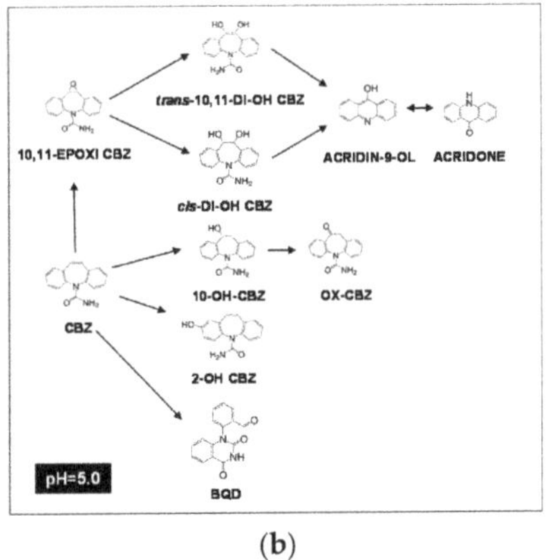

(b)

Figure 3. Equilibrium-species in carbamazepine aqueous solutions oxidized by photo-Fenton treatment. (**a**) Operating at pH = 2.0. (**b**) Operating at pH = 5.0. Experimental conditions: (CBZ) = 50.0 mg/L; (H$_2$O$_2$) = 15.0 mM; (Fe) = 10.0 mg/L; (UV) = 150 W; T = 40 °C.

When carrying out the treatment operating at pH = 4.0, Figure 2a shows that the turbidity potentially increases, reaching maximum values around 16 NTU at 120 min of reaction. However, it is noted that as the oxidation process progresses, the turbidity decreases, reaching values about 5 NTU in the steady state.

To explain this effect, the oxidized water was analyzed at pH = 4.0 (see Table A3), where, from the results obtained, the degradation mechanism shown in Figure 2b is proposed. In a similar way to the rest of the assays, four main degradation pathways are detected, towards the formation of acridon, in this case, through the trans isomer 10,11-DiOH-CBZ, as well as the hydroxylation pathways through the central and lateral benzene ring of CBZ, confirming the formation of BQD and its subsequent oxidation-generating BaQD.

It should be remarked that in all the tests performed during the first 30 min of reaction, a small turbidity peak occurs, whose maximum increases proportionally with the pH. In the case of operating at pH = 4.0, a second turbidity peak appears, with a larger area, which is not observed in the rest of the experiments. This significant increase in turbidity may be due to the formation of ferric species that remain in suspension during the first two hours of the reaction. Afterward, they slowly precipitate until a solution is obtained with turbidity not exceeding 5 NTU.

These results allow us to consider that the iron species, mainly ferric hydroxide, cause turbidity changes when varying the operational pH. Therefore, when applying pH = 4.0, which means that iron is mainly found as a ferric ion, and since the same initial iron catalyst concentration is used ((Fe) = 10.0 mg/L), the final turbidity of the treated water is similar to that of the oxidized samples at pH = 3.0–5.0, which fluctuate around 5 NTU.

The effect of hydrogen peroxide dosage on the formation of turbidity during the oxidation of CBZ was analyzed using a photo-Fenton treatment (see Figure 4a). The tests were performed varying the concentration of oxidant dosed between 2.0 and 15.0 mM, keeping steady the dosage of iron, added as a catalyst in the form of ferrous ion, at 10.0 mg/L and pH = 0. Checking turbidity in the water during the oxidation of CBZ shows three operating ranges that lead to similar turbidity levels. This fact could indicate that the dose of hydrogen peroxide would affect the selectivity of the oxidation pathways of CBZ, leading to the formation of degradation intermediates that cause turbidity.

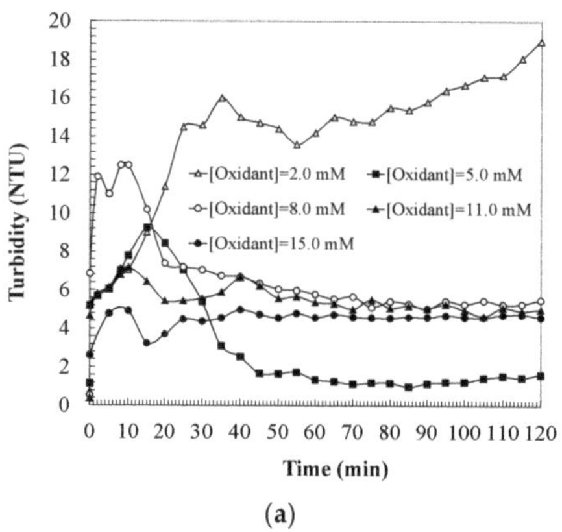

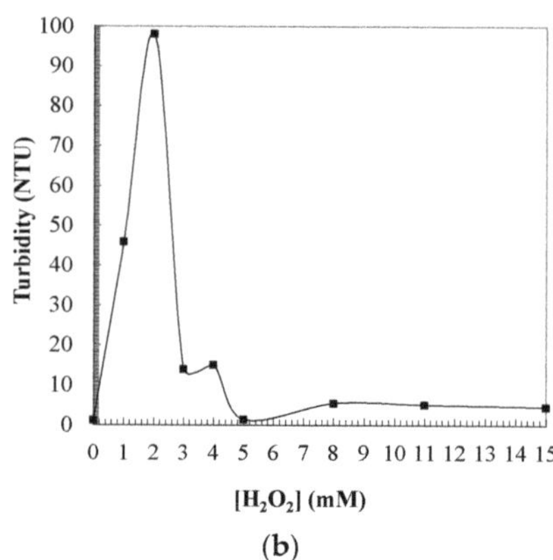

Figure 4. (**a**) Effect of hydrogen peroxide concentration ((Oxidant), mM) on turbidity changes in a photo-Fenton system during the carbamazepine oxidation. (**b**) Effect of hydrogen peroxide on the solutions turbidity once achieved the steady state. Experimental conditions: (CBZ) = 50.0 mg/L; pH = 3.0; (Fe) = 10.0 mg/L; (UV) = 150 W; T = 40 °C.

When applying oxidant concentrations between (H_2O_2) = 8.0 and 15.0 mM, a slight peak of turbidity appears during the first 20 min of the oxidation. It is verified that the maximum turbidity value of the peak is a function of the oxidant concentration. Therefore, that, using (H_2O_2) = 8.0 mM, produces a maximum turbidity of 12.5 NTU. Meanwhile, (H_2O_2) = 1.0 mM produces 7.2 NTU and (H_2O_2) = 15.0 mM creates 5.0 NTU. Once the peak arises, the turbidity evolves according to the kinetics of parallel trend until it coincides in similar values. Besides, it happens that in the steady state (see Figure 4b), the water treated under these conditions presents turbidity around 5 NTU. This result could indicate that the turbidity-causing intermediates formed during the first 20 min, which are dependent on the oxidation degree of the CBZ reached by using different doses of oxidant, are degraded to species of a similar nature.

Experimenting with oxidant concentrations (H_2O_2) = 5.0 mM, the formation of a turbidity peak is observed during the first 60 min of CBZ oxidation. In this case, the pinnacle is of greater area than in the previous interval. Moreover, it is verified that the turbidity evolves to values near 1.5 NTU in the steady state (see Figure 4b). Performing with oxidant concentrations (H_2O_2) = 2.0 mM, the water turbidity undergoes a notable increase during the first 30 min of oxidation of the CBZ, following a linear ratio of 0.34 NTU/min. Subsequently, the turbidity increases over time, but more slowly, at a rate of 0.057 NTU/min, until reaching around 98 NTU in the steady state (see Figure 4b). This result would indicate that CBZ degradation occurs through serial reactions that lead to the formation of species of a different nature, which cause turbidity.

Next, the treated water was analyzed using the oxidant dose (H_2O_2) = 2.0 mM to determine the degradative routes of CBZ towards the formation of species causing turbidity since it creates the highest turbidity in the tests conducted (see Figure 5a). These results allow us to verify that the four general pathways of CBZ degradation observed in the study of the effect of pH also occur here. The oxidation proceeds towards the production of acridones through the formation of the epoxide in the central benzene ring of CBZ, as well as the creation of the epoxide group in the lateral benzene ring, which leads to the development of hydroxylated species 2-OH-CBZ. On the other hand, hydroxylation reactions happen in the central benzene ring of CBZ, with the consequent formation of OX-CBZ and degradation towards the formation of BQD and BaQD.

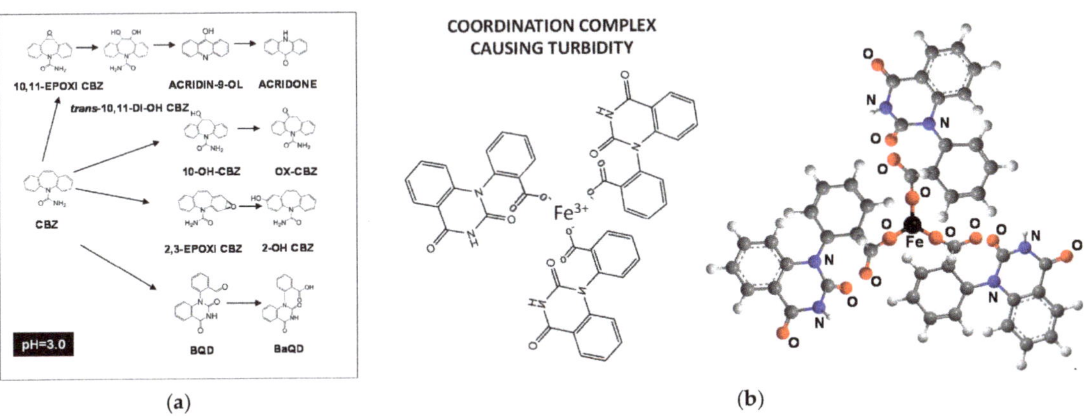

Figure 5. (a) Equilibrium-species in carbamazepine aqueous solutions oxidized by photo-Fenton treatment operating at pH = 3.0. (b) Molecular structure in 2D and 3D of possible coordination iron complex BaQD-Fe(III) causing turbidity in water. Experimental conditions: (CBZ) = 50.0 mg/L; pH = 3.0; (H_2O_2) = 2.0 mM; (Fe) = 10.0 mg/L; (UV) = 150 W; T = 40 °C.

Given the molecular structures of the species detected, it does not seem conceivable that the formation of intermolecular hydrogen bondings generates stable structures of a purely organic nature. In this case, it is contemplated that there are ferric species in the system, since the tests were performed at pH = 3.0, which determines the distribution of the iron species in the solution. Based on this premise, it is plausible that the high levels of turbidity generated in the water when using oxidant concentrations (H_2O_2) =2.0 mM can be caused by the formation of coordination complexes. They consist of the union of three molecules containing substituted carboxylic groups (BaQD), which act as ligands towards an iron atom with oxidation state 3^+, whose molecular structure is shown in Figure 5b.

2.3. Effect of Iron Catalyst

The effect of iron dosage, used as a catalyst, was studied, working with concentrations between (Fe) = 5.0 and 40.0 mg/L (see Figure 6a) and keeping steady the oxidant concentration and pH. The results indicate that the turbidity kinetics analyzed during the oxidation of CBZ show a parallel evolution in all the tests, where the turbidity increases linearly with the iron concentration according to a ratio of 0.616 NTU L/mg Fe (see Equation (1)). These results demonstrate that iron does not affect the CBZ degradation mechanism. Furthermore, by operating at a constant pH, the distribution of ferrous and ferrous species in solution is kept constant. Finally, the concentration of iron species was analyzed, verifying that the catalyst is mainly found as ferric species (see Figure 6b).

$$[NTU] = [NTU]_0 + 0.6159 \,[Fe] \quad (r^2 = 0.9804) \tag{1}$$

Being that:

[NTU]$_0$: turbidity of the aqueous solution of CBZ (=0.2261 NTU);

[NTU]: water turbidity (NTU);

[Fe]: initial iron concentration (mg/L).

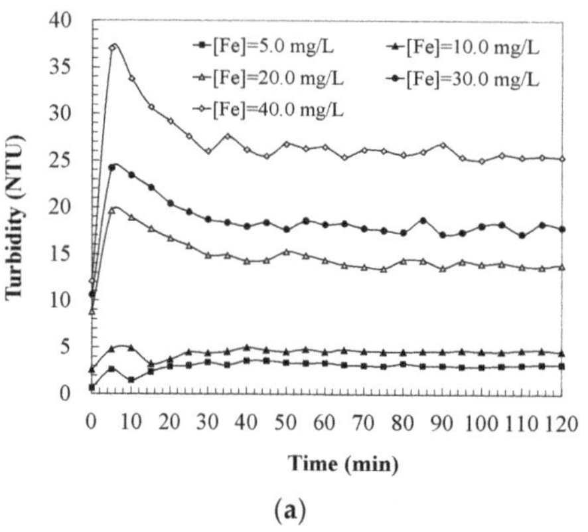

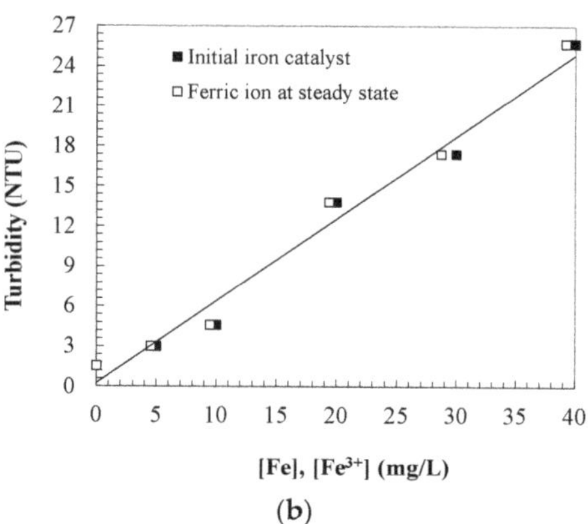

Figure 6. (**a**) Effect of iron catalyst on turbidity changes in a photo-Fenton system during the carbamazepine oxidation. (**b**) Effect of iron concentration on the solutions turbidity once the steady state is achieved. Experimental conditions: (CBZ) = 50.0 mg/L; pH = 3.0; (H_2O_2) = 15.0 mM; (UV) = 150 W; T = 40 °C.

3. Materials and Methods

3.1. Experimental Methods

Samples of carbamazepine aqueous solutions ((CBZ) = 50.0 mg/L, Fagron 99.1%) were studied in a photocatalytic 1.0 L reactor provided with a UV-150W mercury lamp of medium pressure (Heraeus, 95% transmission between 300 and 570 nm). Reactions began adding the iron catalyst as ferrous ion ((Fe), mg/L) operating between (Fe)$_0$ = 5.0–40.0 mg/L (FeSO$_4$ 7H$_2$O, Panreac 99.0%) and the oxidant dosage for each set of experiments, which varied between (H_2O_2) = 0–15.0 mM (Panreac, 30% *w/v*). All the experiments were carried out at around 40 °C in order to simulate real operating conditions, considering the heat absorbed by the water that is in direct contact with the ultraviolet lamp. Assays were performed operating under different initial pH conditions (pH between 2.0 and 5.0) in order to assess the effect of this parameter on color formation during oxidation of carbamazepine aqueous solutions. Acidity was kept constant, adding NaOH and HCl 0.1M.

3.2. Analytical Methods

Turbidity (NTU) was analyzed by a turbidimeter (100Q-Hach) and ferrous ion (Fe^{2+}, mg/L) by the phenanthroline method at λ = 510 nm (Fortune, 1938) using a UV/Vis Spectrophotometer 930-Uvikon, Kontron Instruments (Mazowieckie, Poland). Dissolved oxygen (DO, mg/L) was assessed by a Polarographic Portable Dissolved Oxygen Meter HI9142, Hanna Instruments, S.L. (Eibar, Spain). Total dissolved solids (TDS, mg/L) were analyzed by a TDS Metter Digital and redox potential (V) by a conductimeter Basic 20 Crison, Hach (Derio, Spain).

3.3. Liquid Chromatography-Mass Spectrometry to Elucidate the Intermediates of Carbamazepine Degradation

Samples were stored after receipt under refrigeration. Samples were centrifuged and subsequently diluted before starting analysis. The analysis was carried out with an LC/Q-TOF, with ESI+ Agilent Jet Stream ionization source and the following conditions: column: Kinetex EVO C18 HPLC/UHPLC Core-Shell (100 × 3 mm) 2.6 μm (Phenomenex company, Tianjin, China). Mobile phase 0.1% formic acid (A): acetonitrile with 0.1% of formic acid (B). Gradient: %B: 20; 20; 100; 100; 20 vs. time: 0; 2; 24; 28; 30. Flow: 0.3 mL/min. Column temperature: 35 °C. Injection volume: 5 µL. Ionization: Gas T = 300 °C; drying gas 10 L/min; Nebulizer 20 psig; shealt gas T = 350 °C; shealt gas flow 11 L/min; frag 125 V. Vcap 3500 V.

A screening method was developed to allow the elution and ionization of the greater number of compounds present in the sample. The stabilization of the system, the reproducibility of the signals and the correction of the exact mass were checked before starting the analysis. The compounds were searched using the deconvolution algorithm "Find by molecular features" and subsequent filtering of the proposed compounds based on compounds detected in the blank, background noise and minimum abundance of the compound. The following chromatograms show the major compounds observed for each of the samples (Figure 7). Under the proposed conditions, the following chromatograms were obtained for each of samples at pH=2.0, 3.0, 4.0 and 5.0 (Figures 8–11).

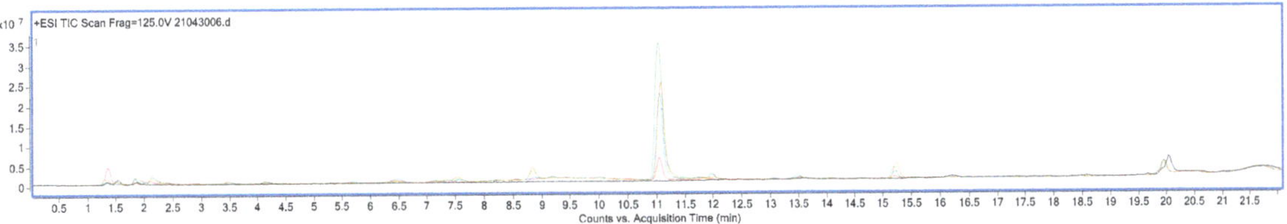

Figure 7. Chromatographic profile of a methanol blank (black line) and samples (red, blue, orange and green line).

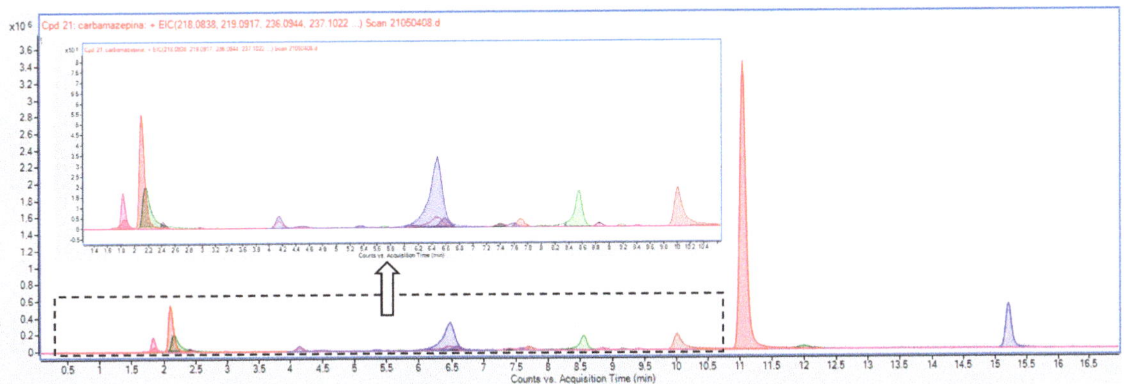

Figure 8. Chromatographic profile of the major compounds detected in the sample of carbamazepine oxidized to pH = 2.0. Experimental conditions: (CBZ) = 50.0 mg/L; (Fe) = 10 mg/L; (H_2O_2) = 15.0 mM; T = 25 °C; (UV) = 150 W.

In order to try to identify the greatest number of compounds, standards of possible carbamazepine degradation compounds were initially prepared to check their retention time and mass spectra. The following commercial standards were used: carbamazepine (CBZ), oxo-carbamazepine (Oxo-CBZ), carbamazepine 10, 12-epoxide (Epoxi-CBZ), 11-dihidro-10-hidroxicarbamazepine (10-OH CBZ), 9-acridanone, acridin-9-ol, 4-aminophenol, malonic acid (Figure 12).

Using the method developed for the screening, the following retention times (Tr) and characteristic ions or mass/charge ratios (m/z) were obtained for each compound (Table 1). Appendix A summarizes the predominant compounds found, as well as their characteristic ions (m/z) and the experimental masses calculated for each sample.

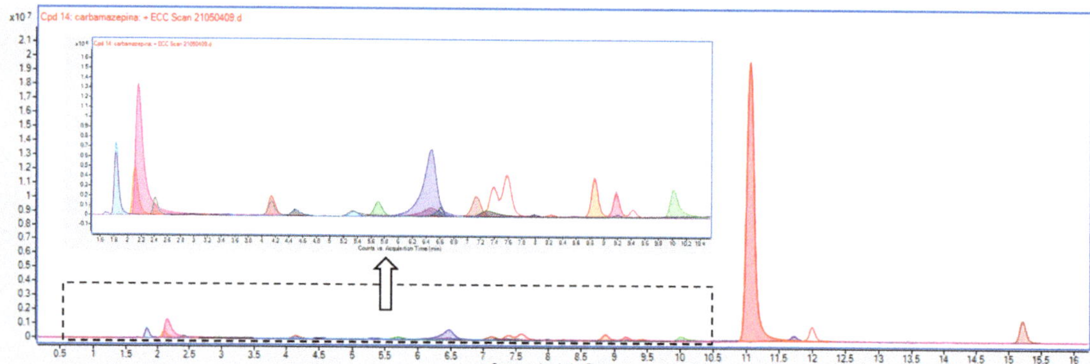

Figure 9. Chromatographic profile of the major compounds detected in the sample of carbamazepine oxidized to pH = 3.0. Experimental conditions: (CBZ) = 50.0 mg/L; (Fe) = 10 mg/L; (H_2O_2) = 15.0 mM; T = 25 °C; (UV) = 150 W.

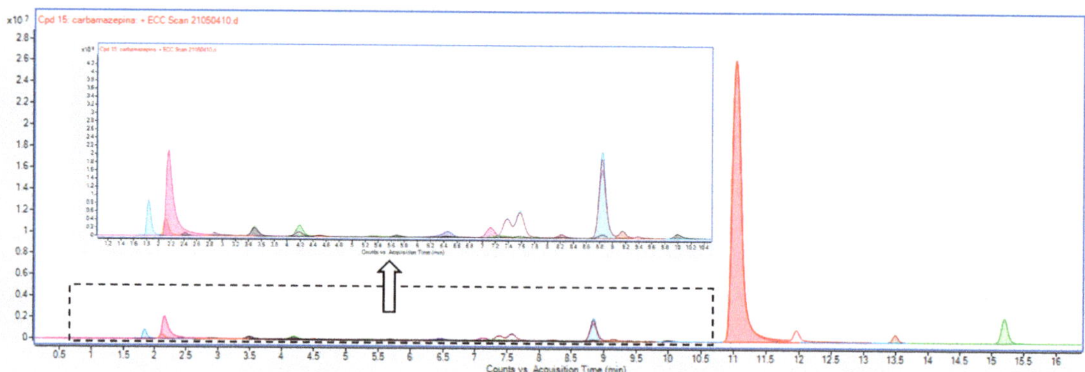

Figure 10. Chromatographic profile of the major compounds detected in the sample of carbamazepine oxidized to pH = 4.0. Experimental conditions: (CBZ) = 50.0 mg/L; (Fe) = 10 mg/L; (H_2O_2) = 15.0 mM; T = 25 °C; (UV) = 150 W.

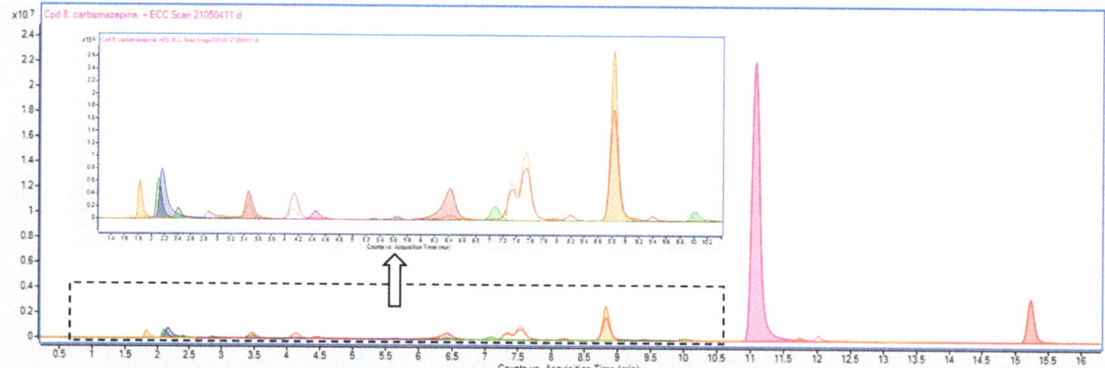

Figure 11. Chromatographic profile of the major compounds detected in the sample of carbamazepine oxidized to pH = 5.0. Experimental conditions: (CBZ) = 50.0 mg/L; (Fe) = 10 mg/L; (H_2O_2) = 15.0 mM; T = 25 °C; (UV) = 150 W.

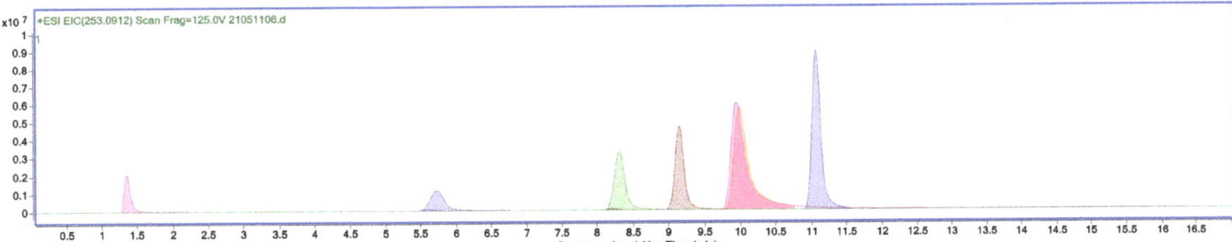

Figure 12. Excerpt chromatogram of standards (EIC) of 1 ppm CBZ, acridone, acridin, Oxo-CBZ, Epoxi-CBZ, 10-OH CBZ, malonic acid and 4-aminophenol.

Table 1. Standards analyzed.

Compound	Tr, min	m/z
4-aminophenol	1.3	110.0600
Malonic acid	1.8	105.0182
10-OH CBZ	5.8	255.1128
Epoxi-CBZ	8.2	253.0972
Oxo-CBZ	9.1	253.0972
9-acridanone	9.9	196.0757
Acridin-9-ol	9.9	196.0757
CBZ	11.2	237.1022

Once the majority of compounds were identified, and in order to determine the concentration of the degradation products in the samples (identified with the commercial standards), calibration was completed. The quantification of the samples was carried out using a calibration at concentrations between 0.001 and 5 µg/mL. The results obtained from the quantitative analysis are shown in Table 2.

Table 2. Results of quantitative analysis, concentrations in µg/mL.

Compound	pH = 2.0	pH = 3.0	pH = 4.0	pH = 5.0
CBZ	1.8	10.8	34.4	17.0
Oxo-CBZ	<LQL	0.25	0.21	0.037
Epoxi-CBZ	-	-	-	-
10-OH CBZ	<LQL	0.065	0.022	0.013
9-acridanone and acridin-9-ol	0.013	0.026	0.004	0.010
4-aminophenol	n.d.	n.d.	n.d.	n.d.
Malonic acid	n.d.	n.d.	n.d.	n.d.

n.d.: not detected, LQL: lower quantification limit (0.002 µg/mL).

Next is explained the validation of the method, wherein Tables 3–7 provide full validation process of the analysis of CBZ and its degradation products with the main validation parameters. The samples were subjected to drastic conditions to acid hydrolysis (1N HCl), basic hydrolysis (1N NaOH), sunlight and temperature (30 °C). Subsequently, the amount recovered was determined in triplicate after 7 days.

Table 3. Limit of quantification values (LOQ) and limit of detection values (LOD) of CBZ and the detected degradation intermediates in water.

Compound	LOQ (ng/mL)	LOD (ng/mL)
CBZ	100	30
Oxo-CBZ	2	0.6
10-OH CBZ	2	0.6
9-acridanone	1	0.3
Acridin-9-ol	1	0.3

Table 4. Linearity values of CBZ and the detected degradation intermediates in water.

Compound	Range (µg/mL)	Regression Equation	R^2
CBZ	0.1–50	y = 3.580 x + 6.340	0.998
Oxo-CBZ	0.002–0.5	y = 2.933 x + 5.480	0.998
10-OH CBZ	0.002–0.5	y = 2.008 x + 3.146	0.993
9-acridanone	0.001–0.5	y = 5.298 x + 9.636	0.998
Acridin-9-ol	0.001–0.5	y = 5.034 x + 8.996	0.998

Table 5. Specificity values of CBZ and the detected degradation intermediates in water.

Compound	Parameter	Amount Added (µg/mL)	Amount Recovered (µg/mL)	Degradation (%)
CBZ	Acidic degradation	25.05	24.58	1.87
	Alkaline degradation	25.05	24.29	3.04
	Solar light	25.05	24.87	0.71
OX-CBZ	Acidic degradation	0.251	0.246	2.03
	Alkaline degradation	0.251	0.241	3.87
	Solar light	0.251	0.249	0.98
10-OH CBZ	Acidic degradation	0.251	0.247	1.54
	Alkaline degradation	0.251	0.240	4.20
	Solar light	0.251	0.249	0.85
Acridanone	Acidic degradation	0.2505	0.2464	1.64
	Alkaline degradation	0.2505	0.2412	3.72
	Solar light	0.2505	0.2480	0.99
Acridin 9-ol	Acidic degradation	0.2505	0.2447	2.31
	Alkaline degradation	0.2505	0.2418	3.47
	Solar light	0.2505	0.2479	1.02

Table 6. Accuracy values of CBZ and the detected degradation intermediates in water.

Compound	Range (µg/mL)	Recovery (Mean $\pm$ % RSD)
CBZ	0.1–50	100.05 $\pm$ 0.023
Oxo-CBZ	0.002–0.5	100.24 $\pm$ 0.030
10-OH CBZ	0.002–0.5	100.56 $\pm$ 0.011
9-acridanone	0.001–0.5	100.08 $\pm$ 0.007
Acridin-9-ol	0.001–0.5	100.91 $\pm$ 0.024

Table 7. Precision values of CBZ and the detected degradation intermediates in water.

Compound	Concentration (µg/mL)	Standard Solution		Sample Solution		Mean	SD	% RSD
		Intraday Precision	Interday Precision	Intraday Precision	Inter-Day Precision			
CBZ	25.05	96.0190	96.9792	98.0188	98.9080	97.4812	98.7352	0.0128
OX-CBZ	0.251	6.2164	6.1542	6.0696	6.0908	6.1328	6.1988	0.0108
10-OH CBZ	0.251	3.6500	3.7595	3.8475	3.7264	3.7459	3.8279	0.0118
Acridanone	0.2505	10.9633	10.6344	10.8027	10.7996	10.8000	10.934	0.0124
Acridin 9-ol	0.2505	10.2570	10.1544	10.0560	10.3688	10.2091	10.3431	0.0131

4. Conclusions

This work checks the effect of the control parameters of photo-Fenton technology applied to CBZ oxidation. Experimental assays show that during the oxidation of aqueous solutions containing CBZ, the water turbidity shows great changes as a function of the operational conditions (pH, hydrogen peroxide and catalyst concentration). The relationship between turbidity and the control parameters of the photo-Fenton reaction would be caused by the degradation intermediates generated in water as a function of the oxidized degree achieved in the treatment. The analysis of treated waters that show the higher turbidity levels allow establishing a general oxidation mechanism, where the CBZ would be oxidized through four main degradation routes. First, the formation of the epoxide (10,11-Epoxy CBZ) leads to the creation of two dihydroxylated isomers (cis and trans 10,11-DiOH-CBZ), which, in turn, degrade, generating acridin-9-ol and acridone. On the other hand, the creation of the epoxide (2,3-Epoxy CBZ) generates hydroxylated benzene rings (3-OH CBZ and 2-OH CBZ). Moreover, the attack of the aromatic ring of CBZ would lead to the production of BQM, where the reaction of BQM with hydroxyl radicals would direct the generation of BQD. Finally, the aldehyde group of BQD could react with the hydroxyl radicals, generating BaQD. Moreover, it has to be considered that in this system, the iron catalyst has the oxidized form Fe^{3+}. Then, the generation of high turbidity would be explained based on the molecular structure of the degradation intermediates detected, where it would be possible to propose the formation of coordination complexes with ferric ions that enhance the turbidity. This would be the case of coordination compounds between a ferric ion atom with three BaQD molecules that consist of stable supramolecular structures that reduce the passage of light through the water, causing turbidity.

Author Contributions: Experimental design, writing, review, conclusions, N.V.; translation, formatting, references, C.F.; experimental development, H.A.Q.; translation, review, J.M.L.; obtaining sponsors, obtaining funding, J.I.L. All authors have read and agreed to the published version of the manuscript.

Funding: Authors are grateful to the University of the Basque Country UPV/EHU for the financial support to carry out this research study through the scholarship Student Movility for Traineeships in the Erasmus + Programme between the Anadolu University in Eskisehir (Turkey) and the Faculty of Engineering Vitoria-Gasteiz (Spain).

Acknowledgments: The authors thank for technical and human support provided by the Central Service of Analysis from Álava—SGIker—UPV/EHU.

Conflicts of Interest: The authors declare no conflict of interest.

Appendix A

Table A1. Results of screening of major compounds. Sample pH = 2.0. Experimental conditions: [CBZ] = 50.0 mg/L; [Fe] = 10.0 mg/L; [H_2O_2] = 15.0 mM; T = 25 °C; [UV] = 150 W.

Label	Tr, min	m/z	Mass	Height	Name	Score	Diff (DB,ppm)	Ions
Comp 1	1.8	224.0718	223.0645	146,262	-	-	-	3
Comp 2	1.8	163.0511	162.0438	41,946	-	-	-	2
Comp 3	2.1	147.0557	146.0485	501,215	-	-	-	2
Comp 4	2.2	180.0811	179.0739	172565	Acridine	99.09	−2.02	2
Comp 5	2.4	271.1054	270.0991	33,111	CIS-d iOH -CBZ	88.39	4.91	8
Comp 6	4.1	271.1092	270.1023	58,811	CIS-diOH-CBZ	94.64	−7.03	7
Comp 7	4.5	267.0781	266.0708	8828	BQD	87.11	−6.51	1
Comp 8	5.4	267.0779	266.0706	8275	BQD	89.7	−5.78	2
Comp 9	5.8	255.1158	254.1093	8104	-	-	-	
Comp 10	6.5	251.0828	250.0755	269,555	T1251	92.29	−5.1	5
Comp 11	6.6	267.0778	266.0706	38,790	BQD	90.62	−5.5	2
Comp 12	7.3	224.0719	223.0639	1553	-	-	-	3
Comp 13	7.4	253.0989	252.0919	12,234	2-OH-CBZ	73.88	−7.86	2
Comp 14	7.6	253.0983	252.0911	16,164	2-OH-CBZ	80.87	−4.92	2
Comp 15	7.7	283.0727	282.0654	32,599	BaQD	93.23	−4.53	2
Comp 16	8.5	267.0779	266.0706	147,319	BQD	90.13	−5.65	3
Comp 17	8.9	253.0994	252.0928	15,599	3-OH-CBZ	77.71	−11.53	3
Comp 18	9.2	253.1034	252.096	3526	-	-	-	2
Comp 19	9.9	196.077	195.0697	184,640	Acridone or acridin-9-ol	96.9	−6.8	10
Comp 20	11.2	259.0863	236.0971	3,433,925	Carbamazepine	94	−9.09	7
Comp 21	12.0	318.2826	317.2753	29,688	-	-	-	2
Comp 22	15.2	226.0881	225.0808	402,491	-	-	-	5

Table A2. Results of screening of major compounds. Sample pH = 3.0. Experimental conditions: [CBZ] = 50.0 mg/L; [Fe] = 10.0 mg/L; [H_2O_2] = 15.0 mM; T = 25 °C; [UV] = 150W. The compounds indicated in grey are the possible species identified by the database.

Label	Tr, min	m/z	Mass	Height	Name	Score	Diff (DB,ppm)	Ions
Comp 1	1.8	224.0716	223.0644	632,743	-	-	-	3
Comp 2	2.1	147.0563	146.0491	436,509	-	-	-	3
Comp 3	2.2	180.0818	179.0745	1,149,862	Acridine	92.86	−5.76	3
Comp 4	2.4	271.1095	270.1017	147,566	TRANS-diOH-CBZ	86.39	−4.87	3
Comp 5	3.5	271.109	270.102	12,320	-	-	-	2
Comp 6	4.1	253.099	270.1023	169,843	TRANS-diOH-CBZ	93.57	−6.99	3
Comp 7	4.5	267.0774	266.0701	56,646	BQD	95.11	−3.92	2
Comp 8	5.3	267.0777	266.0704	45,947	BQD	91.98	−5.07	2
Comp 9	5.7	237.1033	254.1065	131,805	10-OH-CBZ	83.69	−4.02	2
Comp 10	6.5	251.0834	250.0761	544,371	T1251	84	−7.51	5
Comp 11	6.6	267.0776	266.0704	55,876	BQD	93.04	−4.7	2
Comp 12	7.1	269.0934	268.0861	141,392	-	-	-	4
Comp 13	7.3	224.0718	223.0644	52,554	-	-	-	3
Comp 14	7.4	253.0988	252.0916	256,182	2-OH-CBZ	95.43	−6.61	8
Comp 15	7.6	253.0989	252.0916	357,248	2-OH-CBZ	94.24	−6.77	5
Comp 16	7.7	283.0735	282.0662	9024	BaQD	82.12	−7.58	2
Comp 17	7.9	267.0784	266.0709	19,667	BQD	85.55	−6.92	3
Comp 18	8.2	253.0985	252.0914	17,230	EP-CBZ	73.45	−6.07	2
Comp 19	8.6	267.0795	266.0722	8216	-	-	-	2
Comp 20	8.8	253.0994	252.0921	330,535	2-OH-CBZ	90.52	−8.92	4
Comp 21	9.2	253.0988	252.0916	204,534	Oxcarbamazepine	94.32	−6.72	5
Comp 22	10.0	196.0772	195.0699	280,810	Acridone or acridin-9-ol	96.78	−7.92	6
Comp 23	11.2	237.1047	236.0974	10,672,680	Carbamazepine	88.67	−10.08	15
Comp 24	11.7	224.0733	223.066	276,891	-	-	-	5
Comp 25	11.9	473.1992	472.1919	639,344	-	-	-	6
Comp 26	15.2	226.0889	225.0816	1,183,745	-	-	-	5

Table A3. Results of screening of major compounds. Sample pH = 4.0. Experimental conditions: [CBZ] = 50.0 mg/L; [Fe] = 10.0 mg/L; [H_2O_2] = 15.0 mM; T = 25 °C; [UV] = 150W. The compounds indicated in grey are the possible species identified by the database.

Label	Tr, min	m/z	Mass	Height	Name	Score	Diff (DB,ppm)	Ions
Comp 1	1.8	224.0722	223.0649	755,262	-	-	-	3
Comp 2	2.1	147.0566	146.0493	372,649	-	-	-	3
Comp 3	2.2	180.0822	179.0749	1,818,336	Acridine	87.37	−7.78	3
Comp 4	2.4	271.109	270.1018	67,258	TRANS-diOH-CBZ	80.63	−5.01	2
Comp 5	3.5	271.1091	270.1017	178,694	TRANS-diOH-CBZ	96.59	−4.95	3
Comp 6	4.2	253.0988	270.1021	236,040	TRANS-diOH-CBZ	94.47	−6.47	3
Comp 7	4.5	267.0778	266.0705	26,250	BQD	90.8	−5.44	2
Comp 8	5.3	267.078	266.0707	9285	BQD	88.48	−6.13	1
Comp 9	5.7	255.1141	254.1069	31,349	10-OH-CBZ	80.45	−5.52	3
Comp 10	6.5	251.0825	250.0753	119,415	T1251	94.04	−4.46	3
Comp 11	6.6	267.0779	266.0706	13,942	BQD	89.97	−5.7	2
Comp 12	7.1	269.0935	268.0861	166,675	-	-	-	5
Comp 13	7.4	224.0716	223.0643	44,729	-	-	-	2
Comp 14	7.3	253.0985	252.0912	385,655	3-OH CBZ	96.47	−5.18	5
Comp 15	7.6	253.099	252.0917	514,816	3-OH CBZ	93.53	−7.24	8
Comp 16	7.9	267.0781	266.0708	12,768	BDQ	87.89	−6.29	2
Comp 17	8.2	253.099	252.0918	70,044	EP-CBZ	93.04	−7.69	3
Comp 18	8.8	253.0996	252.0923	1,397,381	3-OH-CBZ	89.34	−9.65	11
Comp 19	9.1	253.0989	252.0916	149,274	Oxcarbamazepine	78.9	−6.89	4
Comp 20	9.9	196.0767	195.0696	109,794	Acridone or acridin-9-ol	97.14	−5.92	4
Comp 21	11.2	237.105	236.0973	15,089,995	Carbamazepine	83.8	−9.81	9
Comp 22	11.7	224.0723	223.0653	95,064	-	-	-	2
Comp 23	11.9	473.1994	472.192	688,405	-	-	-	8
Comp 24	13.5	210.0931	209.0859	635,867	-	-	-	3
Comp 25	15.2	226.0889	225.0817	1,793,645	-	-	-	8

Table A4. Results of screening of major compounds. Sample pH = 5.0. Experimental conditions: [CBZ] = 50.0 mg/L; [Fe] = 10.0 mg/L; [H_2O_2] = 15.0 mM; T = 25 °C; [UV] = 150W. The compounds indicated in grey are the possible species identified by the database.

Label	Tr, min	m/z	Mass	Height	Name	Score	Diff (DB,ppm)	Ions
Comp 1	1.8	224.0718	223.0645	520,658	-	-	-	3
Comp 2	2.1	147.0568	146.0496	588,299	-	-	-	3
Comp 3	2.2	180.0822	179.0749	693,527	Acridine	87.18	−7.84	3
Comp 4	2.4	271.1092	270.1019	141,194	TRANS-diOH-CBZ	80.16	−5.49	2
Comp 5	3.5	271.1091	270.1019	432,132	CIS-diOH-CBZ	96.29	−5.38	7
Comp 6	4.1	271.1091	270.1018	409,566	CIS-diOH-CBZ	96.85	−5.13	9
Comp 7	4.5	267.0778	266.0705	114,157	BQD	91.47	−5.23	2
Comp 8	5.3	267.0779	266.0706	13,734	BQD	89.84	−5.73	1
Comp 9	5.7	255.114	254.1069	56,313	10-OH-CBZ	96.03	−5.69	10
Comp 10	6.4	251.0827	250.0755	398,550	T1251	92.37	−5.07	6
Comp 11	6.5	267.0782	266.0709	10,804	BQD	86.14	−6.77	2
Comp 12	7.1	269.0935	268.0862	148,854	-	-	-	8
Comp 13	7.3	224.0721	223.0648	51,741	-	-	-	4
Comp 14	7.3	253.0989	252.0916	473,737	2-OH-CBZ	95.34	−6.62	6
Comp 15	7.5	253.0995	252.0922	802,934	2-OH-CBZ	91.73	−9.17	10
Comp 16	7.9	267.0775	266.0701	34,402	BQD	95.57	−3.73	4
Comp 17	8.2	253.0985	252.0915	81,586	EP-CBZ	97.33	−6.51	8
Comp 18	8.8	253.0996	252.0924	1,722,757	2-OH-CBZ	90.91	−9.73	11
Comp 19	9.1	253.0983	252.0911	37,573	Oxcarbamazepine	80.73	-4.9	2
Comp 20	10.0	196.0773	195.07	158,961	Acridone or acridin-9-ol	95.31	−8.41	6
Comp 21	11.2	237.1048	236.0973	11,711,134	Carbamazepine	89.16	−9.9	15
Comp 22	11.7	224.0731	223.0663	207,462	-	-	-	5
Comp 23	12.0	473.1987	472.1915	239,021	-	-	-	7
Comp 24	15.2	226.0886	225.0814	2,443,394	-	-	-	11

References

1. Lapworth, D.J.; Baran, N.; Stuart, M.E.; Ward, R.S. Emerging Organic Contaminants in Groundwater: A Review of Sources, Fate and Occurrence. *Environ. Pollut.* **2012**, *163*, 287–303. [CrossRef] [PubMed]
2. Hai, F.I.; Yang, S.; Asif, M.B.; Sencadas, V.; Shawkat, S.; Sanderson-Smith, M.; Gorman, J.; Xu, Z.-Q.; Yamamoto, K. Carbamazepine as a Possible Anthropogenic Marker in Water: Occurrences, Toxicological Effects, Regulations and Removal by Wastewater Treatment Technologies. *Water* **2018**, *10*, 107. [CrossRef]
3. Hughes, S.R.; Kay, P.; Brown, L.E. Global Synthesis and Critical Evaluation of Pharmaceutical Data Sets Collected from River Systems. *Environ. Sci. Technol.* **2013**, *47*, 661–677. [CrossRef]
4. Gavrilescu, M.; Demnerová, K.; Aamand, J.; Agathos, S.; Fava, F. Emerging Pollutants in the Environment: Present and Future Challenges in Biomonitoring, Ecological Risks and Bioremediation. *New Biotechnol.* **2015**, *32*, 147–156. [CrossRef]
5. Brack, W.; Dulio, V.; Ågerstrand, M.; Allan, I.; Altenburger, R.; Brinkmann, M.; Bunke, D.; Burgess, R.M.; Cousins, I.; Escher, B.I.; et al. Towards the Review of the European Union Water Framework Directive: Recommendations for More Efficient Assessment and Management of Chemical Contamination in European Surface Water Resources. *Sci. Total Environ.* **2017**, *576*, 720–737. [CrossRef]
6. Kumar, A.; Batley, G.E.; Nidumolu, B.; Hutchinson, T.H. Derivation of Water Quality Guidelines for Priority Pharmaceuticals. *Environ. Toxicol. Chem.* **2016**, *35*, 1815–1824. [CrossRef] [PubMed]
7. Marcelo, V.O.; Salette, R.; Jose, L.F.C.L.; Marcela, A.S. Analytical Features of Diclofenac Evaluation in Water as a Potential Marker of Anthropogenic Pollution. *Curr. Pharm. Anal.* **2016**, *13*, 39–47.
8. Ferrer, I.; Thurman, E.M. Analysis of 100 Pharmaceuticals and Their Degradates in Water Samples by Liquid Chromatography/Quadrupole Time-of-Flight Mass Spectrometry. *J. Chromatogr. A* **2012**, *1259*, 148–157. [CrossRef] [PubMed]
9. Arye, G.; Dror, I.; Berkowitz, B. Fate and Transport of Carbamazepine in Soil Aquifer Treatment (SAT) Infiltration Basin Soils. *Chemosphere* **2011**, *82*, 244–252. [CrossRef]
10. Kasprzyk-Hordern, B.; Dinsdale, R.M.; Guwy, A.J. The Removal of Pharmaceuticals, Personal Care Products, Endocrine Disruptors and Illicit Drugs during Wastewater Treatment and Its Impact on the Quality of Receiving Waters. *Water Res.* **2009**, *43*, 363–380. [CrossRef]
11. Chen, P.; Lin, J.-J.; Lu, C.-S.; Ong, C.-T.; Hsieh, P.F.; Yang, C.-C.; Tai, C.-T.; Wu, S.-L.; Lu, C.-H.; Hsu, Y.-C.; et al. Carbamazepine-Induced Toxic Effects and HLA-B*1502 Screening in Taiwan. *N. Engl. J. Med.* **2011**, *364*, 1126–1133. [CrossRef] [PubMed]
12. Pereira, F.A.; Mudgil, A.V.; Rosmarin, D.M. Toxic Epidermal Necrolysis. *J. Am. Acad. Dermatol.* **2007**, *56*, 181–200. [CrossRef] [PubMed]

13. Jentink, J.; Dolk, H.; Loane, M.A.; Morris, J.K.; Wellesley, D.; Garne, E.; de Jong-van den Berg, L.; EUROCAT Antiepileptic Study Working Group. Intrauterine Exposure to Carbamazepine and Specific Congenital Malformations: Systematic Review and Case-Control Study. *BMJ* **2010**, *341*, c6581. [CrossRef] [PubMed]
14. Cummings, C.; Stewart, M.; Stevenson, M.; Morrow, J.; Nelson, J. Neurodevelopment of Children Exposed in Utero to Lamotrigine, Sodium Valproate and Carbamazepine. *Arch. Dis. Child.* **2011**, *96*, 643–647. [CrossRef] [PubMed]
15. Atkinson, D.E.; Brice-Bennett, S.; D'Souza, S.W. Antiepileptic Medication during Pregnancy: Does Fetal Genotype Affect Outcome? *Pediatr. Res.* **2007**, *62*, 120–127. [CrossRef]
16. Hai, F.I.; Nghiem, L.D.; Khan, S.J.; Price, W.E.; Yamamoto, K. Wastewater Reuse: Removal of Emerging Trace Organic Contaminants. In *Membrane Biological Reactors: Theory, Modeling, Design, Management and Applications to Wastewater Reuse*; IWA Publishing: London, UK, 2014; ISBN 978-1-78040-065-5.
17. Vieno, N.; Tuhkanen, T.; Kronberg, L. Elimination of Pharmaceuticals in Sewage Treatment Plants in Finland. *Water Res.* **2007**, *41*, 1001–1012. [CrossRef]
18. Ying, G.-G.; Kookana, R.S.; Kolpin, D.W. Occurrence and Removal of Pharmaceutically Active Compounds in Sewage Treatment Plants with Different Technologies. *J. Environ. Monit.* **2009**, *11*, 1498–1505. [CrossRef]
19. Tixier, C.; Singer, H.P.; Oellers, S.; Müller, S.R. Occurrence and Fate of Carbamazepine, Clofibric Acid, Diclofenac, Ibuprofen, Ketoprofen, and Naproxen in Surface Waters. *Environ. Sci. Technol.* **2003**, *37*, 1061–1068. [CrossRef]
20. Wijekoon, K.C.; Hai, F.I.; Kang, J.; Price, W.E.; Guo, W.; Ngo, H.H.; Nghiem, L.D. The Fate of Pharmaceuticals, Steroid Hormones, Phytoestrogens, UV-Filters and Pesticides during MBR Treatment. *Bioresour. Technol.* **2013**, *144*, 247–254. [CrossRef]
21. Guo, Y.C.; Krasner, S.W. Occurrence of Primidone, Carbamazepine, Caffeine, and Precursors for N-Nitrosodimethylamine in Drinking Water Sources Impacted by Wastewater1. *JAWRA J. Am. Water Resour. Assoc.* **2009**, *45*, 58–67. [CrossRef]
22. Luo, Y.; Guo, W.; Ngo, H.H.; Nghiem, L.D.; Hai, F.I.; Zhang, J.; Liang, S.; Wang, X.C. A Review on the Occurrence of Micropollutants in the Aquatic Environment and Their Fate and Removal during Wastewater Treatment. *Sci. Total Environ.* **2014**, *473*, 619–641. [CrossRef] [PubMed]
23. Villota, N.; Lombraña, J.I.; Cruz-Alcalde, A.; Marcé, M.; Esplugas, S. Kinetic Study of Colored Species Formation during Paracetamol Removal from Water in a Semicontinuous Ozonation Contactor. *Sci. Total Environ.* **2019**, *649*, 1434–1442. [CrossRef]
24. Mijangos, F.; Varona, F.; Villota, N. Changes in Solution Color During Phenol Oxidation by Fenton Reagent. *Environ. Sci. Technol.* **2006**, *40*, 5538–5543. [CrossRef] [PubMed]
25. Villota, N.; Lomas, J.M.; Camarero, L.M. Study of the Paracetamol Degradation Pathway That Generates Color and Turbidity in Oxidized Wastewaters by Photo-Fenton Technology. *J. Photochem. Photobiol. A Chem.* **2016**, *329*, 113–119. [CrossRef]
26. Villota, N.; Ferreiro, C.; Qulatein, H.A.; Lomas, J.M.; Camarero, L.M.; Lombraña, J.I. Colour Changes during the Carbamazepine Oxidation by Photo-Fenton. *Catalysts* **2021**, *11*, 386. [CrossRef]
27. Sun, S.; Yao, H.; Fu, W.; Liu, F.; Wang, X.; Zhang, W. Enhanced Degradation of Carbamazepine in FeOCl Based Photo-Fenton Reaction. *J. Environ. Chem. Eng.* **2021**, *9*, 104501. [CrossRef]
28. Yao, W.; Qu, Q.; von Gunten, U.; Chen, C.; Yu, G.; Wang, Y. Comparison of Methylisoborneol and Geosmin Abatement in Surface Water by Conventional Ozonation and an Electro-Peroxone Process. *Water Res.* **2017**, *108*, 373–382. [CrossRef]
29. Oturan, M.A.; Aaron, J.-J. Advanced Oxidation Processes in Water/Wastewater Treatment: Principles and Applications. A Review. *Crit. Rev. Environ. Sci. Technol.* **2014**, *44*, 2577–2641. [CrossRef]
30. Ahmed, M.M.; Chiron, S. Solar Photo-Fenton like Using Persulphate for Carbamazepine Removal from Domestic Wastewater. *Water Res.* **2014**, *48*, 229–236. [CrossRef]
31. Expósito, A.J.; Monteagudo, J.M.; Durán, A.; San Martín, I.; González, L. Study of the Intensification of Solar Photo-Fenton Degradation of Carbamazepine with Ferrioxalate Complexes and Ultrasound. *J. Hazard. Mater.* **2018**, *342*, 597–605. [CrossRef] [PubMed]
32. Guo, Q.; Zhu, W.; Yang, D.; Wang, X.; Li, Y.; Gong, C.; Yan, J.; Zhai, J.; Gao, X.; Luo, Y. A Green Solar Photo-Fenton Process for the Degradation of Carbamazepine Using Natural Pyrite and Organic Acid with in-Situ Generated H_2O_2. *Sci. Total Environ.* **2021**, *784*, 147187. [CrossRef] [PubMed]
33. Casierra-Martinez, H.A.; Madera-Parra, C.A.; Vargas-Ramírez, X.M.; Caselles-Osorio, A.; Torres-López, W.A. Diclofenac and Carbamazepine Removal from Domestic Wastewater Using a Constructed Wetland-Solar Photo-Fenton Coupled System. *Ecol. Eng.* **2020**, *153*, 105699. [CrossRef]
34. BOE.Es—BOE-A-2003-3596 Royal Decree 140/2003, of February 7th, Establishing Sanitary Criteria for the Quality of Water for Human Consumption. Available online: https://www.boe.es/buscar/doc.php?id=BOE-A-2003-3596 (accessed on 24 June 2021).
35. BOE.Es—BOE-A-2007-21092 Royal Decree 1620/2007, of December 7th, Establishing the Legal Regime for the Reuse of Purified Waters. Available online: https://www.boe.es/buscar/doc.php?id=BOE-A-2007-21092 (accessed on 24 June 2021).
36. De Laurentiis, E.; Chiron, S.; Kouras-Hadef, S.; Richard, C.; Minella, M.; Maurino, V.; Minero, C.; Vione, D. Photochemical Fate of Carbamazepine in Surface Freshwaters: Laboratory Measures and Modeling. *Environ. Sci. Technol.* **2012**, *46*, 8164–8173. [CrossRef]

37. Krakstrom, M.; Saeid, S.; Tolvanen, P.; Kumar, N.; Salmi, T.; Kronberg, L.; Eklund, P. Ozonation of Carbamazepine and Its Main Transformation Products: Product Determination and Reaction Mechanisms. *Environ. Sci. Pollut. Res.* **2020**, *27*, 23258–23269. [CrossRef] [PubMed]
38. McDowell, D.C.; Huber, M.M.; Wagner, M.; von Gunten, U.; Ternes, T.A. Ozonation of Carbamazepine in Drinking Water: Identification and Kinetic Study of Major Oxidation Products. *Environ. Sci. Technol.* **2005**, *39*, 8014–8022. [CrossRef] [PubMed]

 *catalysts*

Article

Pragmatic Approach toward Catalytic CO Emission Mitigation in Fluid Catalytic Cracking (FCC) Units

Aleksei Vjunov *, Karl C. Kharas, Vasileios Komvokis, Amy Dundee, Claire C. Zhang and **Bilge Yilmaz ***

BASF Corporation, 25 Middlesex/Essex Turnpike, Iselin, NJ 08830, USA; karl.kharas@basf.com (K.C.K.);
vasileios.komvokis@basf.com (V.K.); amy.dundee@basf.com (A.D.); claire.c.zhang@basf.com (C.C.Z.)
* Correspondence: aleksei.vjunov@basf.com (A.V.); bilge.yilmaz@basf.com (B.Y.);
Tel.: +1-732-331-5203 (A.V.); +1-646-413-9090 (B.Y.)

Abstract: The need to mitigate the environmental footprints of refineries in a sustainable and economical way is widely accepted, yet there appears to be a lack of a unilateral pragmatic approach towards CO oxidation to CO_2 among the refining community. In this work we share CO promoter design strategies that can afford a tangible and immediate CO conversion efficiency increase without a need for additional precious metal loading. The key focus is on the support material architecture that is essential to boost the CO conversion and reduce the NO_x generation in the FCC unit. It was demonstrated that the suppression of Pt sintering as well as the enhancement of the oxygen mobility on the catalyst surface can afford an ~40% lower cost of Pt and ~20% lower usage rate compared to current industry-standard designs.

Keywords: Pt-based promoter; CO oxidation; environmental catalysis; refinery compliance; fluid catalytic cracking; FCC

Citation: Vjunov, A.; Kharas, K.C.; Komvokis, V.; Dundee, A.; Zhang, C.C.; Yilmaz, B. Pragmatic Approach toward Catalytic CO Emission Mitigation in Fluid Catalytic Cracking (FCC) Units. *Catalysts* **2021**, *11*, 707. https://doi.org/10.3390/catal11060707

Academic Editors: José Ignacio Lombraña, Héctor Valdés and Cristian Ferreiro

Received: 16 May 2021
Accepted: 1 June 2021
Published: 3 June 2021

Publisher's Note: MDPI stays neutral with regard to jurisdictional claims in published maps and institutional affiliations.

1. Introduction

The ever-more stringent environmental regulations on air pollution, especially that caused by the oil refining industry, are drawing more efforts by refineries to combat these harmful emissions [1]. Some examples of the emissions of concern are CO, NO_x, HCN, and SO_x, to name a few—all of which can cause substantial harm to the environment, and human and animal health.

In this work, we focus on the CO and NO_x emissions as these are common and critical concerns during FCC catalyst regeneration as part of oil cracking. Moreover, the mitigation of such pollutants is now part of most oil refiners' commitment to sustainability [2]. The regenerator and related catalyst logistics are schematically presented in Figure 1. Specifically, incomplete coke combustion in the FCC unit regenerator may lead to CO and O_2 breakthrough into the dilute phase [3,4]. The resulting combustion of CO in the dilute phase, referred to as afterburn, can produce high temperatures, which can damage regenerator internal hardware, and, therefore, cause an FCC unit outage [5]. While seemingly minor, such a sequence of events would cause a major disruption in a refinery operation and would also lead to substantial financial loss. Even without an outage, concerns around afterburn would lead refiners to run their FCC units at suboptimal conditions, hindering them from reaching full potential of their process and catalyst.

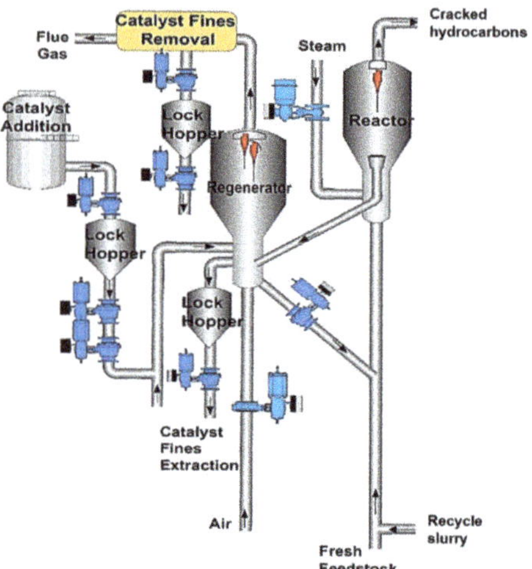

Figure 1. The FCC unit regenerator as well as the associated catalyst logistic streams are schematically shown.

To mitigate CO emissions, refineries typically use CO promoters that catalyze CO oxidation to CO_2 in the dense bed. An additional benefit for CO promoters is improved air utilization. A CO promoter can typically be introduced as part of the FCC catalyst formulation or as a standalone additive. The CO promoter is usually based on highly dispersed supported platinum or palladium that can catalyze the oxidation reaction. The choice of metal is driven by a combination of regulatory certification, current emissions as well as budget for a given FCC unit operation. Typical CO promoter addition rates are 2–5 pounds of additive per short ton fresh FCC catalyst, which often corresponds to ~1–3 ppm platinum group metal (PGM) level in the circulating catalyst inventory [6].

Over the past years there have been numerous contributions to the field of CO oxidation chemistry. These include extensive and fundamental studies using advanced techniques, many summarized in recent review articles [7,8]. In this contribution, however, we focus predominantly on the CO promoter designs relevant to the refining industry with a focus on actual real-life industrial performance rather than fundamental mechanistic insights as to the nature of the occurring chemistry. Pt has been selected as the PGM of choice for this set of experiments. We aim to share the conceptual approach to pragmatic Pt-based CO promoter design as well as the underlying considerations and concepts that are economically viable and can find applications in actual FCC units worldwide.

2. Experimental

2.1. Material Synthesis

A series of reference and experimental catalysts was prepared, the full list is provided in Table 1. In all cases, γ-Al_2O_3 microspheres with a typical D_{50} of 80 μm and a total surface area (TSA) of 90 m^2/g were used as the alumina support. Commercially available cerium nitrate and strontium nitrate, industrial grade in all cases, were used as precursors to form the doping package for the alumina support. An aqueous solution of Pt-nitrate, with a typical Pt content of 10 wt % was used as the PGM source. Both dopants as well as PGM were deposited via incipient wetness impregnation to afford even metal distribution on the support.

Table 1. The sample composition as well as measured CO conversion and NO_x generation values for the catalysts discussed in this contribution are reported.

Promoter	CeO_2 (wt %)	SrO (wt %)	Al_2O_3 (wt %)	Pt (ppm)	CO Conversion (%) [1]	NO_x Generation (%) [1]
Reference-1	0	0	99.97	300	47.6 ± 2.4	127.3 ± 6.4
Reference-2	0	0	99.95	500	43.8 ± 2.2	128 ± 6.4
Reference-3	0	0	99.92	800	47.6 ± 2.4	129 ± 6.5
Catalyst A	3	0	96.97	300	54.4 ± 2.7	117.4 ± 5.9
Catalyst B	6	0	93.97	300	57.5 ± 2.9	115.8 ± 5.8
Catalyst C	10	0	89.97	300	62.1 ± 3.1	112.3 ± 5.6
Catalyst D	18	0	81.97	300	64.2 ± 3.2	113.9 ± 5.7
Catalyst E	10	2.6	87.37	300	64.1 ± 3.2	104.7 ± 5.2
Catalyst F	10	4.7	85.27	300	67.0 ± 3.4	93.0 ± 4.7

[1] Test: 99% spent FCC catalyst + 1% fresh promoter, plug flow reactor operating at 1 L/min flow rate with a feed of 2 vol.% O_2 in N_2 at 700 °C. Values determined by comparing to a promoter-free base case experiment. The method used to calculate CO conversion and NO_x generation is reported in the next chapter. The accuracy of the reported CO and NO_x emissions values was determined by performing a statically relevant number of repetitions. For the chosen reactor setup and under the specific experimental conditions reported in this study, an average variation of ±5% was determined.

Reference catalysts 1–3 were prepared by impregnating the alumina support with the aqueous platinum nitrate solution such that the desired final metal concentration was achieved. Experimental Catalysts A–D were prepared by first impregnating the alumina support with cerium-nitrate such that the desired final CeO_2 concentration was achieved. The support was then calcined at 550 °C for 2 h. Subsequently, the desired amount of the Pt nitrate solution was impregnated onto the ceria/alumina support. Catalysts E-F were prepared by first impregnating the alumina support with an aqueous mixture of cerium- and strontium-nitrates such that the desired final CeO_2 and SrO concentrations were achieved. The support was then calcined at 550 °C for 2 h. Subsequently, the desired amount of the Pt nitrate solution was impregnated onto the ceria/strontia/alumina support. For all presented catalysts, upon Pt deposition, all samples were calcined at 550 °C for 2 h.

2.2. Catalyst Characterization

All catalysts were characterized after synthesis. The surface area measurements were performed using Micromeritics TriStar models 3030/3020. The surface area in all cases was quite similar with the Pt/Al_2O_3 samples having a typical TSA of 90 m^2/g, $Pt/CeO_2/Al_2O_3$ as well as the $Pt/SrO/CeO_2/Al_2O_3$ catalysts having a typical TSA of 80–85 m^2/g. The catalyst composition was determined using the Inductively Coupled Plasma (ICP) test on a Thermo iCAP 7400. The compositions of each presented catalyst are summarized in Table 1.

2.3. Catalyst Testing

Catalysts were tested as-is, also referred to as "fresh" state, as well as after a simulated aging, referred to as "aged". The aging conditions as well as details of the experimental setup are reported in the description of the specific experiment in the results section. The absolute conversion levels were measured at the Chemical Process and Energy Resources Institute (CPERI) located in Thessaloniki, Greece. A spent FCC catalyst was used for the base case tests, while for the evaluation studies mixtures of 1 wt % of the additive and 99 wt % of the spent catalyst were loaded on the reactor. All presented data is reported as the difference compared to the promoter-free base emission. Prior to testing, all catalytic materials were sieved to 63–106 μm. The accuracy of the reported CO and NO_x emissions values was determined by performing a statically relevant number of repetitions. For the chosen reactor setup and under the specific experimental conditions reported in this study, an average variation of $\pm5\%$ was determined. The reaction conditions for the chosen test protocol are reported in Table 2.

Table 2. The CO promoter testing parameters are reported.

Reactor Type	Fluid Bed
Reactor loading	10 gr
Catalyst mixture	99 wt % spent FCC catalyst + 1 wt % CO promoter
Inlet gas flow rate	1 L/min
Inlet gas composition	2 vol.% O_2 in N_2
Reactor bed temperature	700 °C

The CO conversion (%) as well as NO_x generation (%) values were calculated using the formula reported in Equation (1).

$$\text{CO or NO}_x \ (\%) = 100 \times \frac{(\text{CO or NO}_x)_{\text{Base Case}} - (\text{CO or NO}_x)_{\text{Promoter}}}{(\text{CO or NO}_x)_{\text{Base Case}}} \tag{1}$$

Equation (1). The formula used to determine the CO conversion (%) as well as NO_x generation (%) values is reported.

3. Results and Discussion

3.1. Promoter Pt-Loading

The impact of Pt-loading on the activity of a typical CO promoter is demonstrated using the example of the reference catalysts 1–3 (Pt/Al$_2$O$_3$) with different Pt-loadings (see Table 1). Overall, there appears to be essentially no difference between the three chosen Pt-loadings when the overall catalyst design is identical. At first, this result may be somewhat surprising as this observation would suggest that a large fraction of the precious metal is apparently not contributing to the catalytic activity. However, the observation may relate to the chosen synthetic pathway and the resulting relative size of the Pt-clusters on the support surface. From prior reports on spectroscopic studies of precious metal catalysts [9,10], in a 1 nm Pt particle the ratio of surface Pt to Pt forming the core of the particle is ~1:1. With growing particle size, the average Pt coordination number will grow and, as a result, in a 20 nm Pt particle the surface Pt atoms account for approximately 5% of the total metal forming the said particle. Considering the particle size for a typical industrial catalyst, such as those presented here, increasing Pt loading on the support further leads to ever-diminishing returns as the available Pt surface area is essentially unchanged as the diameter of the Pt clusters is increased. Therefore, while some refineries may choose to increase the precious metal levels to combat growing CO emissions, the benefit of such an approach is limited and inherently inefficient.

An example of this is demonstrated using reference catalysts 1–3, which differ only in the amount of Pt deposited onto the support. From Figure 2, within the experimental uncertainty, the efficiency for CO conversion as well as NO_x generation in the fresh state is similar in all cases. The key difference is in the actual cost to the refiner ($/mT) of Pt used, which ranges from ~8700 $/mT to ~23000 $/mT (at 900 $/oz t Pt market price, a reasonable average value in the 2016–2020 timeframe) [11]. Note that the financial impact increases with the growing (USD ~1200/oz t) Pt market price, as has been experienced in late 2020 and early 2021 [11].

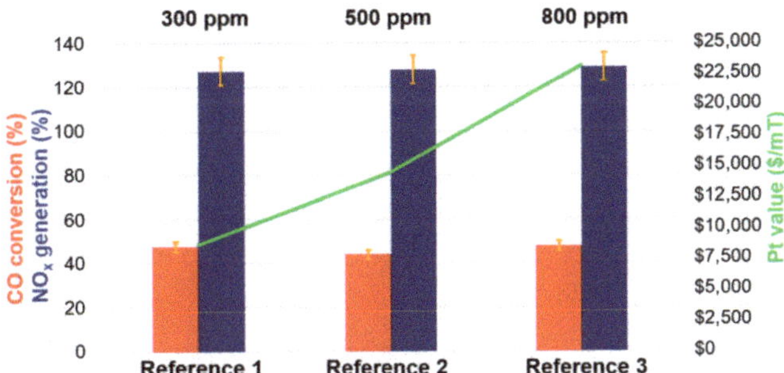

Figure 2. The CO conversion and NO_x generation observed for reference catalysts with varying Pt loadings are reported. The respective value of Pt ($/mT) at a 2016–2020 average market price of 900 $/oz t Pt [11] is also reported for reference.

Therefore, while a refinery may be under the impression that superior performance is obtained through a higher amount of metal, the actual value proposition diminishes substantially with increasing Pt loading. That is, on a relative basis, when the support material is identical in all cases, the oxidation of a CO molecule to CO_2 is ~2.6-times more cost-effective with reference-1 than with reference-3, when the value of the Pt is considered. These considerations, of course, are true when the total concentration of the CO promoter component in an FCC blend is constant and only the PGM amount allocated on this component is increased. An alternative scenario would be for a refinery to increase the total Pt content by increasing the amount of CO promoter in the blend. For example, instead of using 0.25 wt % CO promoter with 1000 ppm Pt, the FCC blend could be formulated using 0.5 wt % CO promoter with 500 ppm Pt. In this case, the benefit of increasing the total Pt in the FCC unit is more pronounced, and in certain cases may lead to the desired improvement of the apparent CO conversion. However, such an approach will typically also lead to a significant NO_x emission increase, a much undesired side-effect that in certain areas with very tight emissions regulations, e.g., California, would not be a viable solution. Beyond the NO_x concern, most of the precious metal in such an FCC blend would still remain inactive since it forms the core of the PGM-particles. Overall, such an approach could be considered a better of the two poor solutions. Indeed, there are many other, cost-efficient, means to consider before the Pt content is increased, as we will discuss in subsequent paragraphs.

We note at this point that the support used for the catalyst adds further, in some cases significant, costs to the overall Pt promoter design. As Pt loading has limited benefit, we hence focus on the maximization of Pt utilization through support tuning, which is a far more efficient and cost-effective way to reduce CO emissions when the total refinery spend on environmental compliance is considered.

3.2. Promoter Doping

To demonstrate the potential of catalyst support modification towards improved activity, we explore the impact of support doping. In this work, the impact of catalyst doping with ceria is demonstrated using the 300 ppm Pt reference-1 as well as a series of catalysts A–D composed of 300 ppm Pt supported on alumina with different levels of CeO_2 (see Table 1 for details). The reference CO promoter achieves a 47.6% conversion of CO to CO_2 under chosen experimental conditions. This will be used as baseline for further assessment of the impact of ceria as an improvement of the CO conversion is observed with increasing amounts of ceria. These results are summarized in Figure 3.

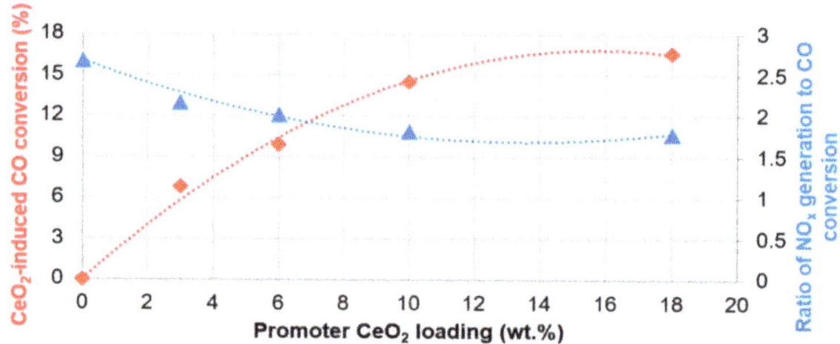

Figure 3. The CeO_2-induced CO conversion as well as the ratio of NO_x generation to CO conversion are reported as a function of promoter CeO_2 content. The lines are intended for trend visualization purposes only. Catalysts were tested by mixing 99% spent FCC catalyst and 1% fresh promoter using a plug flow reactor operating at 1 L/min flow rate with a feed of 2 vol.% O_2 in N_2 at 700 °C. The emissions values determined by comparing to a promoter-free base case experiment. The method used to calculate CO conversion and NO_x generation is reported in the experimental section. The color-coding is reported in the plot.

We observe a nearly linear CO conversion improvement between 0% and ~10% CeO_2 doping of the CO promoter. Beyond ~10%, the impact of ceria is greatly diminished with only minor improvement in ever-increasing CeO_2 loading. There are at least two possible reasons why the addition of ceria is beneficial to the promoter performance when measured in fresh state. First, ceria itself is known as an oxidation catalyst [12,13] that can convert CO to CO_2 at elevated temperatures, either by itself or as a mixture with further oxides, e.g., CuO [14]. Second, the chosen aging and testing conditions may be quite favorable for the doped CO promoter. The testing temperature of 700 °C, when the thermal stability of ceria is considered, is quite mild and does not affect the ability of ceria to serve as an oxygen storage component (OSC) [15]. The addition of OSC, whether in the form of ceria/zirconia or ceria by itself, to PGM is a well-known concept in the field of mobile emissions control [16] and has been shown to allow for optimized surface O_2 concentration on the PGM surface, thus maximizing the metal efficacy for CO oxidation [17].

The diminishing benefit of increasing ceria content above ~10% is explained by several factors: (1) kinetic limitation of the reaction under chosen testing conditions and (2) the way the discussed promoters were prepared. As an example, the degree of the OSC capacity utilization is expected to decrease with increasing CeO_2 content due to mass transfer limitations and the inaccessibility of a portion of ceria in the CeO_2-particle core.

Another important observation for the presented sample set is that the addition of ceria leads to a decrease in NO_x generation by as much as ~10% (Table 1). This feature is important as some refineries, especially those with a high Nelson complexity index (NCI), prefer to view the CO promoter performance as a ratio of CO conversion benefit balanced by the NO_x penalty. That is, one can assess promoter efficiency based on how much NO_x is generated per molecule CO that is oxidized. The ratio of NO_x generation to CO conversion for the studied sample-set is reported in Figure 3. Interestingly, while for every 1% of CO conversion the reference, ceria-free, promoter generates ~2.7% additional NO_x emission, the addition of 10% ceria allows one to decrease the NO_x generation to ~1.8% per 1% CO conversion. Therefore, the addition of ceria has not only allowed boosting the absolute CO conversion levels, but also to decrease the relative NO_x emissions by as much as ~33%, a very significant value for most oil refineries. Again, it should be noted that, similar to CO conversion, the NO_x benefit becomes marginal beyond ~10% CeO_2 content.

While there is a multitude of factors affecting the NO_x generation, there are at least two that we propose relevant to explain the observations in this work. Suggested causes are surmised and are the focus of an ongoing research project outside the frame of the current contribution. First, the addition of ceria is suggested to affect the rates of CO and NO oxidation shifting the balance in favor of CO. Second, under chosen reaction conditions ceria can contribute to the apparently lower NO_x emissions by trapping a portion of the NO_2, a product of NO oxidation, in the form of ceria-nitrate. Examples of NO_x-traps using PGM/ceria components have been previously reported in the field of mobile emissions control [18,19]. Such a mechanism is suggested favorable for a CO promoter geared towards applications in the oil refining industry. As an example, the promoter can trap a portion of the generated NO_x in the regenerator. Subsequently, as the promoter circulates through the FCC unit, cerium nitrate can be regenerated in the riser to ceria while releasing NO_x, which in turn can be converted to ammonia and thus removed at the top of the riser column. The regenerated ceria is then ready for the subsequent regenerator cycle. It is worth noting that while the effect is likely significant to help boost CO conversion, the typical amounts of promoter in the FCC catalyst blends (often around 1 wt %) are generally insufficient to dramatically shift the refinery NO_x emission. Therefore, the proposed NO_x emissions reduction is generally considered less efficient than the installation of a stationary NO_x-scrubber that operates via selective catalytic reduction (SCR) of NO_x with ammonia [20].

3.3. NO_x Emission Suppression

To supplement the NO_x observations discussed in the previous chapter, a series of experiments were performed using formulations that in addition to ceria also incorporated further NO_x-trapping components, similar to those typically reported for Lean NO_x Trap (LNT) diesel applications in mobile emissions control [21]. Two additional samples, catalyst E (catalyst C doped with 2.6% SrO) and catalyst F (catalyst C doped with 4.7% SrO), see Table 1, were prepared and compared in their CO conversion as well as NO_x generation activity to catalyst C, which is one of more active promoter designs discussed in present work. The CO conversion and NO_x generation values for these catalysts are reported in Table 1. Sr, specifically in the form of SrO, has been widely proposed to manage NO_x emissions in the past and is hence a suitable example of such chemistry [22,23]. Further comparison of the impact of SrO on overall activity is shown in Figure 4.

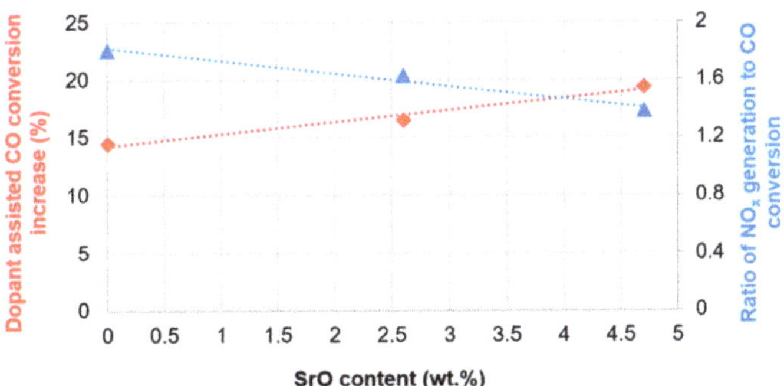

Figure 4. The dopant-assisted CO conversion as well as the ratio of NO_x generation to CO conversion are reported as a function of promoter SrO content. The color-coding is reported in the plot alongside with fitted curves exemplifying the discussed trends. Catalysts were tested by mixing 99% spent FCC catalyst and 1% fresh promoter using a plug flow reactor operating at 1 L/min flow rate with a feed of 2 vol.% O_2 in N_2 at 700 °C. The emissions values determined by comparing to a promoter-free base case experiment. The method used to calculate CO conversion and NO_x generation is reported in the experimental section.

It should be noted that the addition of SrO to catalyst C leads to a near-linear improvement (Figure 4) in both CO conversion increase and NO_x generation reduction, which suggests these reactions are closely coupled. The observation is explained by the ability of SrO to trap NO_2 in the form of strontium nitrate, which can then be decomposed in the riser to yield ammonia and the regenerated SrO, much like the mechanism described for CeO_2 in the previous chapter. The key difference here is that SrO, due to its nature, cannot contribute to CO oxidation. In turn, this means that while SrO selectively stores NO_2, a larger amount of the ceria remains available to oxidize CO rather than be "deactivated" by NO_2 adsorption. This operation is very similar to that of a typical LNT design for diesel emissions system applications, where the NO_x-trapping function is partially decoupled from CO and hydrocarbon oxidation to maximize catalytic efficiency [24].

With the potential NO_x emissions suppression in mind, it is not obvious that the addition of a dedicated NO_x-trapping component is the most practical approach towards NO_x management in a refinery since the necessary amount of such SrO-based trap would need to comprise a substantial amount, e.g., 10–20%, of the FCC formulation, which is not a viable approach towards NO_x mitigation in most existing FCC units.

3.4. Impact of Aging

Two catalysts are chosen for further testing to assess the impact of promoter aging on performance under conditions typical of an FCC unit. Specifically, two designs are compared here, these are reference-2, a 500 ppm Pt containing promoter that is similar to those designs often used in commercial applications, and catalyst C, 300 ppm Pt and 10% CeO_2 containing promoter design that has shown optimal NO_x/CO ratio, as discussed in previous chapter. Additional information on these two catalysts is provided in Table 1. It is important to note that while at first glance the comparison may not seem direct or fair, we focus on a real-life type of application, where a refinery would consider one or two catalysts at most when selecting the CO promoter for their application. Ultimately, what is most critical, is the highest efficiency per unit catalyst, i.e., lb or kg, rather than specific elemental composition, as this approach allows for a minimization of the monthly CO promoter spend at the refinery.

There are two types of aging tests that were developed for this workstream. First, a steam-aging, which is used to simulate the regenerator portion of the FCC operation cycle. Here, we chose a 12 h aging duration with a T_{max} = 787 °C using a sealed steam reactor that is filled with a 90% H_2O-steam/10% N_2 mixture. Second, Lean/Rich (L/R) aging, which is more typical for a mobile emissions control testing setup, e.g., for a very lean diesel engine, that mimics the full FCC unit operation cycle. This aging is also 12 h in duration, reaches a T_{max} = 780 °C and is performed using a flow-through reactor that cycles with 10 min Air, followed by 10 min N_2, followed by 10 min H_2, followed by 10 min N_2, while at a constant 10% H_2O-steam in every step. Due to powder bed penetration difference between air and H_2, the 10 min H_2 reduction is equivalent to ~1 min reduction using hydrocarbons. This point is important as it simulates the relative time a CO promoter spends in the FCC unit, i.e., ~90% of the lifetime in the regenerator and ~10% lifetime in the riser, where the promoter can be regenerated. The comparison of the said two samples in fresh as well as aged form is shown in Figure 5.

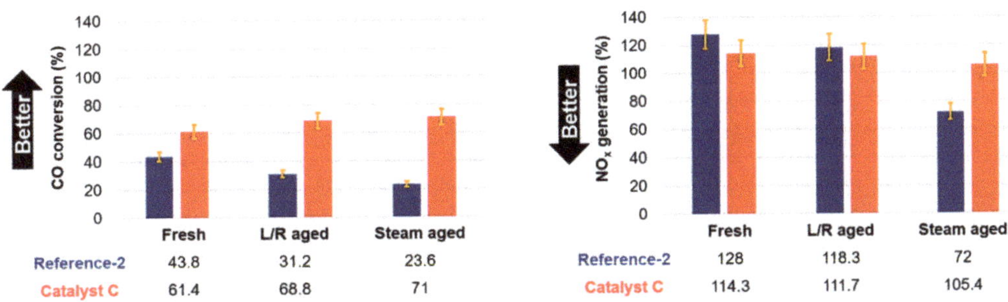

Figure 5. The activity of reference-2 and catalyst C in fresh as well as aged form for CO conversion and NO_x generation are reported. The color-coding as well as the actual measured values are reported in the plots. Catalysts were tested by mixing 99% spent FCC catalyst and 1% fresh promoter using a plug flow reactor operating at 1 L/min flow rate with a feed of 2 vol.% O_2 in N_2 at 700 °C. The emissions values determined by comparing to a promoter-free base case experiment. The method used to calculate CO conversion and NO_x generation is reported in the experimental section.

The superiority of catalyst C to the reference-2, regardless of the lower Pt content in the experimental sample was expected based on fresh testing. However, the extent of the aging on the said catalysts was not obvious prior to testing. For CO conversion, the CO activity of the reference-2 design is decreased by 25–50% depending on the aging type, with steam aging being particularly harmful.

At this point we believe it is necessary to provide a side note on the nature of Pt aging as often observed in mobile emissions control, e.g., for diesel-type emissions mitigation systems. Pt is quite mobile in the oxide form and stable in metallic form when it comes to sintering or Ostwald ripening [25,26]. What this translates to is an elevated degree of mobility on the surface that ultimately leads to a higher extent of agglomeration upon aging in the absence of reductants. Therefore, while both aging protocols in this work had a nominal duration of 12 h, the actual time Pt was subject to oxidation was different, with more favorable conditions in the L/R aging case. Since Pt was offered the occasional regeneration step, that in turn has led to an apparent higher degree of CO activity after the 12 h L/R aging cycle compared to steam aging. The key learning here is that the promoter aging protocol must be chosen such to best suit the actual refinery operation. In turn, this approach not only allows the correct promoter dosing and durability estimate, but also a fair comparison of competing catalyst technologies (e.g., as part of catalyst vendor bids).

With the above in mind, the remarkable stability of catalyst C after both aging protocols deserves a further explanation as the performance appears to improve slightly with aging. There are two key reasons for this observation. First, the CeO_2 that is present in catalyst C will exhibit some CO activity on its own, as discussed previously. The aging conditions chosen in this work are generally below temperatures that are considered harmful to ceria, i.e., cause structural collapse and stop oxygen mobility [27]. Therefore, the CeO_2-contribution to CO oxidation is essentially unaffected. Second, likely even more prominent, ceria when introduced to the promoter design in an appropriate manner, can serve as an anchoring point for Pt as the affinity of Pt to CeO_2 is known to exceed that of Pt to Al_2O_3 [28]. Therefore, as the aging begins, Pt that may be randomly distributed on the surface of the promoter particle, is quickly anchored on the surface of ceria domains. This in turn means that Pt is given little opportunity to find further Pt particles and sinter into larger Pt clusters on the promoter surface. The large number of small anchored Pt particles offers a significantly larger apparent Pt surface area that is available for the CO oxidation reaction. Additionally, ceria can stabilize Pt at atomic dispersion which is active for CO oxidation but not for dehydrogenation chemistry [29,30]. Finally, while it is not possible to completely exclude some degree of Pt encapsulation with ceria, the relatively mild aging temperatures [31], as compared to some of the most aggressive environmental catalysis type aging conditions [32], lead us to surmise that the anchoring of Pt with ceria has a significant net-benefit effect over a simpler Pt/Al_2O_3 type design under operation conditions typical of a refinery FCC unit.

As for NO_x generation, the same general observations and explanations as those discussed for CO conversion are still valid. The decrease of the Pt surface area due to sintering leads to diminishing NO_x emissions. For Catalyst C, the NO_x generation is much more stable, regardless of aging, which serves as an indication of the excellent stability and durability of this promoter design. The latter observation is critical when it comes to real-life applications: refiners are interested in a stable mode of operation with little to no fluctuation of the emissions performance. A promoter such as reference-2 would need to be continuously dosed, either via loader or as part of additional FCC catalyst to maintain average CO conversion activity. In contrast, a design such as catalyst C will allow the refinery to reduce the overall tonnage of the promoter needed monthly. Specifically, for the case exemplified here, one could use ~20% less CO promoter on a weight basis when switching from reference-2 to catalyst C without a sacrifice in CO activity. At the same time, the NO_x emissions would be reduced by ~10–15%. Most importantly, not only would the price per unit promoter ($/kg) be reduced by 40% based on the Pt-loading related value, but the overall cost of CO promotion would be reduced as a function of lower monthly consumption. Furthermore, the improved durability of the promoter would reduce the risk of non-compliance and the amount of effort, especially in regard to monitoring as well as the actual staffing necessary to maintain promoter levels in the unit, thus simplifying the day-to-day operations for the refinery.

4. Conclusions

We have demonstrated that the best strategy to mitigate CO emissions from an FCC unit is to focus on the underlying chemistry rather than PGM dosing. Pt can be effectively anchored on the support by a choice of an appropriate doping package that can suppress Pt sintering. The resulting synergy between the largely retained Pt surface area and the intrinsic activity of the support allows for a substantial increase in promoter efficiency without increasing the Pt content. For instance, CeO_2 is shown to greatly affect CO conversion as well as to have a positive effect on apparent NO_x emission, while at the same time allowing prolonged promoter durability. SrO can be employed to tune the NO_x emissions for refineries operating close to the allowed emissions limit. In turn, the correct choice of more advanced CO promoter systems allows streamlined operations and lower dosage into the FCC unit, which directly contributes to the refinery bottom line.

Author Contributions: Conceptualization, A.V. and K.C.K.; investigation, V.K. and A.D.; writing—original draft preparation, A.V., K.C.K. and C.C.Z.; writing—review and editing, A.V. and B.Y. All authors have read and agreed to the published version of the manuscript.

Funding: This research received no external funding.

Data Availability Statement: The data presented in this study are available on request from the corresponding author. The data are not publicly available due to BASF Corporation Confidentiality policy.

Acknowledgments: The authors would like to recognize the support and high quality of performed CO promoter testing by Angelos Lappas and Eleni Pachatouridou at the Chemical Process and Energy Resources Institute (CPERI) located in Thessaloniki, Greece.

Conflicts of Interest: The authors declare no conflict of interest.

Abbreviations

Platinum group metal (PGM), oxygen storage component (OSC), fluid catalytic cracking (FCC).

References

1. Petroleum Sector (NAICS 324), US EPA. Available online: https://www.epa.gov/regulatory-information-sector/petroleum-sector-naics-324 (accessed on 5 May 2021).
2. Air Emissions -Protecting Air Quality, Chevron Sustainability Roadmap. Available online: https://www.chevron.com/sustainability/environment/air-emissions (accessed on 5 May 2021).
3. Chester, A.W. Chapter 6 CO combustion promoters: Past and present. *Stud. Surf. Sci. Cat.* **2007**, *166*, 67–77. [CrossRef]
4. Stockwell, D.M. Chapter 6 CO combustion promoters: Past and present. *Stud. Surf. Sci. Cat.* **2007**, *166*, 79–102. [CrossRef]
5. Gary, J.H.; Handwerk, G.E. *Petroleum Refining: Technology and Economics*, 4th ed.; CRC Press: New York, NY, USA, 2001; ISBN 0-8247-0482-7.
6. BASF USPTM CO Promoter. Available online: https://catalysts.basf.com/products/ultra-stable-promoter-ups (accessed on 5 May 2021).
7. Lin, J.; Wang, X.; Zhang, T. Recent progress in CO oxidation over Pt-group-metal catalysts at low temperatures. *Chin. J. Catal.* **2016**, *37*, 1805–1813. [CrossRef]
8. Van Spronsen, M.A.; Frenken, J.W.M.; Groot, I.M.N. Surface science under reaction conditions: CO oxidation on Pt and Pd model catalysts. *Chem. Soc. Rev.* **2017**, *46*, 4347–4374. [CrossRef] [PubMed]
9. Jentys, A. Estimation of mean size and shape of small metal particles by EXAFS. *Phys. Chem. Chem. Phys.* **1999**, *1*, 4059–4063. [CrossRef]
10. Frenkel, A.I.; Yevick, A.; Cooper, C.; Vasic, R. Modeling the Structure and Composition of Nanoparticles by Extended X-Ray Absorption Fine-Structure Spectroscopy. *Annu. Rev. Anal. Chem.* **2011**, *4*, 23–39. [CrossRef] [PubMed]
11. APMEX Precious Metals Trading. Available online: https://www.apmex.com/platinum-price (accessed on 5 May 2021).
12. Bunluesin, T.; Putna, E.S.; Gorte, R.J. A comparison of CO oxidation on ceria-supported Pt, Pd, and Rh. *Cat. Lett.* **1996**, *41*, 1–5. [CrossRef]
13. Ferré, G.; Aouine, M.; Bosselet, F.; Burel, L.; Cadete Santos Aires, F.J.; Geantet, C.; Ntais, S.; Maurer, F.; Casapu, M.; Grunwaldt, J.-D.; et al. Exploiting the dynamic properties of Pt on ceria for low-temperature CO oxidation. *Catal. Sci. Technol.* **2020**, *10*, 3904–3917. [CrossRef]
14. Soliman, N.K. Factors affecting CO oxidation reaction over nanosized materials: A review. *J. Mat. Res. Tech.* **2019**, *8*, 2395–2407. [CrossRef]
15. Ozawa, M.; Hattori, M.; Yamaguchi, T. Thermal stability of ceria catalyst on alumina and its surface oxygen storage capacity. *J. Alloy Comp.* **2008**, *451*, 621–623. [CrossRef]
16. Li, J.; Liu, X.; Zhan, W.; Guo, Y.; Guo, Y.; Li, G. Preparation of high oxygen storage capacity and thermally stable ceria–zirconia solid solution. *Catal. Sci. Technol.* **2016**, *6*, 897–907. [CrossRef]
17. Courtois, X.; Bion, N.; Marecot, P.; Duprez, D. Past and Present in DeNO$_x$ Catalysis: From Molecular Modelling to Chemical Engineering. *Stud. Surf. Sci. Cat.* **2007**, *171*, 235–259.
18. Campbell, L.E.; Danzinger, R.; Guth, E.D.; Padron, S. Process for the Reaction and Absorption of Gaseous Air Pollutants, Apparatus Therefor and Method of Making the Same. US Patent 5,451,558, 19 September 1995.
19. Theis, J.; Lambert, C. The Effects of CO, C$_2$H$_4$, and H$_2$O on the NO$_x$ Storage Performance of Low Temperature NO$_x$ Adsorbers for Diesel Applications. *SAE Technical Paper* **2017**. [CrossRef]
20. Han, L.; Cai, S.; Gao, M.; Hasegawa, J.; Wang, P.; Zhang, J.; Shi, L.; Zhang, D. Selective Catalytic Reduction of NO$_x$ with NH$_3$ by Using Novel Catalysts: State of the Art and Future Prospects. *Chem. Rev.* **2019**, *119*, 10916–10976. [CrossRef]
21. Wittka, T.; Holderbaum, B.; Dittmann, P.; Pischinger, S. Experimental Investigation of Combined LNT + SCR Diesel Exhaust Aftertreatment. *Emiss. Contr. Sci. Technol.* **2015**, *1*, 167–182. [CrossRef]
22. Onrubia-Calvo, J.A.; Pereda-Ayo, B.; Caravca, A.; De-La-Torre, U.; Vernoux, P.; Gonzalez-Velasco, J.R. Tailoring perovskite surface composition to design efficient lean NO$_x$ trap Pd–La$_{1-x}$A$_x$CoO$_3$/Al$_2$O$_3$-type catalysts (with A = Sr or Ba). *Appl. Catal. B Enviromental.* **2020**, *266*, 118628. [CrossRef]
23. Zhang, Y.; Liu, D.; Meng, M.; Jiang, Z.; Zhang, S. A Highly Active and Stable Non-Platinic Lean NO$_x$ Trap Catalyst MnO$_x$-K$_2$CO$_3$/K$_2$Ti$_8$O$_{17}$ with Ultra-Low NO$_x$ to N$_2$O Selectivity. *Ind. Eng. Chem. Res.* **2014**, *53*, 8416–8425. [CrossRef]
24. Heck, R.M.; Farrauto, R.J.; Gulati, S.T. *Catalytic Air Pollution Control*; John Wiley and Sons: Hoboken, NJ, USA, 2009; ISBN 9781118397749. [CrossRef]
25. Borgna, A.; Le Normand, F.; Garetto, T.; Apesteguia, C.R.; Moraweck, B. Sintering of Pt/Al$_2$O$_3$ reforming catalysts: EXAFS study of the behavior of metal particles under oxidizing atmosphere. *Catal. Lett.* **1992**, *13*, 175–188. [CrossRef]
26. Hansen, T.W.; DeLaRiva, A.T.; Challa, S.R.; Datye, A.K. Sintering of Catalytic Nanoparticles: Particle Migration or Ostwald Ripening? *Acc. Chem. Res.* **2013**, *46*, 1720–1730. [CrossRef]
27. Lee, J.; Ryou, Y.; Kim, J.; Chan, X.; Kim, T.J.; Kim, D.H. Influence of the Defect Concentration of Ceria on the Pt Dispersion and the CO Oxidation Activity of Pt/CeO$_2$. *J. Phys. Chem. C* **2018**, *122*, 4972–4983. [CrossRef]
28. Nagai, Y.; Hirabayashi, T.; Dohmae, K.; Takagi, N.; Minami, T.; Shinjoh, H.; Matsumoto, S. Sintering inhibition mechanism of platinum supported on ceria-based oxide and Pt-oxide–support interaction. *J. Catal.* **2006**, *242*, 103–109. [CrossRef]
29. Nie, L.; Mei, D.; Xiong, H.; Peng, B.; Ren, Z.; Pereira Hernandez, X.I.; DeLaRiva, A.T.; Wang, M.; Engelhard, M.H.; Kovarik, L.; et al. Activation of surface lattice oxygen in single-atom Pt/CeO$_2$ for low-temperature CO oxidation. *Science* **2017**, *358*, 1419–1423. [CrossRef] [PubMed]

30. Jones, J.; Xiong, H.; DeLaRiva, A.T.; Peterson, E.J.; Pham, H.; Challa, S.R.; Qi, G.; Oh, S.; Wiebenga, M.H.; Pereira Hernández, X.I.; et al. Thermally stable single-atom platinum-on-ceria catalysts via atom trapping. *Science* **2016**, *353*, 150–154. [CrossRef] [PubMed]
31. Yeung, C.M.Y.; Yu, K.M.K.; Fu, Q.J.; Thompsett, D.; Petch, M.I.; Tsang, S.C. Engineering Pt in Ceria for a Maximum Metal−Support Interaction in Catalysis. *J. Am. Chem. Soc.* **2005**, *127*, 18010–18011. [CrossRef] [PubMed]
32. Sala, R.; Bielaczyc, P. Accelerated Ageing Method of Three Way Catalyst Run on Test Bed with Emission Performance and Oxygen Storage Capacity Evaluation. *SAE Technical. Paper* **2020**. [CrossRef]

 *catalysts*

Article

Optimization of Fenton Technology for Recalcitrant Compounds and Bacteria Inactivation

Pablo Salgado [1,2], **José Luis Frontela** [1] and **Gladys Vidal** [1,*]

1   Grupo de Ingeniería y Biotecnología Ambiental, Facultad de Ciencias Ambientales y Centro EULA-Chile, Universidad de Concepción, Concepción 4070386, Chile; pablosalgado@udec.cl (P.S.); jfrontela2018@udec.cl (J.L.F.)
2   Departamento de Ingeniería Civil, Facultad de Ingeniería, Universidad Católica de la Santísima Concepción, Alonso de Ribera 2850, Concepción 4030000, Chile
*   Correspondence: glvidal@udec.cl; Tel.: +56-41-2204067

Received: 11 October 2020; Accepted: 9 December 2020; Published: 19 December 2020

Abstract: In this work, the Fenton technology was applied to decolorize methylene blue (MB) and to inactivate *Escherichia coli* K12, used as recalcitrant compound and bacteria models respectively, in order to provide an approach into single and combinative effects of the main process variables influencing the Fenton technology. First, Box–Behnken design (BBD) was applied to evaluate and optimize the individual and interactive effects of three process parameters, namely Fe^{2+} concentration (6.0×10^{-4}, 8.0×10^{-4} and 1.0×10^{-3} mol/L), molar ratio between H_2O_2 and Fe^{2+} (1:1, 2:1 and 3:1) and pH (3.0, 4.0 and 5.0) for Fenton technology. The responses studied in these models were the degree of MB decolorization ($D_\%^{MB}$), rate constant of MB decolorization (k_{app}^{MB}) and *E. coli* K12 inactivation in uLog units (I_{uLog}^{EC}). According to the results of analysis of variances all of the proposed models were adequate with a high regression coefficient (R^2 from 0.9911 to 0.9994). BBD results suggest that $[H_2O_2]/[Fe^{2+}]$ values had a significant effect only on $D_\%^{MB}$ response, $[Fe^{2+}]$ had a significant effect on all the responses, whereas pH had a significant effect on $D_\%^{MB}$ and I_{uLog}^{EC}. The optimum conditions obtained from response surface methodology for $D_\%^{MB}$ ($[H_2O_2]/[Fe^{2+}] = 2.9$, $[Fe^{2+}] = 1.0 \times 10^{-3}$ mol/L and pH = 3.2), k_{app}^{MB} ($[H_2O_2]/[Fe^{2+}] = 1.7$, $[Fe^{2+}] = 1.0 \times 10^{-3}$ mol/L and PH = 3.7) and I_{uLog}^{EC} ($[H_2O_2]/[Fe^{2+}] = 2.9$, $[Fe^{2+}] = 7.6 \times 10^{-4}$ mol/L and pH= 3.2) were in good agreement with the values predicted by the model.

Keywords: recalcitrant compounds; *E. coli* K12; methylene blue; optimization; Pareto chart; perturbation graph

---

1. Introduction

Some of the effluents produced by industries such as textiles, dyes, tanneries, cosmetics and pulp are colored [1]. In the pulp industry, effluents are colored due to the presence of lignin byproducts and other phenolic compounds formed [2]. These compounds are considered dangerous and recalcitrant because of their low biodegradability and resistant to chemical degradation [1,3]. Besides, recalcitrant compounds with biological activity contained in treated effluent discharges are generating a loss of biodiversity in ecosystems. Even more, some of these compounds with benzyl and phenolic structures are considered endocrine disruptors [4–7]. In addition to the presence of recalcitrant compounds, the bacteria in the effluents of the pulp industry must also be seriously considered. The presence of bacteria in effluents discharged to water bodies are generating humans and animals diseases. In this sense, *Escherichia coli* and other bacteria have been identified in pulp industry effluent [8]. The presence of these bacteria in the effluents of the pulp industry raises an important concern regarding the current technologies (biological treatment) and regulations that govern the discharge of these effluents [9]. The inactivation of a wide range of pathogens in the cellulose industry effluent is effective by chlorination at a relatively low cost [10,11]. However, despite its effectiveness there is a problem to consider: the formation of organochlorine compounds [12]. Tawabini, et al. [13] states that chlorine has a high reactivity that affects the formation of these byproducts (chlorinated organic compounds) when reacting with organic matter. These byproducts are characterized by high toxicity and mutagenic capacity for the environment.

The low effectiveness of conventional water treatments in the destruction of recalcitrant contaminants and the formation of hazardous byproducts has encouraged the search for treatments with a higher oxidative capacity avoiding the formation of harmful byproducts. In this sense, it has been proposed the use of the so-called advanced oxidation processes (AOP) for the elimination of recalcitrant compounds and disinfection of effluents [3,14,15]. Nevertheless, there are differences between bacterial inactivation and decolorization of recalcitrant organic compounds by AOP [16]. Bacteria with the ability to self-repair and grow again after damage are much more complex than recalcitrant compounds [17].

A common point of the vast majority of AOP is the formation of hydroxyl radicals ($\cdot$OH). Furthermore, $\cdot$OH is considered one of the species with the greatest oxidizing power. For example, chlorinated compounds used in conventional effluent treatments such as Cl_2 and ClO_2 have standard reduction potentials of 1.36 and 1.27 V/SHE respectively, while $\cdot$OH has a standard potential of 2.8 V/SHE [18].

Among the AOP, the Fenton technologies has focused a lot of attention for many years [19]. This process involves the reaction of H_2O_2 as an oxidant agent with Fe^{2+} ions as a metal catalyst to produce the degradation agent of $\cdot$OH as illustrated in Equation (1). Fe^{3+} produced by the Fenton reaction can also oxidize H_2O_2 to produce perhydroxyl radicals ($HO_2\cdot$; Equation (2)), named the Fenton-like reaction. The $\cdot$OH and $HO_2\cdot$ produced in Fenton and Fenton-like reactions can participate in parallel reactions to produce singlet oxygen ($^1O_2$; Equations (3) and (4)).

$$Fe^{2+} + H_2O_2 \rightarrow Fe^{3+} + OH^- + \cdot OH \tag{1}$$

$$Fe^{3+} + H_2O_2 \rightarrow Fe^{2+} + HO_2\cdot + H^+ \tag{2}$$

$$HO_2\cdot + HO_2\cdot \rightarrow H_2O_2 + {}^1O_2 \tag{3}$$

$$HO_2\cdot + \cdot OH \rightarrow H_2O + {}^1O_2 \tag{4}$$

Andreozzi, et al. [20] highlight the reactivity of $\cdot$OH, since this species has adequate properties to attack organic compounds, in addition to reacting 10^6–10^{12} times faster than other oxidants. Additionally, the cellular damage produced by $\cdot$OH in the disinfection processes takes place on different macromolecules present in the bacterial membrane causing its inactivation [21].

Many of the parameters that can affect any type of chemical reaction could affect the Fenton reaction, among which the effects of pH and reagent concentration stand out. The pH is one of the most important parameters in the Fenton reaction. However, it is possible to find that the optimum pH varies. One of the reasons why it is possible to find this variety at the optimum pH of the Fenton reaction may be associated with the speciation of Fe^{2+} and Fe^{3+} [22], changes in redox potentials of the main oxidizing species produced [23], or changes in the type of oxidizing species produced depending on the pH [24–26].

Without the presence of Fe^{2+} in the Fenton system there is no formation of $\cdot$OH, so the presence of Fe^{2+} is essential. However, it has been studied that too high Fe^{2+} concentrations can cause the Fenton reaction oxidizing capacity to decrease (Equation (5)) [27].

$$\cdot OH + Fe^{2+} \rightarrow Fe^{3+} + OH^- \tag{5}$$

The H_2O_2 concentration, like Fe^{2+}, is also essential in Fenton systems [28]. However, an excess of H_2O_2 could act as a scavenger of $\cdot$OH [27,28] according to the Equation (6).

$$\cdot OH + H_2O_2 \rightarrow HO_2\cdot + H_2O \tag{6}$$

Accordingly, to minimize Fe^{2+} and H_2O_2 acting as scavengers, but maximizing the production of oxidizing species from these reagents, it is very important to know the optimal $[H_2O_2]/[Fe^{2+}]$ [29].

Therefore, the aim of this work is to evaluate the decolorization of a model recalcitrant compound (methylene blue) and the inactivation of a model bacteria (*E. coli* K12 strain) by Fenton technology considering the operational parameters such as pH, Fe^{2+} concentration ($[Fe^{2+}]$) and the molar ratio between H_2O_2 and Fe^{2+} ($[H_2O_2]/[Fe^{2+}]$), and to reveal the single and combinative effects of the these variables influencing on degree of methylene blue (MB) decolorization ($D_\%{}^{MB}$), rate constant of MB decolorization ($k_{app}{}^{MB}$) and *E. coli* K12 inactivation in uLog units ($I_{uLog}{}^{EC}$).

2. Results and Discussion

2.1. Models and Regression Analysis

Table 1 lists the factors and levels in the experimental design using the following variable: pH, Fe^{2+} concentration ($[Fe^{2+}]$, mol/L) and molar concentration ratio of Fe^{2+} and H_2O_2 ($[H_2O_2]/[Fe^{2+}]$). Also, the experimental degree of MB decolorization ($D_\%{}^{MB}$), the apparent rate constant of MB decolorization ($k_{app}{}^{MB}$) and the inactivation of *E. coli* K12 bacteria in logarithmic units ($I_{uLog}{}^{EC}$) are presented.

Table 1. Actual values and coded levels (in parentheses) of the variables in the Box–Behnken design and experimental values for each response.

	Variables			Responses (Experimental Results)		
#	$[H_2O_2]/[Fe^{2+}]$	$[Fe^{2+}]$ (mol/L)	pH	$D_\%{}^{MB}$ (%)	$k_{app}{}^{MB}$ (min^{-1})	$I_{uLog}{}^{EC}$ (uLog)
1	1:1 (−1)	6.0×10^{-4} (−1)	4.0 (0)	72.8	1.02	0.41
2	3:1 (+1)	6.0×10^{-4} (−1)	4.0 (0)	89.3	1.41	0.84
3	1:1 (−1)	1.0×10^{-3} (+1)	4.0 (0)	81.6	1.94	0.78
4	3:1 (+1)	1.0×10^{-3} (+1)	4.0 (0)	92.9	1.43	0.23
5	1:1 (−1)	8.0×10^{-4} (0)	3.0 (−1)	82.4	1.22	0.33
6	1:1 (−1)	8.0×10^{-4} (0)	5.0 (+1)	96.1	1.63	0.15
7	3:1 (+1)	8.0×10^{-4} (0)	3.0 (−1)	75.8	1.32	0.53
8	3:1 (+1)	8.0×10^{-4} (0)	5.0 (+1)	90.5	1.42	0.34
9	2:1 (0)	6.0×10^{-4} (−1)	3.0 (−1)	89.1	1.20	0.16
10	2:1 (0)	1.0×10^{-3} (+1)	3.0 (−1)	96.7	1.89	0.18
11	2:1 (0)	6.0×10^{-4} (−1)	5.0 (+1)	85.2	1.69	0.15
12	2:1 (0)	1.0×10^{-3} (+1)	5.0 (+1)	89.2	1.78	0.61
13	2:1 (0)	8.0×10^{-4} (0)	4.0 (0)	88.5	1.79	0.15
14	2:1 (0)	8.0×10^{-4} (0)	4.0 (0)	88.0	1.96	0.079
15	2:1 (0)	8.0×10^{-4} (0)	4.0 (0)	88.3	1.82	0.36

Using Design Expert software (version 10), experimental data in Table 1 were analyzed by a second-order linear polynomial regression model (Equation (7)).

$$\eta = \gamma_0 + \gamma_1 A + \gamma_2 B + \gamma_3 C + \gamma_{12} AB + \gamma_{13} AC + \gamma_{23} BC + \gamma_{11} A^2 + \gamma_{22} B^2 + \gamma_{33} C^2 \tag{7}$$

in which η is the dependent factor (response), γ_0 is the intercept; A ($[H_2O_2]/[Fe^{2+}]$), B ($[Fe^{2+}]$) and C (pH) are the independent variables; γ_1, γ_2 and γ_3 are the coefficients of the linear part of the predicted model; γ_{12}, γ_{13} and γ_{23} are the interaction coefficients and γ_{11}, γ_{22} and γ_{33} are the quadratic coefficients. Interaction and quadratic coefficients refer to the effects of the interaction among independent variables.

Analysis of variances (ANOVAs) and significant test results for the quadratic regression equations are shown in Table 2.

Table 2. ANOVA of the regression model for the prediction of degree of methylene blue (MB) decolorization, rate constant of MB decolorization and *E. coli* K12 inactivation.

Source	Sum of Squares	Df	Mean Square	*F*-Value	*p*-Value	Observations
Degree of MB decolorization ($D_{\%}{}^{MB}$)						
Model	627.24	9	69.69	975.81	<0.0001	significant
A-[H_2O_2]/[Fe^{2+}]	397.41	1	397.41	5564.29	<0.0001	-
B-[Fe^{2+}]	72.17	1	72.17	1010.43	<0.0001	-
C-pH	69.91	1	69.91	978.81	<0.0001	-
AB	6.64	1	6.64	92.99	0.0002	-
AC	0.22	1	0.22	3.14	0.1365	-
BC	3.52	1	3.52	49.26	0.0009	-
A^2	58.93	1	58.93	825.15	<0.0001	-
B^2	0.055	1	0.055	0.76	0.4223	-
C^2	13.62	1	13.62	190.72	<0.0001	-
Residual	0.36	5	0.071	-	-	-
Lack of Fit	0.21	3	0.069	0.92	0.5584	not significant
Pure Error	0.15	2	0.075	-	-	-
Cor Total	627.60	14	-	-	-	-
Rate constant of MB decolorization ($k_{app}{}^{MB}$)						
Model	1.17	9	0.13	6.47	0.0267	significant
A-[H_2O_2]/[Fe^{2+}]	0.019	1	0.019	0.94	0.3768	-
B-[Fe^{2+}]	0.37	1	0.37	18.64	0.0076	-
C-pH	8.98×10^{-3}	1	8.98×10^{-3}	0.45	0.5328	-
AB	0.20	1	0.20	10.09	0.0246	-
AC	0.024	1	0.024	1.18	0.3275	-
BC	0.089	1	0.089	4.45	0.0887	-
A^2	0.40	1	0.40	19.76	0.0067	-
B^2	0.025	1	0.025	1.25	0.3141	-
C^2	0.069	1	0.069	3.43	0.1232	-
Residual	0.10	5	0.020	-	-	-
Lack of Fit	0.083	3	0.028	3.27	0.2430	not significant
Pure Error	0.017	2	8.48×10^{-3}	-	-	-
Cor Total	1.27	14	-	-	-	-
E. coli K12 inactivation in uLog units ($I_{uLog}{}^{EC}$)						
Model	0.80	9	0.089	62.10	0.0001	significant
A-[H_2O_2]/[Fe^{2+}]	2.03×10^{-3}	1	2.03×10^{-3}	1.42	0.2870	-
B-[Fe^{2+}]	0.082	1	0.082	57.42	0.0006	-
C-pH	0.17	1	0.17	119.79	0.0001	-
AB	0.29	1	0.29	205.03	<0.0001	-
AC	3.57×10^{-3}	1	3.57×10^{-3}	2.50	0.1745	-
BC	0.056	1	0.056	39.45	0.0015	-
A^2	0.011	1	0.011	7.53	0.0406	-
B^2	0.16	1	0.16	114.39	0.0001	-
C^2	0.033	1	0.033	23.33	0.0048	-
Residual	7.14×10^{-3}	5	1.43×10^{-3}	-	-	-
Lack of Fit	6.78×10^{-3}	3	2.26×10^{-3}	12.54	0.0747	not significant
Pure Error	3.60×10^{-4}	2	1.80×10^{-4}	-	-	-
Cor Total	0.81	14	-	-	-	-

Df: degrees of freedom. Parameter "A" represents the [H_2O_2]/[Fe^{2+}], "B" represents the [Fe^{2+}] and "C" represent the pH. AC, AC, BC, A^2, B^2 and C^2 represent the interactions of A, B and C parameters on the responses.

Table 2 listed the results of variance analysis for the MB decolorization and *E. coli* K12 removal using the Fenton process. The values of the sum of squares demonstrate the contribution of independent variables on responses [30]. The mean squares, which are the sums of squares divided by the degree of freedom. Adequacy of the model parameters in the present study for response variables ($D_\%^{MB}$, k_{app}^{MB} and I_{uLog}^{EC}) was determined by the Fisher value (*F*-value), obtained by dividing the mean squares of each effect by the mean squares of error [31]. The probability critical level (*p*-value) of 0.05 was considered to reflect the statistical significance of the parameters of the proposed model. The F-values > 0.001 (975.81, 6.47 and 62.10) and *p*-values < 0.05 obtained for $D_\%^{MB}$, k_{app}^{MB} and I_{uLog}^{EC} responses confirming the qualification of the model to predict the decolorization of MB ($D_\%^{MB}$ and k_{app}^{MB}) and the inactivation of *E. coli* K12 (I_{uLog}^{EC}) by the Fenton reaction. In addition, the validity of the model is confirmed by the *p*-value of the lack of fit with values greater than the lowest limit of fit as recommended (>0.05) [32]. As a result, the models developed in this work for predicting the $D_\%^{MB}$, k_{app}^{MB} and I_{uLog}^{EC} by the Fenton process were considered adequate. These models can be described as shown in Table 3 with coded three factors.

Table 3. Statistical results of the proposed models in terms of the coded factors.

Response	Proposed Quadratic Model	R^2	R_{adj}^2
$D_\%^{MB}$	$88.3 + 7.05A + 3.00B - 2.96C - 1.29AB + 0.24AC - 0.94BC - 4.00A^2 - 0.12B^2 + 1.92C^2$	0.9994	0.9984
k_{app}^{MB}	$1.86 + 0.048A + 0.22B + 0.034C - 0.22AB - 0.077AC - 0.15BC - 0.33A^2 - 0.082B^2 - 0.14C^2$	0.9210	0.7787
I_{uLog}^{EC}	$0.20 - 0.061A + 0.030B + 0.10C - 0.24AB - 0.003AC + 0.11BC + 0.22A^2 + 0.15B^2 - 0.075C^2$	0.9911	0.9752

The ANOVA results of three parameters ($D_\%^{MB}$, k_{app}^{MB} and I_{uLog}^{EC}) showed that the significant ($p < 0.05$) response surface models with high R^2 value (0.9210–0.9994) were obtained as shown in Table 3, ensuring a satisfactory adjustment of the quadratic models to the experimental data. The R_{adj}^2 values (0.7787–0.9984) obtained suggests that the three proposed models had an adequate predictive capacity. Even more, plots comparing the experimental and predicted values for $D_\%^{MB}$, k_{app}^{MB} and I_{uLog}^{EC} indicated a good agreement between experimental and predicted data from the model (Figure 1). Therefore, this finding indicates high correlation and adequacy of the proposed model to predict performance of the Fenton process ($D_\%^{MB}$, k_{app}^{MB} and I_{uLog}^{EC}).

2.2. Effect of Variables on MB Decolorization ($D_\%^{MB}$)

Figure 2a showed the standardized effects of the components and their contribution to the $D_\%^{MB}$ in a Pareto chart. The sign of standardized effects in Pareto chart, + (favorable effect) or − (unfavorable effect), along the length of the bars provided the physical meaning of model terms. In this Pareto chart, we saw that A, B, C, AB, BC, AA and CC crossed the reference line ($p = 0.05$). It is evident that the most important model term was A ($[H_2O_2]/[Fe^{2+}]$), followed by linear terms B and C ($[Fe^{2+}]$ and pH respectively), quadratic terms corresponding to AA, etc. Thus, e.g., it can be concluded that larger A value, i.e., higher $[H_2O_2]/[Fe^{2+}]$ values, would result in an increase in the $D_\%^{MB}$.

The perturbation plots (Figure 2b) illustrates the effect of all parameters on the $D_\%^{MB}$. The positive effect means that if the effect factor level increases then the response value increases. On the other side, the negative effect means that if the effect factor level increases then the response value decreases. In other words, steep slope or curvature in a factor shows that the response is sensitive to that variable, while a relatively flat line indicates a low sensitivity of response to change with that particular variable. It was observed that the $[H_2O_2]/[Fe^{2+}]$ (A) and $[Fe^{2+}]$ values (B) had significant positive effects on $D_\%^{MB}$, while the initial pH (C) had a negative effect on this response. Gulkaya, Surucu and Dilek [29] demonstrated that $[H_2O_2/Fe^{2+}]$ is a critical parameter for improving the Fenton technology as a treatment of a carpet dyeing wastewater. Otherwise, Babuponnusami and Muthukumar [33] and Chen, et al. [34] demonstrated the positive effect of $[Fe^{2+}]$ on the degradation of phenol and Acridine Orange by Fenton technologies. In both publications it was established that an increase in the $[Fe^{2+}]$ values leads to an increase in the percentage of degradation of phenol and Acridine Orange by the Fenton reaction, in line with what has been demonstrated in the present investigation. Regards to pH in a large part of the experiments carried out by Fenton technologies exhibit an optimal pH close to

3 [24,33,34]. In these research, at pH less or greater than 3 the efficiency of Fenton technology decreases, as observed in this publication.

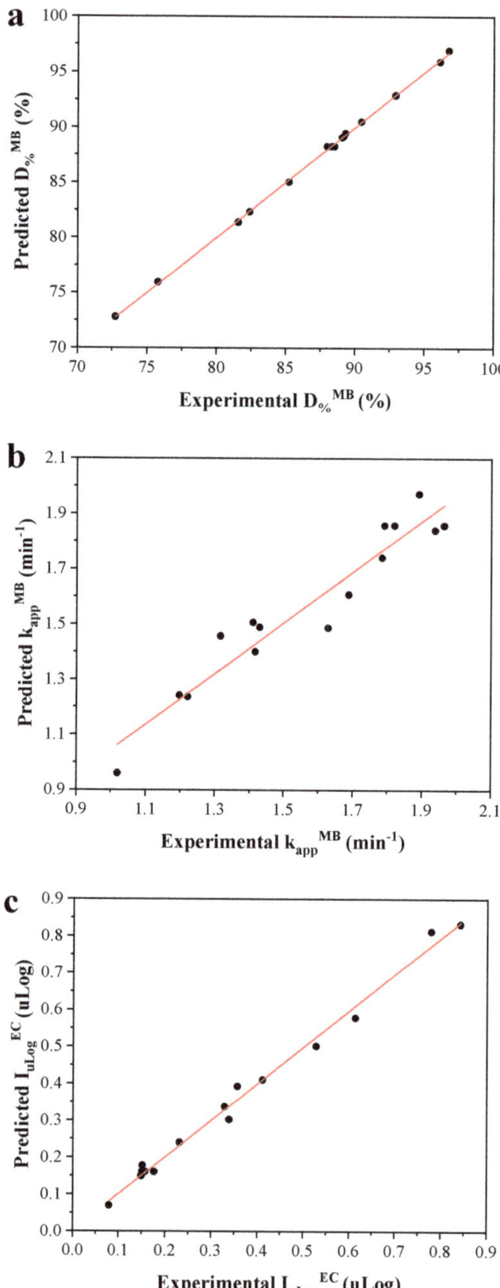

Figure 1. Correlations between the experimental and predicted values of (**a**) $D_\%^{MB}$ values, (**b**) k_{app}^{MB} values and (**c**) I_{uLog}^{EC} values.

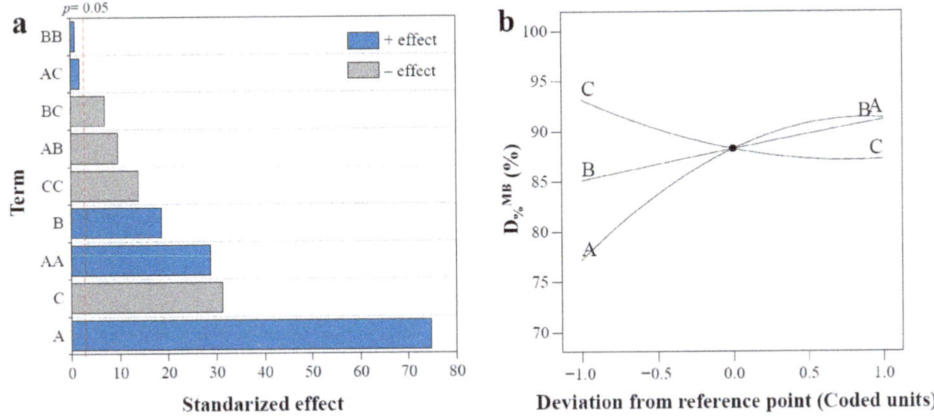

Figure 2. (**a**) Pareto chart showing the standardized effects of variables (first order, quadratic and interaction terms) on $D_\%^{MB}$ (vertical line represents the 95% confidence interval), and (**b**) Perturbation graphs for $D_\%^{MB}$ (A-[H$_2$O$_2$]/[Fe^{2+}], B-[Fe^{2+}], C-pH).

The 3D surface and contour plots in Figure S1 show the individual effects of the process variables and their interactions on the $D_\%^{MB}$. Optimum conditions determination of different variables is the main objective of the response surface methodology (RSM) study, which can affect the $D_\%^{MB}$. By considering the predicted response, [H$_2$O$_2$]/[Fe^{2+}] = 2.9, [Fe^{2+}] = 1.0 × 10^{-3} mol/L and pH = 3.2 of Fenton process were the optimum condition for $D_\%^{MB}$ (94.57%).

2.3. Effect of Variables on the MB Decolorization Rate Constant (k_{app}^{MB})

Figure 3a shows the standardized effects of the components and their contribution to the k_{app}^{MB} in a Pareto chart. In this Pareto chart, we saw that B, AA and AB crossed the reference line ($p = 0.05$). It is evident that the most important model terms are AA and B, followed by interaction term AB ([H$_2$O$_2$]/[Fe^{2+}] and [Fe^{2+}] interaction). Thus, e.g., the AA term implies that k_{app}^{MB} were not influenced by [H$_2$O$_2$]/[Fe^{2+}] in a linear level, but strongly influenced by this parameter in a quadratic level, i.e., a slight variation in [H$_2$O$_2$]/[Fe^{2+}] will result in an increase in the k_{app}^{MB}.

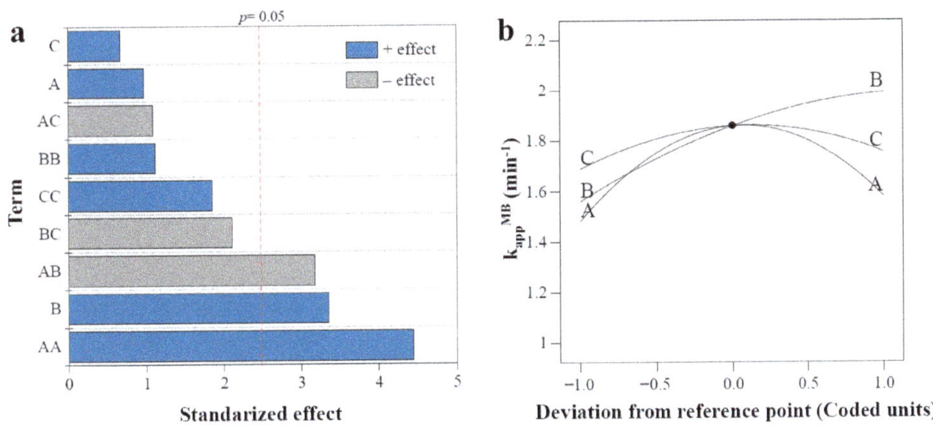

Figure 3. (**a**) Pareto chart showing the standardized effects of variables (first order, quadratic and interaction terms) on k_{app}^{MB} (vertical line represents the 95% confidence interval), and (**b**) Perturbation graphs for k_{app}^{MB} (A-[H$_2$O$_2$]/[Fe^{2+}], B-[Fe^{2+}], C-pH).

Figure 3b shows the perturbation plot of the effect of the parameters on the k_{app}^{MB}. It was observed that the $[Fe^{2+}]$ values (B) had a significant positive effect on k_{app}^{MB}, while the $[H_2O_2]/[Fe^{2+}]$ (B) and pH (C) had an insignificant effect on the response. It has also been reported the main role of $[Fe^{2+}]$ on the rate constants of discoloration of other dyes. For example, Tunç, et al. [35] indicated that the rate constant of acid orange 8 decolorization increased almost 10 times (0.0027–0.0267 min^{-1}) if $[Fe^{2+}]$ values change from 5.0×10^{-6} to 2.5×10^{-5} mol/L. In the same publication the rate constant of acid red 44 decolorization increased almost 4 times (0.0085–0.0331 min^{-1}) if $[Fe^{2+}]$ values incremented from 2.5×10^{-6} to 2.5×10^{-5} mol/L. Melgoza, et al. [36] reported also that the decolorization rate constant of MB increased 2.5 times (0.0014–0.0035 min^{-1}) if $[Fe^{2+}]$ values change from 1.0×10^{-3} to 2.0×10^{-3} mol/L.

Figure S2 show the individual effects of the process variables and their interactions on the k_{app}^{MB} in the 3D surface and contour plots. By considering the predicted response, $[H_2O_2]/[Fe^{2+}] = 1.7$, $[Fe^{2+}] = 1.0 \times 10^{-3}$ mol/L and pH = 3.7 of the Fenton process were the optimum condition providing k_{app}^{MB} (2.08 min^{-1}).

2.4. Effect of Variables on E. coli K12 Removal (I_{uLog}^{EC})

Figure 4a showed the standardized effects of the components and their contribution to the I_{uLog}^{EC} in a Pareto chart. In this Pareto chart, we saw that B, C, AA, BB, CC, AB and BC crossed the reference line ($p = 0.05$). It is evident that the most important model terms was AB, followed by linear term C (pH) and quadratic term BB ($[Fe^{2+}]^2$). Thus, e.g., it can be concluded that smaller C value, i.e., lower pH values, would result in an increase in the I_{uLog}^{EC}.

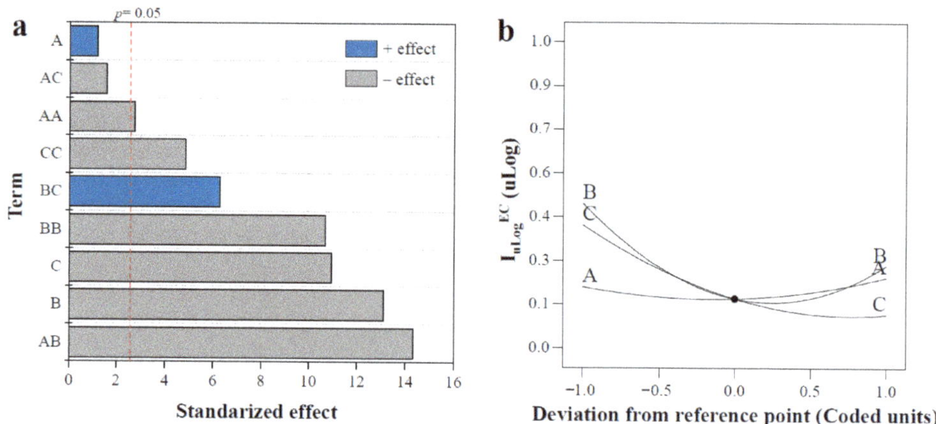

Figure 4. (a) Pareto chart showing the standardized effects of variables (first order, quadratic and interaction terms) on I_{uLog}^{EC} (vertical line represents the 95% confidence interval), and (b) Perturbation graphs for I_{uLog}^{EC} (A-$[H_2O_2]/[Fe^{2+}]$, B-$[Fe^{2+}]$, C-pH).

The perturbation plots (Figure 4b) illustrates the effect of all parameters on the I_{uLog}^{EC}. It was observed that the $[Fe^{2+}]$ (B) and pH (A) values had a significant negative effects on I_{uLog}^{EC}, while the $[H_2O_2]/[Fe^{2+}]$ (C) values did not have a statistically significant effect on this response. Asad, et al. [37] also reported that the inactivation of *E. coli* was mostly affected by Fenton technologies if the $[Fe^{2+}]$ was low. This is evident when considering Equation (5), since at high $[Fe^{2+}]$ values the activity of the ·OH formed could be inhibited.

On the other hand, it is known that the ·OH production by Fenton technology is benefited at acidic pH close to 3. Some examples of Fenton technologies applied to the inactivation of a few bacteria are presented in Table 4. Although the experiments do not have the same conditions of $[H_2O_2]/[Fe^{2+}]$

and reaction time, it is possible to identify that at lower pH values and at lower values of $[Fe^{2+}]$ the bacterial elimination efficiencies tend to increase.

Table 4. Examples of Fenton technologies applied to the bacteria inactivation.

$[Fe^{2+}]$ (mol/L)	$[H_2O_2]/[Fe^{2+}]$	pH	Time (min)	Bacteria	Reduction (uLog)	Ref.
7.8×10^{-4}	62.2	3.0	300	*E. coli*	2.12	Blanco, et al. [38]
7.2×10^{-5}	20.4	3.5	100	*E. faecalis*	1.0	Ortega-Gómez, et al. [39]
1.0×10^{-3}	20	7.0	25	*E. coli*	0.108	Ahmad and Iranzo [40]
5.0×10^{-3}	10	8.5	1440	*E. coli*	0.36	Cengiz, et al. [41]

Figure S3 show the individual effects of the process variables and their interactions on the I_{uLog}^{EC} in the 3D surface and contour plots. By considering the predicted response, $[H_2O_2]/[Fe^{2+}] = 2.9$, $[Fe^{2+}] = 7.6 \times 10^{-4}$ mol/L and pH = 3.2 of the Fenton process were the optimum condition providing I_{uLog}^{EC} (0.89 uLog).

2.5. Analysis of Optimization and Model Validation

The optimal conditions obtained for MB decolorization and *E. coli* K12 inactivation are different for each of the responses studied. These results indicate that although some authors have suggested that it is possible to analyze the bacteria inactivation of AOP by extrapolating from dye decolorization [42], these processes have differences. The results in the present study (Table 1) show that if for example experiments #1 ($[H_2O_2]/[Fe^{2+}] = 1:1$, $[Fe^{2+}] = 6.0 \times 10^{-4}$ mol/L and pH = 4.0) and #10 ($[H_2O_2]/[Fe^{2+}] = 2:1$, $[Fe^{2+}] = 1.0 \times 10^{-3}$ mol/L and pH = 3.0) are compared, the MB decolorization reaches 72.8% and 96.7% respectively, while in the same experiments the *E. coli* K12 inactivation reaches 61.3% (0.412 uLog) and 33.4% (0.176 uLog), i.e., when the MB decolorization is high the *E. coli* K12 inactivation tends to be low, and vice versa. Similar compartments can be observed when comparing (Table 1), for example, experiments #3 ($[H_2O_2]/[Fe^{2+}] = 1:1$, $[Fe^{2+}] = 1.0 \times 10^{-3}$ mol/L and pH = 4.0) and #14 ($[H_2O_2]/[Fe^{2+}] = 2:1$, $[Fe^{2+}] = 8.0 \times 10^{-4}$ mol/L and pH = 4.0). These observations support what has been observed in other studies on the complexity involved in the bacteria inactivation compared to the recalcitrant compounds elimination [43].

To validate the model obtained by the Box–Behnken optimization technique, experiments were carried out with the suggested optimum values of independent variables. Table 5 shows the optimal conditions predicted by the models, the predicted response value and the response value obtained experimentally (Table 5).

Table 5. Results of validation experiments under optimized conditions.

Conditions	$[H_2O_2]/[Fe^{2+}]$	$[Fe^{2+}]$ (mol/L)	pH	Experimental Value	Predicted Value	Error
$D_\%^{MB}$	2.9	1.0×10^{-3}	3.2	94.57	97.12	−2.55
k_{app}^{MB}	1.7	1.0×10^{-3}	3.7	2.08	2.02	0.06
I_{uLog}^{EC}	2.9	7.6×10^{-4}	3.2	0.89	0.92	−0.03

The result obtained from experiments for all response parameters was in agreement with the model prediction. Low errors showed the model and parameters could accurately reflect on the three responses analyzed.

2.6. Effect of $[H_2O_2]/[Fe^{2+}]$ on Responses

The increasing of $[H_2O_2]/[Fe^{2+}]$ value (studied at molar ratio 1:1, 2:1 and 3:1) showed a significant positive effect only on the values of $D_\%^{MB}$. Gulkaya, Surucu and Dilek [29] studied the treatment of wastewater from carpet dyes by Fenton technologies, also finding that $[H_2O_2]/[Fe^{2+}]$ plays a crucial role in the removal of dyes. Considering the above, it is possible to indicate that it is necessary to maintain a considerable concentration of H_2O_2, to ensure that the $\cdot$OH production is maintained for a longer time, obtaining a high efficiency in the discoloration of MB. However, the values of $[H_2O_2]/[Fe^{2+}]$ do not seem to significantly affect the k_{app}^{MB} and I_{uLog}^{EC} values.

2.7. Effect of [Fe²⁺] on Responses

Statistical analyses identified $[Fe^{2+}]$ as an important factor for the three types of responses studied ($D_{\%}^{MB}$, k_{app}^{MB} and I_{uLog}^{EC}). For both the $D_{\%}^{MB}$ analysis and for k_{app}^{MB} the increasing of $[Fe^{2+}]$ value (studied at 6.0×10^{-4}, 8.0×10^{-4} and 1.0×10^{-3} mol/L) exhibited a significant positive effect on these responses, while for the I_{uLog}^{EC} analysis this parameter showed a significant negative effect. In Equation (1) it is observed that the formation of ·OH due to the oxidation of Fe^{2+} to Fe^{3+} in the presence of H_2O_2 will increase with a higher concentration of Fe^{2+}. A single ·OH attack on the MB structure leads to its decolorization ($D_{\%}^{MB}$ and k_{app}^{MB}), which seems to agree with the importance of $[Fe^{2+}]$ found in the present research. However, if the ·OH also had a preponderant role in the inactivation of *E. coli* K12, then the $[Fe^{2+}]$ values should not have a significant negative effect on the I_{uLog}^{EC} response. Some authors [44–47] suggest that although ·OH oxidize most simple organic compounds (such as recalcitrant compounds), the inactivation of a bacterium (such as *E. coli* K12) is not directly affected by the production of these radicals. These researchers indicate that the formation of singlet oxygen ($^1O_2$) is responsible for the inactivation of bacteria. Consequently, the I_{uLog}^{EC} decreases with increasing values of $[Fe^{2+}]$, since excess Fe^{2+} could react with the ·OH formed (Equation (5)), decreasing the possibility that the ·OH react to form $^1O_2$ (Equations (3) and (4)).

2.8. Effect of pH on Responses

The pH value (studied at pH 3.0, 4.0 and 5.0) showed a significant effect on the $D_{\%}^{MB}$ and I_{uLog}^{EC} responses but did not show a significant effect on the k_{app}^{MB} response. The Fenton reaction (Fe^{2+} and H_2O_2), with a rate constant 76 L·mol^{-1}s^{-1} [48], form ·OH quickly, consume Fe^{2+} and produce Fe^{3+}. The Fenton-like reaction (Fe^{3+} and H_2O_2) has a much slower rate constant (0.01 L·mol^{-1}s^{-1}) than the Fenton reaction [48] and it only produces $O_2 \cdot^-$, a much less reactive radical than ·OH. Additionally it has been established that the species of Fe(II) that prevails in the working pH range (3.0–5.0) is Fe^{2+} [49], while in the same pH range the speciation of Fe(III) demonstrates the formation of $Fe(OH)^{2+}$ and $Fe(OH)_2^+$ species, species that are less reactive than Fe^{3+} [22]. Based on this information, it is expected that the rate of ·OH formation from the Fenton reaction, at least in the first minutes of reaction that directly influence the determination of k_{app}^{MB}, will not be greatly altered when changing system pH between 3.0 and 5.0. However, $D_{\%}^{MB}$ and I_{uLog}^{EC}, which are obtained in a final time of 15 min, will be influenced by both the Fenton reaction and the subsequent Fenton-like reaction. Considering this, the participation of the Fenton-like reaction implies that the pH and its effect on Fe(III) speciation have a greater influence on the $D_{\%}^{MB}$ and I_{uLog}^{EC} responses, as observed in this investigation.

3. Materials and Methods

3.1. Chemicals and Materials

Iron sulfate heptahydrate ($FeSO_4 \cdot 7H_2O$), hydrogen peroxide (H_2O_2, 30%), sodium hydroxide (NaOH), hydrochloric acid (HCl), Luria Bertani (LB) medium and methylene blue ($C_{16}H_{18}N_3SCl \cdot 3H_2O$) were purchased from Merck S.A. (Santiago, Chile).

3.2. Fenton Experiments

Methylene blue, a dye that does not generate toxic byproducts when reacting with ·OH [50–52], was used as a model of recalcitrant compound, while *E. coli* K12, a non-pathogenic *E. coli* [53], was used as a model of bacteria. Experiments were performed in 20 mL glass reactors containing the MB solution (5.0×10^{-5} mol/L) or *E. coli* K12 (10^6 CFU), kept under magnetic stirring at room temperature (25 °C) [43,44]. First, $FeSO_4 \cdot 7H_2O$ solution was added to each sample according to the experimental design. The pH of each sample was adjusted by using NaOH (0.25 mol/L) or HCl (0.10 mol/L) solutions. Reactions were started by adding an aliquot of H_2O_2 solution. After the experimental time elapsed (15 min), for *E. coli* K12 analysis, 0.2 mL of each sample was collected for its enumeration. The decolorization of MB was studied by determining its kinetic constants of color decay and the degree of decolorization. After 2 min of maintaining the reaction under constant agitation, samples (3.0 mL) were withdrawn, and immediately injected into a cuvette for analysis at time intervals of 3, 6, 9, 12 and 15 min. The analyses in samples were performed spectrophotometrically by UV–Vis spectrophotometry (Shimadzu UV-1800, Shimadzu Inc., Kyoto, Japan) at 668 nm using quartz cells with path lengths of 1 cm. A calibration curve was constructed (5.30×10^{-7}–1.30×10^{-5} mol/L; $R^2 = 0.999$). Fitting decolorization kinetics and the rate constant was obtained by Sigma Plot 11.0 software (Systat Software, Inc., San Jose, CA, USA).

3.3. Detection and Enumeration of E. coli K12

Strain samples were stored in cryo-vials containing 20% glycerol at −20 °C. To prepare the bacterial pellet for the experiments, one colony was picked from the precultures and loop-inoculated into a 50 mL sterile PE eppendorf flask containing the Luria Bertani (LB) medium. The flask was then incubated aerobically at 37 °C and 150 rpm in a shaker incubator (Gerhardt THO500, Gerhardt GmbH & Co., Königswinter, Germany) until the stationary physiological phase was reached. After 24 h, cells were centrifuged (SIGMA 2-16P, Sigma Laborzentrifugen GmbH, Steinheim, Germany) and diluted until optical density 0.5 a.u. (i.e., 10^6 CFU/mL) at 600 nm [43,44]. Component of LB medium included sodium chloride (10 g), tryptone (10 g) and yeast extract (5 g) in 1 L of deionized water; this solution was then sterilized by autoclaving for 20 min at 121 °C. The bacterial pellet was resuspended and washed three times with a saline solution (NaCl/KCl). The final pellet was resuspended in saline solution. This procedure resulted in a cell density of approximately 10^9 colony forming units (CFU) per milliliter. The pH of the solution was adjusted to 7.0 and the solution was then sterilized by autoclaving for 30 min at 121 °C. The bacterial solution was diluted in reactors to the required cell density corresponding to 10^6 CFU/mL [43,44].

CFU were performed by plating on plates (PCA method). Of the samples 0.2 mL was withdrawn. Samples were diluted (10% *v/v*) and 0.1 mL poured on plates. Plates were aerobic incubated for 24 h at 37 °C (Heraeus B6, Kendro, Langenselbold, Germany) and the CFU were counted manually. All experiments were performed in triplicates. The enumeration of colonies was expressed as CFU (colony forming units) per 100 mL of sample. These concentrations were transformed to $\log_{10}$ and the removal of bacteria, uLog = $\log(N_t/N_0)$, was calculated from the initial bacteria concentration (N_0) and the remaining bacteria population at "t" time (N_t).

3.4. Experimental Design

To determine the optimal experimental conditions for the decolorization of MB and the inactivation of *E. coli* K12 by Fenton technology, a Box–Behnken design was performed. pH, Fe^{2+} concentration ($[Fe^{2+}]$, mol/L) and molar concentration ratio of Fe^{2+} and H_2O_2 ($[H_2O_2]/[Fe^{2+}]$) were selected as independent variables in the experimental design (Table 6).

Table 6. Independent variables and levels used in the Box–Behnken design for Fenton technology.

Variable	Coded	Coded Factor Level		
		−1	**0**	**1**
$[H_2O_2]/[Fe^{2+}]$ (A)	X_1	1:1	2:1	3:1
$[Fe^{2+}]$ (mol/L) (B)	X_2	6.0×10^{-4}	8.0×10^{-4}	1.0×10^{-3}
pH (C)	X_3	3.0	4.0	5.0

Three replicates were performed at the central point, with 15 runs performed for each study. The chosen levels of the independent variables were based on literature reports [54]. The experimental responses were the degree of MB decolorization ($D_\%^{MB}$), the apparent kinetic constant of MB decolorization (k_{app}^{MB}), and removal of bacteria in uLog units (I_{uLog}^{EC}) for variables showed in Table 6.

A second-order linear polynomial regression model (Equation (7)) was obtained to analyze the data. Data were statistically evaluated and an analysis of variance (ANOVA) was applied at with a confidence level of 95% using software Design Expert version 10 (Stat-Ease Inc., Minneapolis, MN, USA). Responses of the experimental tests were compared to the estimated values, and the fit of model was assessed. Experimental tests, performed under optimal conditions, were performed to achieve maximal $D_\%^{MB}$, k_{app}^{MB} and I_{uLog}^{EC}.

4. Conclusions

The present study provided a comprehensive description regarding the application of the Fenton technology as a process for MB decolorization and *E. coli* K12 inactivation in aqueous solutions at different $[H_2O_2]/[Fe^{2+}]$ values (1.0, 2.0 and 3.0), $[Fe^{2+}]$ values (6.0×10^{-4}, 8.0×10^{-4} and 1.0×10^{-3} mol/L) and pH values (3.0, 4.0 and 5.0) up to 15 min of reaction. It was found that the Box–Behnken model could effectively predict and optimize the performance of Fenton technology for MB decolorization and *E. coli* K12 inactivation.

The maximum $D_\%^{MB}$ of 94.57% was predicted at $[H_2O_2]/[Fe^{2+}] = 2.9$, $[Fe^{2+}] = 1.0 \times 10^{-3}$ mol/L and pH = 3.2; for k_{app}^{MB} the maximum of 2.08 min^{-1} was predicted at $[H_2O_2]/[Fe^{2+}] = 1.7$, $[Fe^{2+}] = 1.0 \times 10^{-3}$ mol/L and pH = 3.7 and the maximum I_{uLog}^{EC} of 0.89 uLog was predicted at $[H_2O_2]/[Fe^{2+}] = 2.9$, $[Fe^{2+}] = 7.6 \times 10^{-4}$ mol/L and pH = 3.2. This analysis revealed good agreement between experimental results and the RSM predictions, further illustrating that RSM is a suitable approach to optimize the MB decolorization and *E. coli* K12 inactivation.

The Pareto and perturbation analysis of the model terms showed that all parameters analyzed have different effects on the responses. The $[H_2O_2]/[Fe^{2+}]$ values show a significant positive effect only on $D_\%^{MB}$. The pH values show a significant negative effect on $D_\%^{MB}$ and I_{uLog}^{EC}, which could involve the main role of speciation of Fe(II) and Fe(III) species in the total process of MB decolorization and *E. coli* K12 inactivation by Fenton technology. The positive and negative significant effect of the $[Fe^{2+}]$ values on the MB decolorization ($D_\%^{MB}$ and k_{app}^{MB}) and *E. coli* K12 inactivation (I_{uLog}^{EC}) respectively, suggest that different oxidizing species are involved in these processes.

Thus, considering that bacteria are larger than dye molecules, the complex self-repair mechanisms of bacteria and the different external structures of bacteria compared to the dyes structure, the *E. coli* inactivation proved to be less effective than MB decolorization by Fenton processes.

Supplementary Materials: The following are available online at http://www.mdpi.com/2073-4344/10/12/1483/s1, Figure S1: Three-dimensional response surface plots (a, c and e) and their corresponding contour plots (b, d and f) representing de modeled $D_\%^{MB}$ as a function of: $[H_2O_2]/[Fe^{2+}]$ and $[Fe^{2+}]$ (a, b), pH and $[H_2O_2]/[Fe^{2+}]$ (c, d), $[Fe^{2+}]$ and pH (e, f) at central point values of other parameters, Figure S2: Three-dimensional response surface plots (a, c and e) and their corresponding contour plots (b, d and f) representing de modeled k_{app}^{MB} as a function of: $[H_2O_2]/[Fe^{2+}]$ and $[Fe^{2+}]$ (a, b), pH and $[H_2O_2]/[Fe^{2+}]$ (c, d), $[Fe^{2+}]$ and pH (e, f) at central point values of other parameters, Figure S3: Three-dimensional response surface plots (a, c and e) and their corresponding contour plots (b, d and f) representing de modeled I_{ulog}^{EC} as a function of: $[H_2O_2]/[Fe^{2+}]$ and $[Fe^{2+}]$ (a, b), pH and $[H_2O_2]/[Fe^{2+}]$ (c, d), $[Fe^{2+}]$ and pH (e, f) at central point values of other parameters.

Author Contributions: Conceptualization, P.S. and G.V.; methodology, P.S. and G.V.; software, P.S. and J.L.F.; validation, P.S. and J.L.F.; formal analysis, P.S. and G.V.; investigation, J.L.F.; writing—original draft preparation, P.S. and G.V.; writing—review and editing, P.S. and G.V.; supervision, G.V.; project administration, G.V.; funding acquisition, P.S. and G.V. All authors have read and agreed to the published version of the manuscript.

Funding: This research was funded by ANID/FONDAP/15130015 and ANID FONDECYT/Postdoctorado 3180566.

Acknowledgments: José Luis Frontela thanks ERASMUS mobility program.

Conflicts of Interest: The authors declare no conflict of interest.

References

1. Lizama, C.; Freer, J.; Baeza, J.; Mansilla, H.D. Optimized photodegradation of Reactive Blue 19 on TiO$_2$ and ZnO suspensions. *Catal. Today* **2002**, *76*, 235–246. [CrossRef]
2. Vidal, G.; Videla, S.; Diez, M.C. Molecular weight distribution of *Pinus radiata* kraft mill wastewater treated by anaerobic digestion. *Bioresour. Technol.* **2001**, *77*, 183–191. [CrossRef]
3. Yeber, M.C.; Oñate, K.P.; Vidal, G. Decolorization of Kraft Bleaching Effluent by Advanced Oxidation Processes Using Copper (II) as Electron Acceptor. *Environ. Sci. Technol.* **2007**, *41*, 2510–2514. [CrossRef] [PubMed]
4. Chamorro, S.; Hernández, V.; Monsalvez, E.; Becerra, J.; Mondaca, M.A.; Piña, B.; Vidal, G. Detection of Estrogenic Activity from Kraft Mill Effluents by the Yeast Estrogen Screen. *Bull. Environ. Contam. Toxicol.* **2010**, *84*, 165–169. [CrossRef] [PubMed]
5. Chamorro, S.; Hernández, V.; Matamoros, V.; Domínguez, C.; Becerra, J.; Vidal, G.; Piña, B.; Bayona, J.M. Chemical characterization of organic microcontaminant sources and biological effects in riverine sediments impacted by urban sewage and pulp mill discharges. *Chemosphere* **2013**, *90*, 611–619. [CrossRef]
6. Chamorro, S.; Monsalvez, E.; Piña, B.; Olivares, A.; Hernández, V.; Becerra, J.; Vidal, G. Analysis of aryl hydrocarbon receptor ligands in kraft mill effluents by a combination of yeast bioassays and CG-MS chemical determinations. *J. Environ. Sci. Health A* **2013**, *48*, 145–151. [CrossRef]
7. Orrego, R.; Guchardi, J.; Hernandez, V.; Krause, R.; Roti, L.; Armour, J.; Ganeshakumar, M.; Holdway, D. Pulp and paper mill effluent treatments have differential endocrine-disrupting effects on rainbow trout. *Environ. Toxicol. Chem.* **2009**, *28*, 181–188. [CrossRef]
8. Singh, A.K.; Chandra, R. Pollutants released from the pulp paper industry: Aquatic toxicity and their health hazards. *Aquatic Toxicol.* **2019**, *211*, 202–216. [CrossRef]
9. Xavier, C.; Chamorro, S.; Vidal, G. Chronic Effects of Kraft Mill Effluents and Endocrine Active Chemicals on *Daphnia magna*. *Bull. Environ. Contam. Toxicol.* **2005**, *75*, 670–676. [CrossRef]
10. Wang, Y.; Zhu, G.; Yang, Z. Analysis of water quality characteristic for water distribution systems. *J. Water Reuse Desalinat.* **2018**, *9*, 152–162. [CrossRef]
11. Kulkarni, P.; Olson, N.D.; Paulson, J.N.; Pop, M.; Maddox, C.; Claye, E.; Rosenberg Goldstein, R.E.; Sharma, M.; Gibbs, S.G.; Mongodin, E.F.; et al. Conventional wastewater treatment and reuse site practices modify bacterial community structure but do not eliminate some opportunistic pathogens in reclaimed water. *Sci. Total Environ.* **2018**, *639*, 1126–1137. [CrossRef] [PubMed]
12. Pantelaki, I.; Voutsa, D. Formation of iodinated THMs during chlorination of water and wastewater in the presence of different iodine sources. *Sci. Total Environ.* **2018**, *613–614*, 389–397. [CrossRef] [PubMed]
13. Tawabini, B.; Al-Mutair, M.; Bukhari, A. Formation potential of trihalomethanes (THMs) in blended water treated with chlorine. *J. Water Reuse Desalinat.* **2011**, *1*, 172–178. [CrossRef]
14. Selvabharathi, G.; Adishkumar, S.; Jenefa, S.; Ginni, G.; Rajesh Banu, J.; Tae Yeom, I. Combined homogeneous and heterogeneous advanced oxidation process for the treatment of tannery wastewaters. *J. Water Reuse Desalinat.* **2015**, *6*, 59–71. [CrossRef]
15. Gandhi, K.; Lari, S.; Tripathi, D.; Kanade, G. Advanced oxidation processes for the treatment of chlorpyrifos, dimethoate and phorate in aqueous solution. *J. Water Reuse Desalinat.* **2015**, *6*, 195–203. [CrossRef]
16. Yeber, M.C.; Soto, C.; Riveros, R.; Navarrete, J.; Vidal, G. Optimization by factorial design of copper (II) and toxicity removal using a photocatalytic process with TiO$_2$ as semiconductor. *Chem. Eng. J.* **2009**, *152*, 14–19. [CrossRef]
17. Wang, J.; Li, C.; Zhuang, H.; Zhang, J. Photocatalytic degradation of methylene blue and inactivation of Gram-negative bacteria by TiO$_2$ nanoparticles in aqueous suspension. *Food Control* **2013**, *34*, 372–377. [CrossRef]
18. Sirés, I.; Brillas, E.; Oturan, M.A.; Rodrigo, M.A.; Panizza, M. Electrochemical advanced oxidation processes: Today and tomorrow. A review. *Environ. Sci. Pollut. Res.* **2014**, *21*, 8336–8367. [CrossRef]
19. Miklos, D.B.; Remy, C.; Jekel, M.; Linden, K.G.; Drewes, J.E.; Hübner, U. Evaluation of advanced oxidation processes for water and wastewater treatment—A critical review. *Water Res.* **2018**, *139*, 118–131. [CrossRef]
20. Andreozzi, R.; Caprio, V.; Insola, A.; Marotta, R. Advanced oxidation processes (AOP) for water purification and recovery. *Catal. Today* **1999**, *53*, 51–59. [CrossRef]

21. Gao, M.; An, T.; Li, G.; Nie, X.; Yip, H.-Y.; Zhao, H.; Wong, P.-K. Genetic studies of the role of fatty acid and coenzyme A in photocatalytic inactivation of *Escherichia coli*. *Water Res.* **2012**, *46*, 3951–3957. [CrossRef] [PubMed]

22. Salgado, P.; Melin, V.; Contreras, D.; Moreno, Y.; Mansilla, H.D. Fenton reaction driven by iron ligands. *J. Chil. Chem. Soc.* **2013**, *58*, 2096–2101. [CrossRef]

23. Sawyer, D.T. The Chemistry and Activation of Dioxygen Species (O_2, O_2^-, and HOOH) in Biology. In *Oxygen Complexes and Oxygen Activation by Transition Metals*; Martell, A.E., Sawyer, D.T., Eds.; Springer: Boston, MA, USA, 1988; pp. 131–148. [CrossRef]

24. Salgado, P.; Melin, V.; Albornoz, M.; Mansilla, H.; Vidal, G.; Contreras, D. Effects of pH and substituted 1,2-dihydroxybenzenes on the reaction pathway of Fenton-like systems. *Appl. Catal. B Environ.* **2018**, *226*, 93–102. [CrossRef]

25. Lee, H.; Lee, H.-J.; Sedlak, D.L.; Lee, C. pH-Dependent reactivity of oxidants formed by iron and copper-catalyzed decomposition of hydrogen peroxide. *Chemosphere* **2013**, *92*, 652–658. [CrossRef]

26. Bataineh, H.; Pestovsky, O.; Bakac, A. pH-induced mechanistic changeover from hydroxyl radicals to iron(iv) in the Fenton reaction. *Chem. Sci.* **2012**, *3*, 1594–1599. [CrossRef]

27. Mansoorian, H.J.; Bazrafshan, E.; Yari, A.; Alizadeh, M. Removal of azo dyes from aqueous solution using Fenton and modified Fenton processes. *Health Scope* **2014**, *3*, 1–9. [CrossRef]

28. Mohajeri, S.; Aziz, H.A.; Isa, M.H.; Bashir, M.J.K.; Mohajeri, L.; Adlan, M.N. Influence of Fenton reagent oxidation on mineralization and decolorization of municipal landfill leachate. *J. Environ. Sci. Health A* **2010**, *45*, 692–698. [CrossRef]

29. Gulkaya, İ.; Surucu, G.A.; Dilek, F.B. Importance of H_2O_2/Fe^{2+} ratio in Fenton's treatment of a carpet dyeing wastewater. *J. Hazard. Mater.* **2006**, *136*, 763–769. [CrossRef]

30. Ahmadian-Kouchaksaraie, Z.; Niazmand, R.; Najafi, M.N. Optimization of the subcritical water extraction of phenolic antioxidants from *Crocus sativus* petals of saffron industry residues: Box-Behnken design and principal component analysis. *Innov. Food Sci. Emerg. Technol.* **2016**, *36*, 234–244. [CrossRef]

31. Peydayesh, M.; Bagheri, M.; Mohammadi, T.; Bakhtiari, O. Fabrication optimization of polyethersulfone (PES)/polyvinylpyrrolidone (PVP) nanofiltration membranes using Box-Behnken RSM method. *RSC Adv.* **2017**, *7*, 24995–25008. [CrossRef]

32. Al-Musawi, T.J.; Kamani, H.; Bazrafshan, E.; Panahi, A.H.; Silva, M.F.; Abi, G. Optimization the Effects of Physicochemical Parameters on the Degradation of Cephalexin in Sono-Fenton Reactor by Using Box-Behnken Response Surface Methodology. *Catal. Lett.* **2019**, *149*, 1186–1196. [CrossRef]

33. Babuponnusami, A.; Muthukumar, K. Degradation of Phenol in Aqueous Solution by Fenton, Sono-Fenton and Sono-photo-Fenton Methods. *Clean-Soil Air Water* **2011**, *39*, 142–147. [CrossRef]

34. Chen, C.-C.; Wu, R.-J.; Tzeng, Y.-Y.; Lu, C.-S. Chemical Oxidative Degradation of Acridine Orange Dye in Aqueous Solution by Fenton's Reagent. *J. Chin. Chem. Soc.* **2009**, *56*, 1147–1155. [CrossRef]

35. Tunç, S.; Duman, O.; Gürkan, T. Monitoring the Decolorization of Acid Orange 8 and Acid Red 44 from Aqueous Solution Using Fenton's Reagents by Online Spectrophotometric Method: Effect of Operation Parameters and Kinetic Study. *Ind. Eng. Chem. Res.* **2013**, *52*, 1414–1425. [CrossRef]

36. Melgoza, D.; Hernández-Ramírez, A.; Peralta-Hernández, J.M. Comparative efficiencies of the decolourisation of Methylene Blue using Fenton's and photo-Fenton's reactions. *Photochem. Photobiol. Sci.* **2009**, *8*, 596–599. [CrossRef]

37. Asad, L.M.B.O.; Asad, N.R.; Silva, A.B.; de Almeida, C.E.B.; Leitão, A.C. Role of SOS and OxyR systems in the repair of *Escherichia coli* submitted to hydrogen peroxide under low iron conditions. *Biochimie* **1997**, *79*, 359–364. [CrossRef]

38. Blanco, J.; Torrades, F.; De la Varga, M.; García-Montaño, J. Fenton and biological-Fenton coupled processes for textile wastewater treatment and reuse. *Desalination* **2012**, *286*, 394–399. [CrossRef]

39. Ortega-Gómez, E.; Esteban García, B.; Ballesteros Martín, M.M.; Fernández Ibáñez, P.; Sánchez Pérez, J.A. Inactivation of *Enterococcus faecalis* in simulated wastewater treatment plant effluent by solar photo-Fenton at initial neutral pH. *Catal. Today* **2013**, *209*, 195–200. [CrossRef]

40. Ahmad, S.I.; Iranzo, O.G. Treatment of post-burns bacterial infections by Fenton reagent, particularly the ubiquitous multiple drug resistant *Pseudomonas* spp. *Med. Hypotheses* **2003**, *61*, 431–434. [CrossRef]

41. Cengiz, M.; Uslu, M.O.; Balcioglu, I. Treatment of *E. coli* HB101 and the *tet*M gene by Fenton's reagent and ozone in cow manure. *J. Environ. Manag.* **2010**, *91*, 2590–2593. [CrossRef]

42. Chen, F.; Yang, X.; Xu, F.; Wu, Q.; Zhang, Y. Correlation of Photocatalytic Bactericidal Effect and Organic Matter Degradation of TiO_2 Part I: Observation of Phenomena. *Environ. Sci. Technol.* **2009**, *43*, 1180–1184. [CrossRef] [PubMed]

43. Marugán, J.; van Grieken, R.; Pablos, C.; Sordo, C. Analogies and differences between photocatalytic oxidation of chemicals and photocatalytic inactivation of microorganisms. *Water Res.* **2010**, *44*, 789–796. [CrossRef] [PubMed]

44. Fisher, M.B.; Keane, D.A.; Fernández-Ibáñez, P.; Colreavy, J.; Hinder, S.J.; McGuigan, K.G.; Pillai, S.C. Nitrogen and copper doped solar light active TiO_2 photocatalysts for water decontamination. *Appl. Catal. B Environ.* **2013**, *130–131*, 8–13. [CrossRef]

45. Rengifo-Herrera, J.A.; Pulgarin, C. Photocatalytic activity of N, S co-doped and N-doped commercial anatase TiO_2 powders towards phenol oxidation and *E. coli* inactivation under simulated solar light irradiation. *Sol. Energy* **2010**, *84*, 37–43. [CrossRef]

46. Keane, D.A.; McGuigan, K.G.; Ibáñez, P.F.; Polo-López, M.I.; Byrne, J.A.; Dunlop, P.S.M.; O'Shea, K.; Dionysiou, D.D.; Pillai, S.C. Solar photocatalysis for water disinfection: Materials and reactor design. *Catal. Sci. Technol.* **2014**, *4*, 1211–1226. [CrossRef]

47. Luo, Y.; Han, Z.; Chin, S.M.; Linn, S. Three chemically distinct types of oxidants formed by iron-mediated Fenton reactions in the presence of DNA. *Proc. Natl. Acad. Sci. USA* **1994**, *91*, 12438–12442. [CrossRef]

48. Salgado, P.; Melin, V.; Durán, Y.; Mansilla, H.; Contreras, D. The Reactivity and Reaction Pathway of Fenton Reactions Driven by Substituted 1,2-Dihydroxybenzenes. *Environ. Sci. Technol.* **2017**, *51*, 3687–3693. [CrossRef]

49. Fischbacher, A.; von Sonntag, C.; Schmidt, T.C. Hydroxyl radical yields in the Fenton process under various pH, ligand concentrations and hydrogen peroxide/Fe(II) ratios. *Chemosphere* **2017**, *182*, 738–744. [CrossRef]

50. Liu, Y.; Jin, W.; Zhao, Y.; Zhang, G.; Zhang, W. Enhanced catalytic degradation of methylene blue by α-Fe_2O_3/graphene oxide via heterogeneous photo-Fenton reactions. *Appl. Catal. B Environ.* **2017**, *206*, 642–652. [CrossRef]

51. Su, S.; Liu, Y.; Liu, X.; Jin, W.; Zhao, Y. Transformation pathway and degradation mechanism of methylene blue through β-FeOOH@GO catalyzed photo-Fenton-like system. *Chemosphere* **2019**, *218*, 83–92. [CrossRef]

52. Biswas, S.; Pal, A. Visible light assisted Fenton type degradation of methylene blue by admicelle anchored alumina supported rod shaped manganese oxide. *J. Water Process Eng.* **2020**, *36*, 101272. [CrossRef]

53. Leech, J.; Golub, S.; Allan, W.; Simmons, M.J.H.; Overton, T.W. Non-pathogenic *Escherichia coli* biofilms: Effects of growth conditions and surface properties on structure and curli gene expression. *Arch. Microbiol.* **2020**, *202*, 1517–1527. [CrossRef] [PubMed]

54. Gao, A.; Gao, H.; Zhu, Z.; Jiao, Z. Application of response surface methodology to optimize the treatment of cephems pharmaceutical wastewater by ultrasound/Fenton process. *Desalinat. Water Treat.* **2016**, *57*, 10866–10877. [CrossRef]

Publisher's Note: MDPI stays neutral with regard to jurisdictional claims in published maps and institutional affiliations.

catalysts

Review

Photocatalysis for Organic Wastewater Treatment: From the Basis to Current Challenges for Society

Salma Izati Sinar Mashuri [1], Mohd Lokman Ibrahim [1,2,*], Muhd Firdaus Kasim [2], Mohd Sufri Mastuli [2], Umer Rashid [3,*], Abdul Halim Abdullah [4], Aminul Islam [5], Nurul Asikin Mijan [6], Yie Hua Tan [7], Nasar Mansir [8,9], Noor Haida Mohd Kaus [10] and Taufiq-Yap Yun Hin [11]

[1] School of Chemistry and Environment, Faculty of Applied Sciences, Universiti Teknologi MARA, Shah Alam 40450, Malaysia; salmaizati.work@gmail.com

[2] Centre for Functional Materials and Nanotechnology, Institute of Science, Universiti Teknologi MARA, Shah Alam 40450, Malaysia; muhdfir@uitm.edu.my (M.F.K.); mohdsufri@uitm.edu.my (M.S.M.)

[3] Institute of Advanced Technology, Universiti Putra Malaysia (UPM), Serdang 43400, Malaysia

[4] Department of Chemistry, Faculty of Science, Universiti Putra Malaysia (UPM), Serdang 43400, Malaysia; halim@upm.edu.my

[5] Department of Petroleum and Mining Engineering, Jessore University of Science and Technology, Jashore Sadar 7408, Bangladesh; aminul03211@yahoo.com

[6] Department of Chemical Sciences, Faculty of Science and Technology, Universiti Kebangsaan Malaysia, Bangi 43600, Malaysia; ckin_mijan@yahoo.com

[7] Department of Chemical Engineering, Faculty of Engineering and Science, Curtin University Malaysia, CDT 250, Miri 98009, Malaysia; tanyiehua@curtin.edu.my

[8] Catalysis Science and Technology Research Centre, Faculty of Science, Universiti Putra Malaysia, Serdang 43400, Malaysia; nmansir09@yahoo.com

[9] Department of Chemistry, Federal University, Dutse 7156, Nigeria

[10] Nano|Hybrid|Materials Research Group, School of Chemical Sciences, Universiti Sains Malaysia, Penang 11800, Malaysia; noorhaida@usm.my

[11] Faculty of Science and Natural Resources, Universiti Malaysia Sabah, Kota Kinabalu 88400, Malaysia; taufiq@upm.edu.my

* Correspondence: mohd_lokman@uitm.edu.my (M.L.I.); umer.rashid@upm.edu.my (U.R.); Tel.: +6-03-5521-1763 (M.L.I.); +60-3-8946-7393 (U.R.)

Received: 14 September 2020; Accepted: 25 October 2020; Published: 30 October 2020

Abstract: Organic pollutants such as dyes, antibiotics, analgesics, herbicides, pesticides, and stimulants become major sources of water pollution. Several treatments such as absorptions, coagulation, filtration, and oxidations were introduced and experimentally carried out to overcome these problems. Nowadays, an advanced technique by photocatalytic degradation attracts the attention of most researchers due to its interesting and promising mechanism that allows spontaneous and non-spontaneous reactions as they utilized light energy to initiate the reaction. However, only a few numbers of photocatalysts reported were able to completely degrade organic pollutants. In the past decade, the number of preparation techniques of photocatalyst such as doping, morphology manipulation, metal loading, and coupling heterojunction were studied and tested. Thus, in this paper, we reviewed details on the fundamentals, common photocatalyst preparation for coupling heterojunction, morphological effect, and photocatalyst's characterization techniques. The important variables such as catalyst dosage, pH, and initial concentration of sample pollution, irradiation time by light, temperature system, durability, and stability of the catalyst that potentially affect the efficiency of the process were also discussed. Overall, this paper offers an in-depth perspective of photocatalytic degradation of sample pollutions and its future direction.

Keywords: photocatalysis; organic wastewater; preparation method; degradation; characterizations; hybridization

1. Introduction

Nowadays, about 300 to 400 million tons of untreated organic pollutants are produced annually, which leads to water pollution problems, especially near industrial areas [1]. To overcome this issue, the majority of countries introduced strict regulations to control environmental pollution. Moreover, it attracts scientists' attention to intensively study the best technology in this scope of research with the hope to control environmental pollution and improve the wellbeing of the environment.

One of the attractive solutions to degrade organic pollutants is by using photocatalysts. This is because of its promising, effective, and efficient degradation activity of pollutants, which occur by allowing both spontaneous and non-spontaneous reactions to optimize the whole process. Photocatalysis was introduced in 1972 by Fujishima and Honda in their research on the electrochemical photolysis of water at a semiconductor electrode and published in Nature [2]. The process is defined as a chronological succession of advanced oxidation processes (AOPs), which improve their drawbacks such as high cost, incomplete mineralization, and require high hydroxyl radical [3]. The photocatalysis process requires light energy to activate the photocatalyst, hence this shows interest because the reaction can be controlled using the light or photon sources.

Currently, the most studied photocatalysts are titanium dioxide (TiO_2) [4–6] and zinc oxide (ZnO) [7–9]. However, they suffer from the recombination of the electrons (e^-) and holes (h^+), which are considered as the major drawback to these materials. Improving the efficiency and stability of photocatalyst become the target goal among researchers.

In this review, we focused on the hybridization type, preparation technique, effect of parameters, and current challenges of photocatalysis sciences in organic wastewater treatment. As has been mentioned before, the recombination of the electrons (e^-) and holes (h^+) are common problems for the photocatalyst. Hence, the hybridization technique was proposed by several researchers to overcome this problem. Examples of such techniques are doping, coupling heterojunction, and supporting materials. Further mechanisms and explanations on the effect of hybridization type on the performance of catalysts are discussed in this review.

Several techniques were also proposed by researchers to synthesize photocatalyst, however, there is no single study that reviewed the performance of each method and quality of produced photocatalyst. In this paper, there are two types of green and simple techniques identified that could affectively affect the performance of the photocatalyst such as physical- and chemical-based techniques. Moreover, this paper also discussed extensively the effect of operating parameters involved during the photocatalysis process. For example, the effect of catalyst amount, pH of pollutant, irradiation intensity, the temperature of pollutant, initial concentration, and effect of size and shape of photocatalyst toward the rate of degradation of organic pollutants.

2. Photocatalysis

Catalysis is the study involving synthesis, modification, and mechanism of a substance that can increase or accelerate the rate of a chemical reaction due to the participation of a substance called a catalyst, which remains unchanged at the end of the reaction. The reaction will occur faster with the catalyst because they require less activation energy than that of a normal reaction. The activation energy (E_a) is threshold energy, which must be overcome for a reaction to occur as illustrated in Figure 1. Despite the normal reaction pathway, the presence of the catalyst opens up an alternative approach with lower activation energy, thus, the rate of reaction will increase. Nevertheless, the result and the overall thermodynamics are the same [10].

Nowadays, photocatalysis shows high potential in reclaiming the environmental practice and green technology. It drives researchers to enhance this technology by looking forward to the best type of photocatalyst and the reactors. Currently, it has been applied in many applications such as coating technology [11,12], environmental pollution [13,14], and air pollution treatment [15,16].

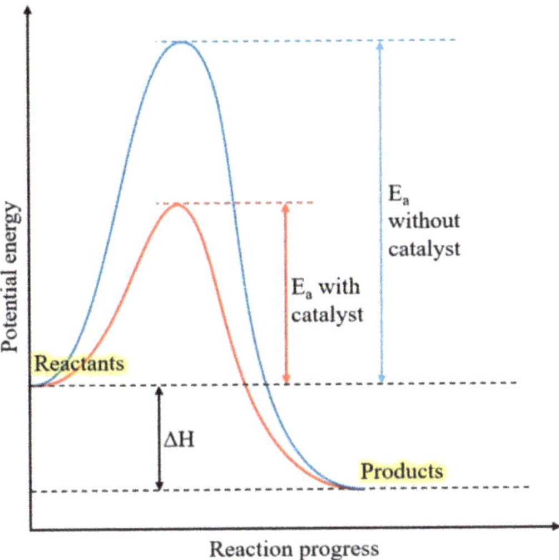

Figure 1. Effect of catalyst on energy diagram profile.

Theoretically, photocatalysis requires light to activate the photocatalyst to initiate the reaction. According to Ohtani [17], the difference between photocatalytic and catalytic reaction depends on the preferential crystal facets of the photocatalyst. Therefore, the intensity of irradiated light will affect the kinetics of the reaction, to produce electron–hole pairs, which makes the reaction happens, whereas the catalytic reaction depends on the active sites for the reaction to occur [18–20]. In catalysis, the density of active sites concurrent with the reaction kinetic [21,22] as illustrated in Figure 2. Generally, the catalytic reaction is limited to spontaneous reaction only, in which the Gibbs free energy is more than zero. However, photocatalyst allows both spontaneous and non-spontaneous reactions, which relied on the photo-absorption ability of the material that can provide energy source and turn it into chemical energy [17].

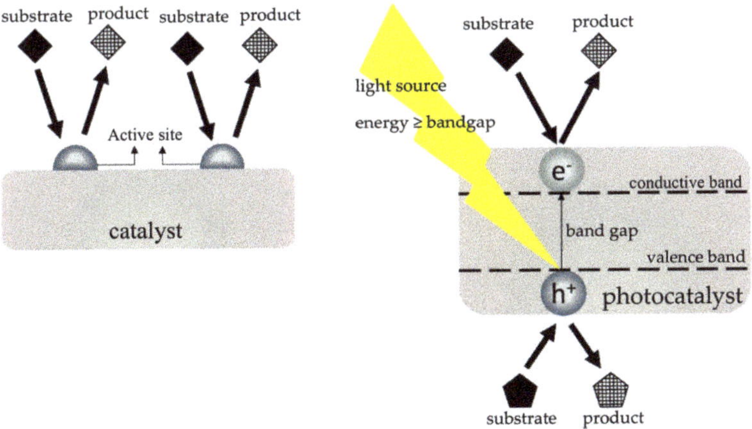

Figure 2. The illustration of catalytic and photocatalytic reaction processes.

As illustrated in Figure 3, the activity of photocatalyst (i.e., semiconductor) depends on the ability to create e^- and h^+ pairs to generate free radicals, which are needed to initiate the reaction. An electron from the valence band (VB) will be excited to the conductive band (CB) by absorption of the light energy equally or more than its band gap, which is an energy difference between VB and CB in the semiconductor.

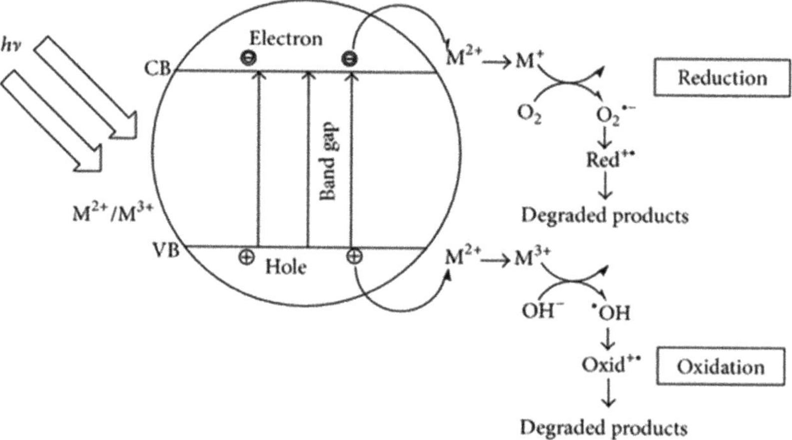

Figure 3. The basic mechanism of photocatalysis [23].

Theoretically, the reaction starts when enough photons ($h\nu$) from the light source hit the e^- on VB. The e^- excites the CB, leaving the h^+ on VB. Both e^- and h^+ will migrate to the surface of the photocatalyst. Simultaneously on the surface, h^+ will oxidize the water to form hydroxyl radicals (OH·) that initiate the chain reaction to oxidize the organic pollutants. While e^- will donate to the electron acceptor such as oxygen (O_2) extending the formation of superoxide or metal ion that is reduced to its lower valence state and deposited on the photocatalyst surface [23]. Both oxidation and reduction processes produce a degraded product of organic pollutant that is more environmentally friendly.

Table 1 shows recent publications on the catalytic study of the photocatalyst in the degradation of (i) dyes like methyl orange and methylene blue, (ii) antibiotics such as doxycycline, levofloxacin, tetracycline hydrochloride, and tetracycline, (iii) analgesics such as acetaminophen, (iv) herbicides such as atrazine, (v) pesticides such as naphthalene, and (vi) stimulants such as caffeine—all of which could benefit from further research and applications. According to Gogate and Pandit [24] and Low et al. [25], organic contaminants from industries were found to be the major source of water pollution. Globally, around 15 g pharmaceutical products were consumed per capita and 3 to 10 times higher in developed countries [26]. The remaining untreated refractory organic contaminants could stimulate microbial growth, leading to oxygen depletion and disturb the entire water bodies' ecosystem [27]. Numerous research reported in this area is scientifically important. For example, Cheshme et al. [28] and Vaiano et al. [4] in their research, succeeded in completely degrading the methyl orange within 50 min and paracetamol sample in 120 min using Cu doped ZnO/Al$_2$O$_3$ and TiO$_2$-graphite composite photocatalyst, respectively.

Table 1. Research on photocatalysis; parameter condition and results reported from 2016 to 2018.

Type of Organic Pollutant	Treat	Photocatalyst	Light Source	Band Gap	Degradation Rate (%)	Ref.
Analgesics	Acetaminophen	CdS sub-microspheres	50 W LED visible light λ = 455 nm	2.16 eV	Acetaminophen: 85% Levofloxacin: 70% in 240 min	[29]
	Paracetamol	TiO_2-graphite composites	UV lamp 8 W λ = 365 nm	3.24 eV	UV: 100% in 120 min TOC: 88% in 180 min	[30]
Antibiotics	Doxycycline	$BiOBr/FeWO_4$	300 W Xenon lamp with a 400-nm cutoff filter	2.46 eV	90.4% in 60 min	[31]
	Levofloxacin	CdS sub-microspheres	50 W LED visible light λ = 455 nm	2.16 eV	Levofloxacin: 70% in 240 min	[29]
	Tetracycline hydrochloride	$CdTe/TiO_2$	400 W halogen lamp equipped with a cutoff filter (λ > 400 nm)	1.39 eV	78% in 30 min	[32]
	tetracycline	Fe-based metal–organic frameworks	300 W Xenon lamp visible lamp λ > 420 nm	1.88 eV	96.6% in 3 h	[33]
	Tetracycline hydrochloride	$ZnFe_2O_4$ porous hollow cube	300 W Xe Lamp equipped with 350 nm–780 nm reflection filter and 420 nm cutoff filter (irradiation wavelength of 420 nm–780 nm)	1.5 eV	84.08% in 60 min	[34]
	Tetracycline	$ZnWO_{4-x}$ nanorods	UV lamp mercury 300 W Xenon lamp 300 W	3.1 eV	91% in 80 min	[35]
	Nitrofurantoin	$Nd_2Mo_3O_9$	300 W tungsten incandescent lamp lamp intensity is 150 mW/cm^2	2.82 eV	99% in 45 min	[36]
Dyes	Rhodamine B	polycaprolactone/TiO_2 nanofibrous	25 W of 254 nm UV light	-	100% in 300 min	[37]
	Methyl orange	Cu-doped ZnO/Al_2O_3	Visible light 400 W high-pressure mercury-vapor lamp λ = 546.8 nm	2.18 eV	100% in 50 min	[28]
	Orange G	Sepiolite-TiO_2 nanocomposites	300 W Xe lamp	-	98.8% in 150 min	[38]
	Nitroblue tetrazolium Methylene blue	Ternary g-$C_3N_4/Al_2O_3/ZnO$	Visible light 300 W xenon lamp with λ > 420 nm cut-off filter	ZnO 3.20 eV Al_2O_3 4.86 eV g-C_3N_4 2.76 eV	85% in 50 min	[39]
	Methyl orange	Tungsten doped Al_2O_3/ZnO coating aluminum	Simulate solar irradiation 300 W	-	95% in 10 h	[40]
Herbicide	atrazine	Cu-BiOCl	Mercury UV lamp Λ = 254 nm	3.0 eV	35% in 30 min	[41]
Pesticide	naphthalene	ZnO	254 nm irradiation under 50 W mercury lamp	2.98 eV	70% in 2 h	[42]
Stimulant	Caffeine	Mg doped $ZnO-Al_2O_3$	UV mercury lamp 400 W	-	89.18% in 70 min	[43]

3. Types of Hybridization Photocatalyst

The single semiconductor was used as a photocatalyst, facing the problem of recombination of e^- and h^+ due to limited diffusion length and large band gap [44]. In the past decade, researchers put effort into hybridizing the photocatalyst to enhance the light absorption in the range of visible sunlight. Therefore, several types of hybridization of photocatalysts such as doping, metal loading, and coupling heterojunction have been explored to solve the problem.

3.1. Metal-Doped Photocatalyst

Doping is a method used to add other substance that has energy levels almost the same as the valence band or conducting band edge of the main semiconductor, that enables it to enhance the concentration of charge carriers either by donating or accepting electrons. However, the opposite effect has been observed, which could promote e^- and h^+ recombination [45]. Thus, morphology manipulation is important to produce nano-sized photocatalyst to reduce travel distance from the e^- to the surface hence creating h^+ in the bulk phase [46]. Nevertheless, it also has a side effect, which leads to an increase in the bandgap due to the particle-in-box model [46]. According to Ge et al. [47], metal loading requires a balanced amount of loading to favor the effective reaction to occur effectively. When an excess of metal is loaded, it will cover the active site on the surface, yet it decreases the separation capacity: it may also act as the center of recombination.

Theoretically, the introduction of metal dopant on photocatalyst could improve both adsorption and photocatalytic degradation efficiencies. As been reported by Cao et al. [48] in Table 2, the addition of 1 wt.% to 8 wt.% of Co $Zr_6O_4(OH_4)BDC_{12}$ in photocatalyst improved the adsorption and degradation process from 9.9% to 78.5%. This is because of the photocatalyst's crystallinity level, which improves the light adsorption and increases the efficiency of charge separation. Fundamentally, the introduced dopant or metal on photocatalyst has introduced a sub-energy level below the conductive band. Moreover, it acts as a trapping site for excitation and delay of the recombination of the excited electron as shown in Figure 4.

Table 2. Effect of different loading of dopant on the adsorption and photocatalytic degradation.

Catalysts	Quantity of Dopant	Morphological Characterization	Adsorption Capacity (wt.% Dopant: % Adsorption)	Photocatalytic Degradation (wt.% Dopant: % Degradation)	Ref.
Cu-doped ZnO/Al_2O_3	Minimum dopant loading	Undoped ZnO/Al_2O_3 observed as spherical morphology	-	0.0 wt.%: 6.4% 2.5 wt.%: 39.5% 5.0 wt.%: ~93.0%	[28]
	Optimum dopant loading	After doping, the surface became lamellar morphology and XRD peaks relative intensity showed slightly decreased	-	7.5 wt.%: 100%	
	Maximum dopant loading	-	-	10.0 wt.%: ~93.0%	
Mg-doped $ZnO-Al_2O_3$	Minimum dopant loading	Crystallite size 21 nm; Observable porosity on surface	0.0 wt.%: 7.0%	0.0 wt.%: 89.2%	[43]
	Optimum dopant loading	Crystallite size 8 nm; The surface increase in grain size	1.0 wt.%: 11.1%	1.0 wt.%: 98.9%	
	Maximum dopant loading	Crystallite size ≤ 35 nm	3.0 wt.%: 6.7% 5.0 wt.%: 1.2%	3.0 wt.%: <98.9% 5.0 wt.%: <98.9%	
Co-doped $Zr_6O_4(OH_4)BDC_{12}$	Minimum dopant loading	Agglomerated cubic morphology with diameter of 230 nm; surface area was 584 m^2g^{-1}	0.0 wt.%: 9.9%	0.0 wt.%: <78.5%	[48]
	Optimum dopant loading	Dispersive and uniform cubic with diameter of 170 nm; increase surface area to 815 m^2g^{-1}	1.0 wt.%: 68.1%	1.0 wt.%: 78.5%	
	Maximum dopant loading	The higher amount of Co doped, the higher surface area observed	2.0 wt.%: 61.3% 4.0 wt.%: 58.6% 8.0 wt.%: 55.4%	2.0 wt.%: <78.5% 4.0 wt.%: <78.5% 8.0 wt.%: <78.5%	
Ag-doped ZnS	Minimum dopant loading	Average diameter was 3.0–5.0 nm; surface area was 78 m^2g^{-1}	-	No dopant: 79.7%	[49]
	Optimum dopant loading	Average diameter was 3.0–5.3 nm; increase surface area to 89 m^2g^{-1}	-	With dopant: 92.8%	
	Maximum dopant loading	-	-	-	

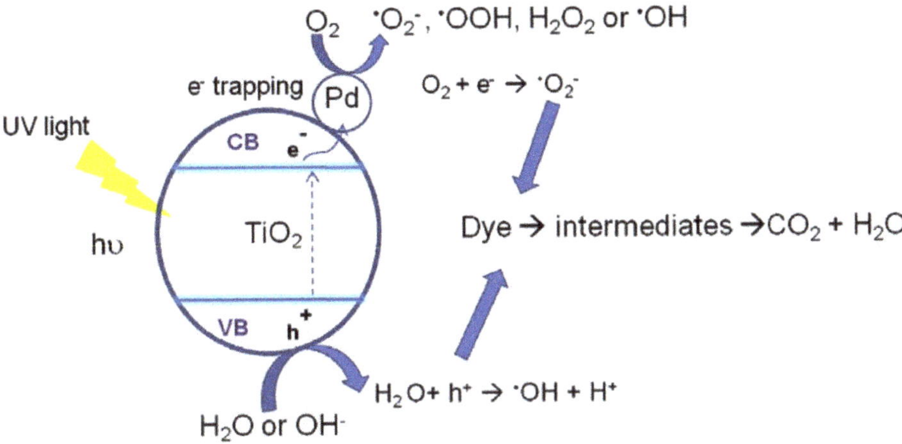

Figure 4. Example of doping as an electron trapping site of photocatalyst [50].

Furthermore, the synergy between dopant and photocatalyst is also an important criterion that must be observed. Based on the reported studies presented in Table 2, where Mg-doped ZnO-Al$_2$O$_3$ without or with minimum metal dopant showing low photocatalytic degradation. Further addition of dopant to the optimum level (1 wt.%) improves the photocatalysis degradation up to 98.9%. However, the addition of dopant (3–5 wt.%) exceeded the optimum level causes the decrease of the photocatalysis degradation lower than 98.9%. The decrease of the photocatalytic degradation might be due to several factors: (1) the addition of excess dopant, increases the crystallinity of catalyst, which hinder their active site on the surface [43] and (2) the excess of dopant tends to act as the center of recombination, which accelerates the recombination between e$^-$ and h$^+$ [28].

3.2. Coupling Heterojunction Photocatalyst

A coupling heterojunction is a combination of semiconductor to other semiconductor(s). It creates variation towards interfacial interactions between the semiconductors, which resulted in a new unique photocatalyst. Based on the band alignment, the heterojunctions are categorized into three types as shown in Figure 5a.

In semiconductor/semiconductor heterojunction, the e$^-$ and h$^+$ flow towards less negative potential and less positive potential, respectively. Type I shows the CB in the first semiconductor (SC-1), which is more negative than the second semiconductor (SC-2) while VB is more positive. Therefore, both e$^-$ and h$^+$ are accumulated in SC-2 and make them potentially recombine and reduce photocatalytic degradation. Subsequently, the type II band alignment is the most preferred because e$^-$ in CB of SC-1 will flow to SC-2, while h$^+$ in SC-2 will flow to SC-1 and both will transfer to the surface to undergo redox reaction. On the other hand, type III shows no heterojunction because SC-1 and SC-2 will work as a single semiconductor. In each semiconductor, the e$^-$ tends to recombine to its own h$^+$ [51].

Meanwhile, Figure 5b–d show the binary, ternary, and quaternary coupling heterojunctions. The same concept was applied as in Figure 5a, which depends on the level of band gaps for the coupled metals which determine the types of band alignment.

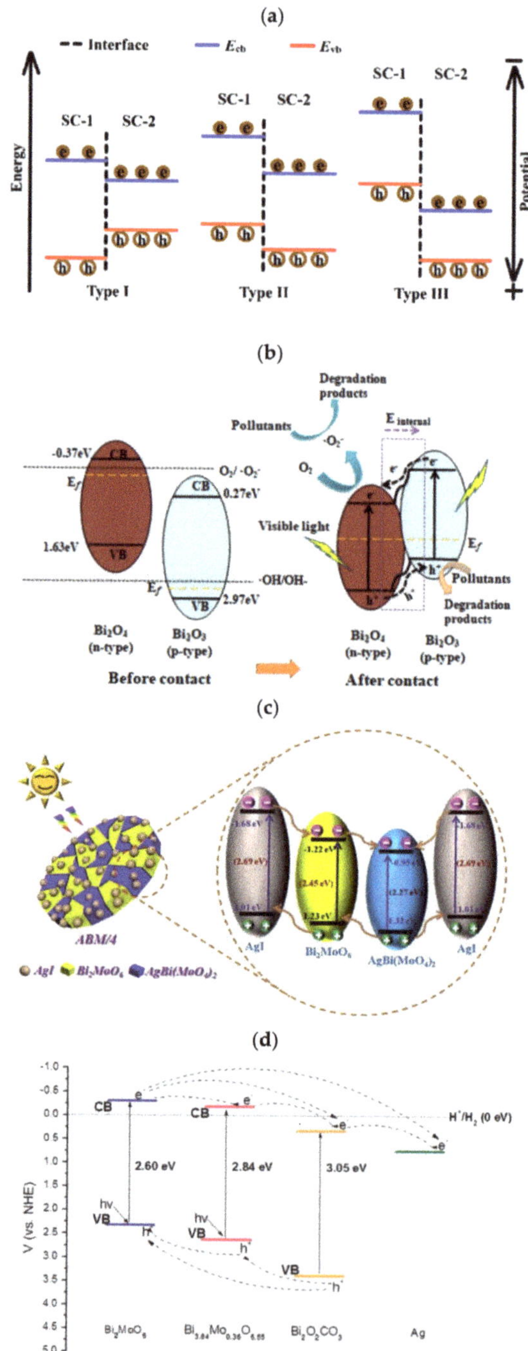

Figure 5. Illustrated (**a**) type of band alignment between two semiconductors in a heterojunction [51], (**b**) binary [52], (**c**) ternary [53], and (**d**) quaternary coupling heterojunctions [54].

3.3. Supported Material Photocatalyst

Currently, the trend in the development of active catalyst's support has attracted much attention due to its ability to disperse the active site on the catalyst surface. The support could also enhance the process due to its responsibility to provide a high surface area for depositing the primary photocatalyst test solutions [55,56]. The support's material can be inert or active during the photocatalytic process, which could act as a co-catalyst or secondary catalyst [57]. It is a requisite to overcome the difficulty in separating the catalyst usually in the form of powder, microcrystalline, and nanocrystalline after mixing with an aqueous sample and make it a milky dispersion [58]. To minimize the inconvenient, researchers discovered that the support material such as polymer membrane [59,60], silica [61,62], metal oxide [63,64], graphene [65,66], zeolite [67], carbon nanotube (CNT) [68], ceramics [69] and aluminum oxide [70] play key roles in maximizing the efficiency of the catalyst.

Several reported research papers studied the efficiency of metal oxides like the catalyst's support. For example, titanium dioxide [71,72], zinc oxide [73,74], silicon dioxide [75,76], and aluminum oxide [77,78]. However, most of the introduced supports suffer from having a low surface area and a lack of physical and chemical stabilities. Besides, some of these oxides offer good support; these include aluminum oxide, which offers very high surface area, high thermal conductivity, as well as promising chemical and physical stabilities based on the previous research by Shi et al. [78], Larimi and Khorasheh [77], and Sun et al. [79].

The chemical illustration of support materials of graphene, carbon nanotube, and aluminum oxide are presented in Figure 6. From this Figure, it is observed that the support material is one of the most important materials used to increase the absorption of the sample molecules, which allow the dispersion of photocatalyst on their exposed surface. Hence, it is potentially increasing the performance of the degradation process.

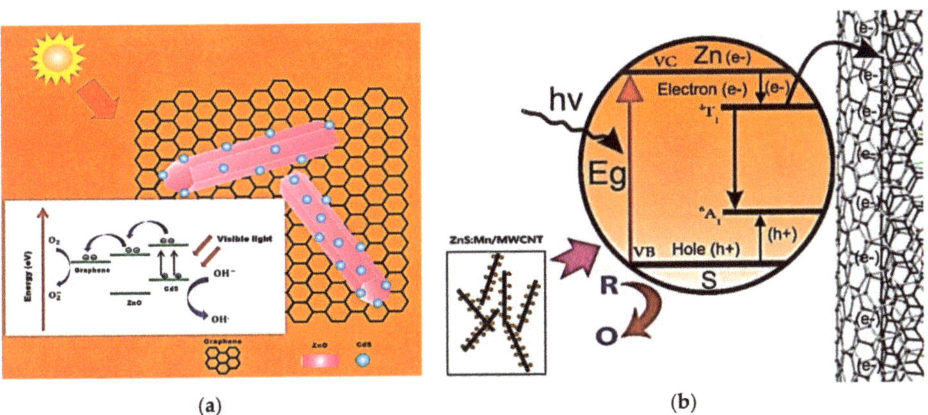

Figure 6. *Cont.*

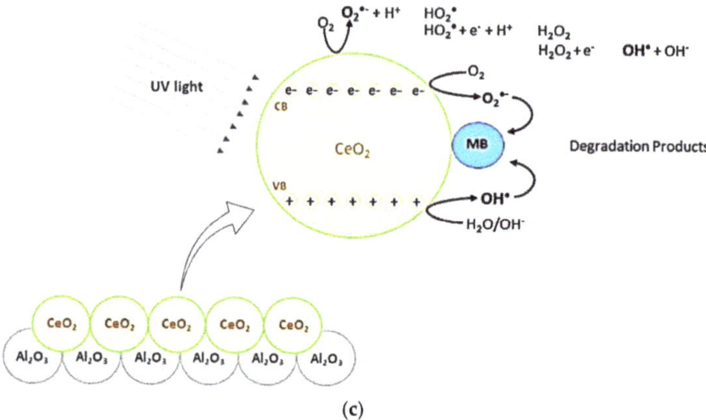

(c)

Figure 6. Example of recent studies on supported material photocatalyst such as (**a**) graphene [65], (**b**) carbon nanotube [68], and (**c**) aluminum oxide [70].

4. Preparation Techniques of Photocatalyst

In the past few years, various approaches of catalyst preparation techniques such as hydrothermal, electrospinning, impregnation, co-precipitation, solid-state, sonication, microemulsion, thermal evaporation, and sol-gel have been introduced, as shown in Table 3. These methods can be divided into solid-based and solution-based methods.

Table 3. Published preparation methods for photocatalysts.

Type of Method	Method	Photocatalyst	Reference
Solid-based method	Sonication	Cu-doped ZnO/Al$_2$O$_3$	[28]
		Ce(MoO$_4$)$_2$	[80]
		SnO/g-C$_3$N$_4$	[81]
	Solid-state	Mg doped ZnO-Al$_2$O$_3$	[43]
		ZnO	[82]
		CuO/Al$_2$O$_3$/TiO$_2$	[83]
	Thermal evaporation	SnO$_2$	[84]
		ZnO	[85]
		ZnO	[86]
Solution-based method	Hydrothermal	Sepiolite-TiO$_2$	[38]
		CdS	[29]
		CdTe/TiO$_2$	[32]
	Electrospinning	ZnO	[42]
		polycaprolactone/TiO$_2$	[37]
		Ag/LaFeO$_3$	[87]
	Impregnation	Mg-ZnO/Al$_2$O$_3$	[43]
		Cu-BiOCl	[41]
		Se-ZnS	[88]
	Precipitation	TiO$_2$-graphite	[30]
		γ-Al$_2$O$_3$	[89]
		ZnWO$_4$	[90]
	Co-precipitation	g-C$_3$N$_4$/Al$_2$O$_3$/ZnO	[39]
		Mg doped ZnO-Al$_2$O$_3$	[43]
		ZnFe$_2$O$_4$	[34]
		Cu-doped ZnO/Al$_2$O$_3$	[28]
		ZnO	[91]
	Microemulsion	Fe$_2$O$_3$	[92]
		ZnO	[93]
		Si doped TiO$_2$	[94]
	Sol-gel	Nd$_2$Mo$_3$O$_9$	[36]
		TiO$_2$	[94]
		Bi$_2$Mo$_3$O$_{12}$	[95]

4.1. Preparation of Photocatalyst by Physical Techniques

The preparation of photocatalyst using the physical technique is commonly based on the machine, force, energy, and pressure to mix or to modify the structure of the materials such as the sonication [28], solid-state [43], and thermal evaporation [85]. These methods offer several benefits such as simple technical aspects and a high yield of purity compared to chemical-based techniques. However, the composition, shape, and size of the catalyst are difficult to control [96].

A Sonication bath is one of the instruments that can be used to synthesize the photocatalyst using sound energy. The ultrasonic probe delivers the ultrasonic waves that come from different power capacities, pressure, and structural form as shown in Figure 7. According to Warner et al. [97] and Taylor [98], different products will be obtained with different frequency, amplitude, power supply, and processing time. It was also reported that this product has a high potential to produce a pure product with high yield and minimal waste from the process.

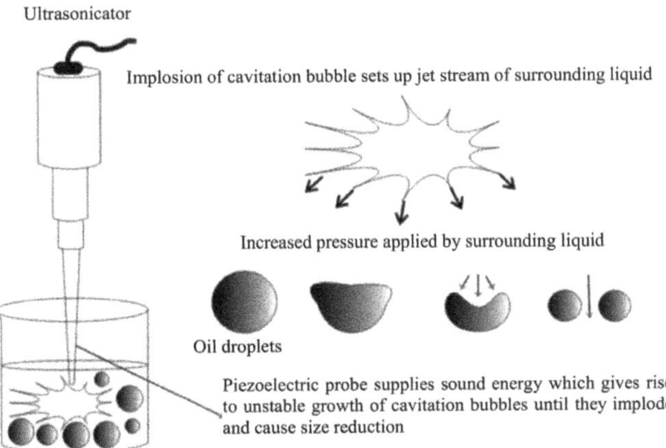

Figure 7. Illustration of sonication effect on the shape and size of the catalyst [99].

According to Sobhani-Nasab et al. [80], the nanosheet of $Ce(MoO_4)_2$ photocatalyst as shown in Figure 8a has been successfully synthesized by the initial reactant of $Ce(NO_3)_3 \cdot 6H_2O$ and $(NH_4)_6Mo_7O_{24} \cdot 4H_2O$ solution with the addition of glucose as surfactant using sonication. The surfactant played a key role in modeling the photocatalyst into a nanosheet. The research reported that the product was of high purity based on the X-ray diffraction (XRD) and energy dispersive spectrometer (EDS) analysis results. However, it was found that the increase of ultrasonic power could increase the nanosheet size of the catalyst. Therefore, the power needs to be analyzed so that to be maintained on the nano size scale. Similarly, Liang et al. [81], reported that ultrasonic power was used in the bonding of g-C_3N_4 and SnO. At a lower amount of g-C_3N_4, the observed morphology was in the form of a nanosheet as shown in Figure 8b; however, at a higher amount of g-C_3N_4, the morphology became irregular due to excess g-C_3N_4, which inhibited the formation of nanosheet and reduced the specific surface area of the photocatalysts.

Figure 8. SEM images of (**a**) nanosheet Ce(MoO$_4$)$_2$ [80] and (**b**) g-C$_3$N$_4$/SnO photocatalyst [81].

Meanwhile, the solid-state is a facile additive-free method that is promising in synthesizing the product in a large quantity [100]. This simple technique is normally employed for commercial production in the industry. The temperature is usually applied until the unwanted substances decompose and produce the product [82]. Without contact with any additive, the product produces a high yield with high purity. Despite that, the promising benefit of this method is that it is a simple and easy technique to operate. Figure 9 shows the illustration of the solid-state technique for the preparation of the photocatalyst.

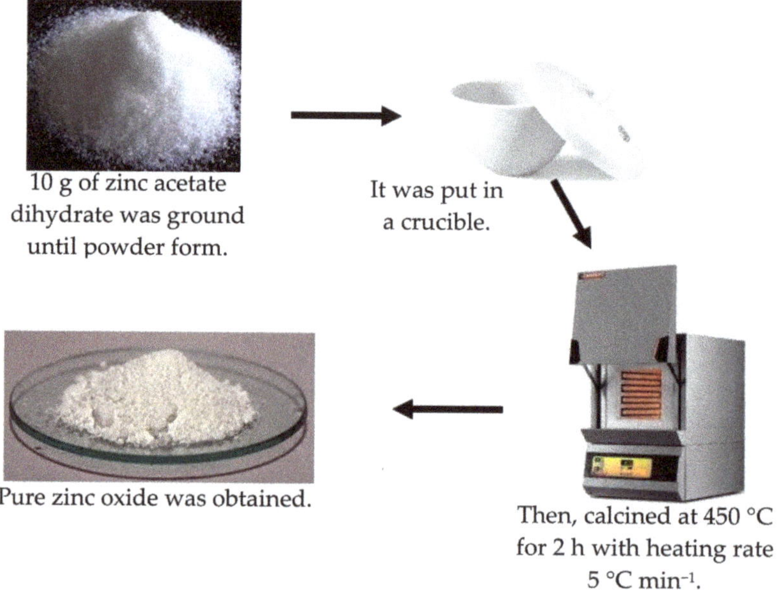

Figure 9. The solid-state technique for the preparation of ZnO.

Elhalil et al. [43] in their report, found that the solid-state method produced an Mg doped ZnO-Al$_2$O$_3$ photocatalyst when the calcination process was able to remove the carbonate group based on their FTIR result, thus, coupling the semiconductors. Their findings also showed that there are no impurities involved during the process, which was confirmed by the Joint Committee on Powder Diffraction Standards (JCPDS) data of XRD analysis. The porosity of the photocatalyst is presented in Figure 10a. Similarly, according to Rokesh et al. [82], the ZnO produced using the solid-state method was found to be of high purity as well as high crystallinity as proved by XRD data and energy dispersive x-ray (EDX) spectrum. The calcination process made the structure of microcrystal folding as shown in Figure 10b, thereby producing the rough surface and large quantity of surface defect. It also indicated the high surface area of the ZnO. Consequently, this method is strongly convinced to be good considering it is simplicity and ability to produce products with high purity and crystallinity.

Figure 10. SEM images of (**a**) Mg doped ZnO-Al$_2$O$_3$ photocatalyst [43] and (**b**) cone-like ZnO shape [82].

4.2. Preparation of Photocatalyst by Chemical Techiques

The solution-based method is the most common technique used for the preparation of photocatalysts because it offers numerous advantages such as being environmentally friendly, having affordable reagents, and requires incredibly low energy input. From the previous study, the efficiency of the produced photocatalyst can be controlled by manipulating the operating parameters. The variables involved during the process design could result in different compositions, shapes, and sizes, which potentially affects the performance of photocatalyst. Therefore, this method is preferable compared to a solid-based method. The common examples of these methods are sol-gel [36], co-precipitation [34], electrospinning [37], and hydrothermal [29], see Table 3.

4.2.1. Electrospinning Technique

Electrospinning is the process that requires the precursors in the form of solution or suspension, in order to be transferred into the syringe pump, so that the spinning tip will eject the sample drop by drop. The droplets are aided by the electrical field, which will be charged. The surface tension will subsequently be overtaken by electrostatic repulsion. The droplet repulse and elongate until it is deposited on a collector plate or drum [51]. The synthesized product by this technique is usually in nanofibers form. Figure 11 shows an illustration of the electrospinning technique.

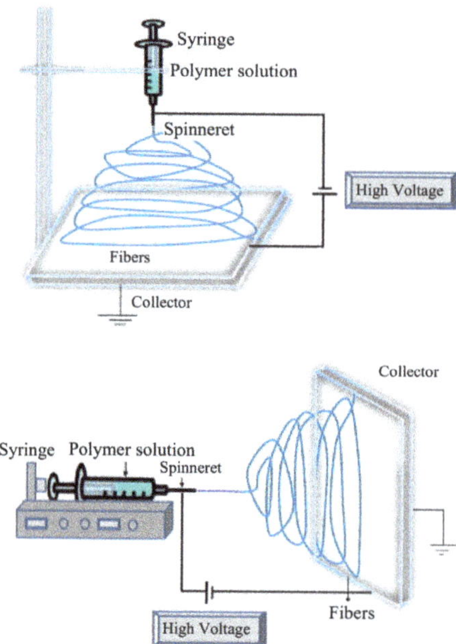

Figure 11. Visual of the formation of nanofibers by high voltage and spinneret injection of the solution [101].

According to Sekar et al. [102], the product obtained was smooth and has a uniform surface of nanofibers as shown in Figure 12a. Therefore, additional calcination was compulsorily needed to increase the surface area and its porosity. Meanwhile, Li et al. [87] reported that the electrospinning method creates a lack of uniformity of product diameter due to the interference of the inner needle. Hence, the best viscosity of precursor is necessary to avoid wrinkled nanofibers as shown in Figure 12b.

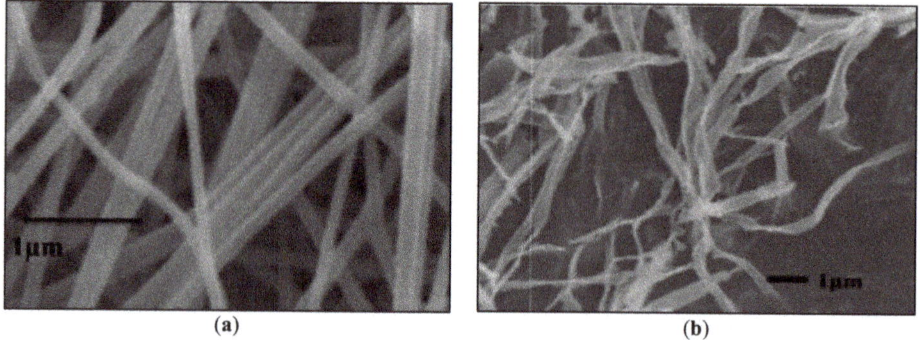

(a)   (b)

Figure 12. SEM images of (**a**) Fe-ZnO/PVA nanofibers [102]; (**b**) the wrinkled Ag/LaFeO$_3$ nanofibers [87].

4.2.2. Sol-Gel Technique

There are few steps involved in the sol-gel technique, which include hydrolysis, condensation, and the drying process [8]. Generally, a metal precursor undergoes a rapid hydrolysis process to form the metal hydroxide. Rapid condensation is then applied to produce the gel followed by a drying process to solidify the gel [103]. Figure 13 shows the illustrations of the sol-gel technique in the preparation ZnO [104].

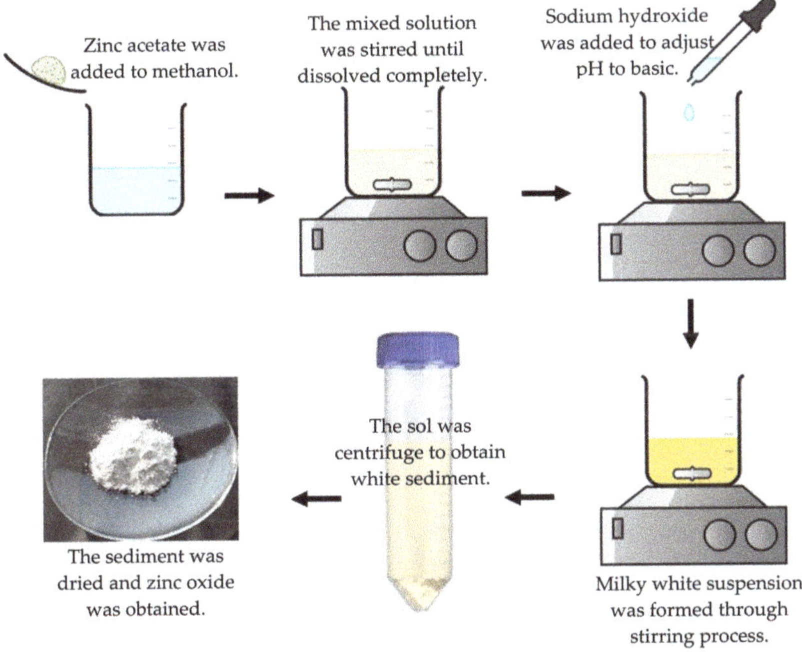

Figure 13. Illustration of ZnO photocatalyst preparation using a sol-gel technique.

Research conducted by Li et al. [95] reported that high purity of $Bi_2Mo_3O_{12}$ photocatalyst was produced in the Nano foam shaped using a sol-gel technique. Figure 14a presents the uniform and compacted size nanoparticles. The nanoparticles provide a large surface area in the catalyst, which is beneficial for catalytic activity. Meanwhile, Vinoth et al. [36] claimed that they obtained high purity crystal of $Nd_2Mo_3O_9$ without any redundant impurities. In addition, the flower-like photocatalyst shown in Figure 14b below contains a fairly uniform distribution of Nd, Mo, and O elements. Thus, the sol-gel method is believed to produce high purity and uniformly size photocatalyst. Moreover, the steps involved are quite simple and support sustainable green technology.

Figure 14. SEM images of (**a**) $Bi_2Mo_3O_{12}$ photocatalyst [95] (**b**) uniformly size and shape of $Nd_2Mo_3O_9$ [36].

4.2.3. Co-Precipitation Technique

The co-precipitation method is used to design different morphologies of photocatalysts based on the operating parameters. Both anionic and cationic solutions are mixed and stirred together to form a uniform mixture. The nucleation occurs in the formed mixture when one ion is replaced by another ion, forming the crystal lattice. The growing process will continue until the precipitation agent is added to agglomerate and formed the stable colloid suspension or precipitant. The obtained product from this method can be control either by solution concentration, pH, washing medium, and calcination temperature [96]. Figure 15 shows the illustration of the co-precipitation technique of Fe_3O_4 [105].

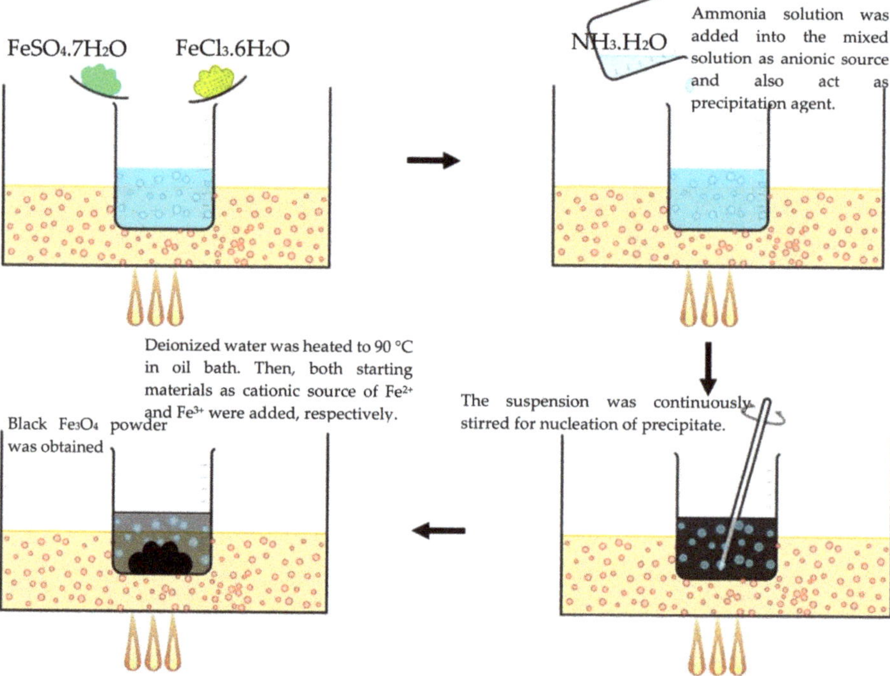

Figure 15. Illustration of the preparation of Fe_3O_4 by the co-precipitation technique.

As reported by Pan et al. [91], the ZnO photocatalyst obtained has a flower-like shape as shown in Figure 16a. They control the morphology of the catalyst by manipulating the concentration of Zn^{2+} and OH^-. Therefore, the concentration of aqueous $ZnCl_2$ and NaOH is increased to obtain a high reactant concentration. In line with Cao et al. [34], the porous hollow cube shape of $ZnFe_2O_4$ shown in Figure 16b was controlled by the synthesis of Prussian Blue precursor. During nucleation and growth, polyvinylpyrrolidone (PVP) was added as a capping agent, to produce the cube shape. In this process, calcination is needed to decompose PVP and cyanide ligands, while gas was also diffused to produce the small voids.

Figure 16. SEM images of (**a**) flower-like ZnO [91] (**b**) the cubic shaped of porous hollow $ZnFe_2O_4$ [34].

5. Degradation of Various Source of Pollutants

Generally, the sources of organic pollutants in water originally comes from domestic sewage, municipal sewage, and industrial wastewater, which commonly contained organic and inorganic types of pollutants that partially biodegrade. Moreover, even at low concentrations, it tends to poison aquatic organisms [106]. Therefore, chemical degradation process such as photocatalysis is required to remove the organic pollutant into more environmentally friendly substances. The pollutants that were reportedly studied were dye-based, antibiotic-based, and other-based pollutants such as herbicides and toxic substances.

Truong et al. [5], reported having used graphene@Fe-Ti binary oxide composites under the exposure of 350 W Xenon sunlight simulated lamp to degrade 10 mg/L methylene blue dye solution. About 6 mg photocatalyst was used and successfully degraded by about 100% for 20 min. According to Cheshme et al. [28], the Cu doped ZnO/Al_2O_3 under visible light of 400 W and a high-pressure mercury-vapor lamp with a wavelength of 546.8 nm, could degrade 100% of methyl orange (15 ppm) within 50 min with 0.6 g/L photocatalyst.

Based on Cao et al. [34], in their work, they successfully degraded about 84.08% of 500 ppm tetracycline hydrochloride using $ZnFe_2O_4$ photocatalyst within 50 min. 300 W Xenon lamp was used to supply the light to 100 mL solution with a 40 mg/L photocatalyst dosage. In keeping with Osotsi et al. [35], $ZnWO_{4-x}$ photocatalyst was used to degrade 20 ppm of tetracycline solution. About 2 g/L photocatalyst dosage was used under the exposure of 300 W of a ultra violet (UV) lamp. Therefore, 91% of degradation was achieved within 80 min.

As reported by Sekar et al. [102], 40 ppm of naphthalene was successfully degraded under a 16 W UV lamp using Fe-ZnO/PVA photocatalyst. About 96% of naphthalene solution was also degraded in 4 h with 0.06 g/L of Fe-ZnO/PVA. Moreover, Moyet et al. [41], used 0.1 g/L of Cu-BiOCl photocatalyst under the exposure of mercury UV lamp with 254 nm wavelength, degrading 10 ppm of atrazine solution in 30 min. Other examples are shown in Table 4.

Table 4. Example of a current study of the photocatalyst in degradation from a different source of organic pollutants.

Photocatalyst	Type of Light Used	Source of Pollution	Condition Set				Degradation	Ref.
			Catalyst Loading	Initial Concentration	Reaction Time	pH		
$BiVO_4/CHCOO(BiO)$	Visible light using 300 W Xenon lamp	Sulfamethoxazole	1 g/L	10 mg/L	5 h	6	85%	[107]
		Bisphenol A	1 g/L	10 mg/L	5 h	6	99%	
		4-aminoantipyrine	1 g/L	10 mg/L	5 h	6	46%	
		Ibuprofen	1 g/L	10 mg/L	5 h	6	65%	
CuO/ZnO	500 W visible lamp	Methylene blue	1 g/L	10 mg/L	25 min	-	96.57%	[108]
$BiVO_4$/carbon	350 W Xenon lamp with 400 nm cut off filter	Methylene blue	1.0 g/L	0.0001 mol/L	180 min	-	95%	[109]
		Rhodamine B	1.0 g/L	1×10^{-5} mol/L	180 min	-	80%	
		Phenol	1.0 g/L	5 mg/L	5 h	-	50.13%	
CdS-reduced graphene oxide	300 W Xenon lamp with UV cut off filter	Rhodamine B	0.4 g/L	20 mg/L	60 min	-	97.2%	[110]
		Acid chrome blue K	0.4 g/L	20 mg/L	60 min	-	65.7%	
$BiVO_4/Bi_4V_2O_{10}$	300 W Xenon lamp with 400 nm cut off filter	Rhodamine B	1 g/L	20 mg/L	15 min	-	100%	[111]
		Methylene blue	250 g/L	20 mg/L	60 min	-	75%	
		Phenol	1 g/L	30 mg/L	60 min	-	95%	
$H_3PW_{12}O_{40}/Ag_3PO_4$	300 W Xe lamp λ > 420 nm	4-fluorophenol	3 g/L	10 mg/L	720 s	7	100%	[112]
		Methyl orange	3 g/L	10 mg/L	720 s	7	100%	
ZnS:Mn/MWCNT	Low-pressure mercury lamp	AR18 dye	0.1 g/L	20 mg/L	180 min	-	70%	[68]
CdS/TiO_2	300 W xenon lamp with 420 nm cut off filter	Hexavalent chromium	2 g/L	10 mg/L	5 h	3.5	100%	[113]
		Phenol	1 g/L	10 mg/L	3 h	-	78%	
		Rhodamine B	1 g/L	10 mg/L	60 min	-	83%	
Zr/TiO_2	300 W Xenon lamp with 320 nm cut off filter	Chloridazon	0.1 g/L	0.005 mM	4 h	5	100%	[114]
		Phenol	0.1 g/L	0.001 mM	4 h	5.45	90%	
		4-chlorophenol	0.1 g/L	0.001 mM	4 h	5.53	95%	
WO_3	400 W metal halide lamp. Light intensity ≈ 86,800 lx	Rhodamine B	1 g/L	20 mg/L	3 h	-	95%	[115]
$RP-MoS_2/rGO$	300 W Xenon lamp 420 nm cut off filter	Rhodamine B	0.4 g/L	20 mg/L	30 min	-	99.3%	[116]
		Hexavalent chromium	0.4 g/L	40 mg/L	30 min	-	98%	

6. Effect of Parameters on the Efficiency of Photocatalytic Degradation Process

Several variables can affect the performance of the photocatalysis process. The variables include catalyst dosage, pH, irradiation time, temperature, and initial concentration. These variables might be affecting the photodegradation of pollutants. Therefore, an analysis with a series of experiments is needed to get the optimum operational parameters for the process.

6.1. Photocatalyst Dosage

Catalyst dosage has a major impact on the photocatalysis process as described by many researchers. According to Cheshme et al. [28], the dosage range of 0.005 to 0.04 g of Al_2O_3/ZnO:Cu photocatalyst was used for the degradation of 15 mg/L of methyl orange. However, the increment in photodegradation of methyl orange solution occurs with the addition of only 0.005 to 0.03 g of catalyst dosage. However, the addition of 0.04 g of catalyst exceeded the optimum amount of the amount of catalyst used and the abundance of catalysts will only reduce its performance. Therefore, the optimum photocatalyst loading was found to be 0.03 g. According to Elhalil et al. [43], about 0.1 to 0.3 g/L showed an increment in degradation efficiency of 20 mg/L caffeine from 69.42% to 98.9%. As the amount of photocatalyst increased to more than 0.3 g/L, the photocatalytic degradation showed a slight decrease. This was due to the excess amount of photocatalyst, which scattered the light and reduced its penetration into the solution. Therefore, their optimum photocatalyst dosage was found to be 0.3 g with 98.9% caffeine solution degradation.

6.2. pH of Wastewater Sample

The pH plays a major role in photocatalytic degradation. This is because it has an important role to play on the surface charge of the material in the aqueous media, especially in adsorption studies. If the surface charge of the material is opposing the adsorption due to the fact of having the same charge as the adsorbate, which needs to be modified and find the pH conditions that show the best adsorption. Therefore, there is a need to find out the pH at which the surface charge of the material is zero in the aqueous media or the pH_{pzc} (pH point of zero charge) of the adsorbent material [117,118].

Elhalil et al. [43], discovered the optimum pH value of 9.5 for Mg-ZnO-Al_2O_3 photocatalyst to degrade 20 mg/L caffeine solution. The dramatically decreased was found with a pH of 3.5. The researchers reported that the pH solution affects the surface charge and ionization of caffeine molecules, as a result, it enhanced hydroxyl radical formation. Compared to a pH of 3.5, many factors contribute simultaneously such as non-favorable adsorption, dissolution, and decomposition of the photocatalyst. However, Subash et al. [119], reported that the optimum pH of 9 was recorded for the higher adsorption efficiency for photodegradation of Acid Black 1 dye solution recording up to 90.1% using ZnO photocatalyst. At acidic pH level, the removal efficiency was less due to the dissolution of the photocatalyst. Therefore, the solution of pH must be tested as the photocatalyst surface charge is inverse to the solution charge.

6.3. Irradiation Intensity

Irradiation of light intensity affects the photocatalysis process via the production of more hydroxyl radicals as intensity increase [120]. Tekin and Saygi [121] reported that at a higher intensity (132 W/m^2), 25 mg/L of acid black 1 dye solution was degraded completely for 40 min reactions. Compared to low intensity (44 W/m^2), which has only an 80% degradation rate. Bhatia et al. [122] also reported the degradation of atenolol solution (25 mg/L) with 1.5 g/L TiO_2-graphene photocatalyst under 1000 mW/cm^2, the degradation almost completed in 60 min compared to under irradiation of 250 mW/cm^2, the solution degraded to 20 mg/L only. An increase in the irradiation light intensity can produce more photons to interact with a photocatalyst, thus, releasing more hydroxyl radicals and superoxide ions. However, the effective cost should be calculated to determine the best light source intensity and high efficiency of photodegradation of pollutants.

6.4. Temperature of Wastewater Sample

Usually, the elevation of temperature will increase the photodegradation within a certain range. It is agreeable to Tambat et al. [123] findings, which reported that at room temperature and high temperature (60 °C), the methyl orange solutions were completely degraded using CeO_2 photocatalyst. In contrary to low temperature (20 °C), the solution was not degraded completely. Moreover, according to Ateş et al. [124], the degradation of methyl orange solution using Al_2O_3-NP/SnO_2 photocatalyst increased from 60.90% to 93.95% as the temperature increased from 10 to 55 °C. However, beyond that, the rate of reaction temperature could give a negative effect when electron–hole recombine and cause a decrease in degradation reaction rate [121]. Thus, it is crucial to control the temperature in the range of reaction temperature.

6.5. Initial Concentration of Wastewater Sample

It is important to specify the optimum initial concentration of solution in the photocatalytic process due to the synergistic effect between photocatalyst, the formation of the oxidizing agent, and the pollutant. Subash et al. [119] reported a decrease in removal efficiency when they increased the Acid Black 1 dye concentration. The report claimed that the path length of the photons from the light source decreases, thus, it did not fully penetrate the solution. On the contrary, Elhalil et al. [43] increased the initial concentration of caffeine solution, which lead to an increase in the degradation rate. This was due to the higher ability of collision between hydroxyl radicals and that of caffeine molecules. Therefore, the initial concentration is one of the variables that can influence the photocatalysis.

6.6. Stability and Durability of Photocatalyst

The stability and durability of the photocatalyst are important to ensure the efficiency and the quality of the synthesized catalyst. Based on the previous studies, the morphological characteristics affect the stability and durability of the photocatalyst. The reduction of performance of the photocatalyst is known as the deactivation process. This phenomenon is also depending on the type of reactions, solvents, and the mechanical process. Most of the photocatalyst deactivation is discovered after several cycles of the reaction process. Several factors lead to the deactivation of the photocatalyst. These include (1) loss of photocatalyst mass especially during the washing/purification process, for example, the research reported by Taddesse et al. [125], where the Rod-like ZnO stacking on Cu_2O/Ag_3PO_4, which has lost some quantity during filtration; (2) leaching of dopants is also a common phenomenon that normally happens during the reaction due to photo-etching as experienced by Zhang et al. [126], when the synthesized ZnO catalyst undergo photodecomposition after three cycles in degradation of Rhodamine B; (3) residual pollutant and/or organic intermediates that adsorbed on the surface of photocatalyst. This normally occurs for the nanosized-photocatalyst with high porosity surface. For example, the nanocomposite of WO_3-ZnO has uniform surface morphology with a concomitant irregular distribution with an average thickness of 27 nm, which showed a gradual decrease in degradation until the fifth cycle. The researchers discovered that the methylene blue of organic intermediates adsorbed on the photocatalyst surface even on every cycle [127,128]. In summary, the strength, durability, and stability of the catalyst towards a harsh environment need to be analyzed and carefully studied. This is to ensure that the prepared catalyst has high reusability and recycle the ability to minimize the cost of the process.

7. Future Direction

Currently, several photocatalysts were developed and introduced, especially for water treatment technologies. This is due to their inexpensive cost, efficiency, and being environmentally friendly. However, every technique introduced in the reported papers in past decades have their own advantages and weaknesses in terms of catalyst stability, efficiency, cost production, structure, and performance of the catalyst. In this review, the recent reported common techniques used to prepare the catalyst to provide a better understanding to the readers and researchers were analyzed.

For example, the most common technique used was co-precipitation, which is easy to control the morphology of the catalyst. However, this technique involves several problems such as being easy to be contaminated and complex purification process. Tremendous attention has been put on the improvement of the discussed techniques. The factors that need to focus to improve the techniques such as (1) bandgap of the semi-conductors; (2) type of carrier transport; (3) crystallinity of the materials; (4) surface area; (5) stability of photocatalyst, which could be controlled using the proper technique of preparation. All these factors potentially affect the electron–hole recombination, synergic between pollutant and photocatalyst, purities of the materials, availability of active sites, and reusability of the photocatalyst, respectively. These factors are the clear paths of directions for the next researchers who studied the sciences behind photocatalysis technology.

The hurdles, limitations, challenges, and research gap in this evolution of studies become an opportunity for researchers to explore and find a clear direction for them. Figure 17 shows the key factors of different types of hybridization of photocatalyst, which might be the main guide or reference for other researchers to study the effect of phase structure towards the catalytic activity.

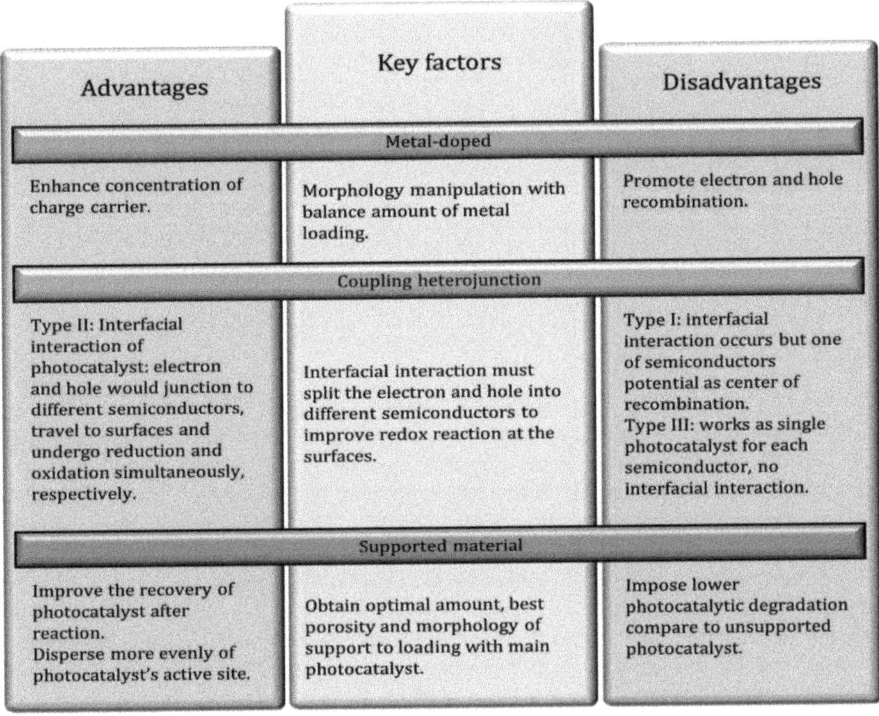

Figure 17. Different types of hybridization photocatalyst and the key factor to focus.

This review offers a comprehensive summary and novel insight into photocatalysis: These insights include (i) hybridization of photocatalyst; (ii) preparation techniques; (iii) degradation of various sources; (iv) effects of parameter towards photocatalysis reaction, thus aiming the future study to design and develop improved photocatalysts that are functional and hence facilitating the photocatalysis process for industrial scale. Figure 18 shows the importance of optimum reaction parameters during photodegradation processes. This is crucial to ensure the maximum capacity and performance of the catalyst–reactant relationship.

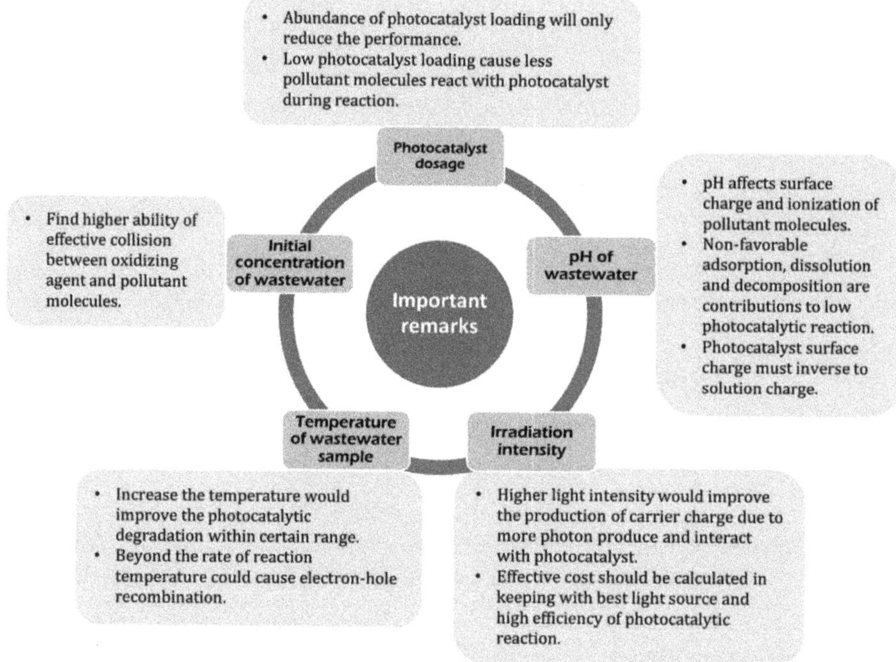

Figure 18. Important remarks: Parameter effects on photocatalytic degradation efficiency.

8. Summary

Photocatalyst shows promising techniques to degrade organic pollutants to environmentally friendly substances by allowing both spontaneous and non-spontaneous reactions. From the previous studies, the single semiconductor normally facing the recombination problem of electron–hole pairs. The hybridization photocatalyst was proposed to overcome the problems. However, it has slight weaknesses such as promoting the center of recombination, absence of interfacial interaction, and imposing lower photocatalytic degradation. Consequently, the proper preparation and hybridization techniques are important to ensure the maximum potential of the catalyst, which was discussed in this review. Adequate methods were proposed by researchers to synthesize the photocatalyst, however, to the best of our knowledge, no study compares the methods and quality of photocatalyst production. Therefore, this manuscript provides detailed observation, comparison, and conclusion on the previously reported papers working on photocatalysis, especially for organic wastewater treatment in recent years. Moreover, clear direction to overcome the challenges to the researchers and society from the basis of key factors and parameters were highlighted and discussed.

Author Contributions: Writing—original draft preparation, S.I.S.M.; conceptualization, S.I.S.M. and U.R.; visualization, S.I.S.M.; methodology, S.I.S.M. and M.F.K.; writing—review and editing, M.L.I., N.A.M., and A.I.; investigation, S.I.S.M. and Y.H.T.; validation, M.L.I., M.F.K, M.S.M., N.M., and N.H.M.K.; supervision, M.L.I., A.H.A., and T.-Y.Y.H. All authors have read and agreed to the published version of the manuscript.

Funding: This research was funded by Malaysia Higher Education for the FRGS research fund File No. FRGS/1/2018/STG07/UITM/03/5.

Acknowledgments: The authors would acknowledge the Malaysia Higher Education and Universiti Teknologi MARA (UiTM) for the FRGS research fund File No. FRGS/1/2018/STG07/UITM/03/5. Special thanks to the Institute of Science (IOS), Universiti Teknologi MARA for all the facilities provided throughout this work.

Conflicts of Interest: The authors declare no conflict of interest.

References

1. Palaniappan, M.; Gleick, P.H.; Allen, L.; Cohen, M.J.; Christian-Smith, J.; Smith, C. References. In *Clearing the Waters. A Focus on Water Quality Solutions*; Ross, N., Ed.; Pacific Institute: Nairobi, Kenya, 2010; pp. 1–88. ISBN 978-92-807-3074-6.
2. Fujushima, A.; Honda, K. Electrochemical photolysis of water at a semiconductor electrode. *Nature* **1972**, *238*, 37–38. [CrossRef] [PubMed]
3. Ong, C.B.; Ng, L.Y.; Mohammad, A.W. A review of ZnO nanoparticles as solar photocatalysts: Synthesis, mechanisms and applications. *Renew. Sustain. Energy Rev.* **2018**, *81*, 536–551. [CrossRef]
4. Vaiano, V.; Sacco, O.; Matarangolo, M. Photocatalytic degradation of paracetamol under UV irradiation using TiO_2-graphite composites. *Catal. Today* **2018**, *315*, 230–236. [CrossRef]
5. Truong, N.T.; Thi, H.P.N.; Ninh, H.D.; Phung, X.T.; Van Tran, C.; Nguyen, T.T.; Pham, T.D.; Dang, T.D.; Chang, S.W.; Rene, E.R.; et al. Facile fabrication of graphene@Fe-Ti binary oxide nanocomposite from ilmenite ore: An effective photocatalyst for dye degradation under visible light irradiation. *J. Water Process Eng.* **2020**, *37*, 101474. [CrossRef]
6. He, Y.; Sutton, N.B.; Rijnaarts, H.H.H.; Langenhoff, A.A.M. Degradation of pharmaceuticals in wastewater using immobilized TiO_2 photocatalysis under simulated solar irradiation. *Appl. Catal. B Environ.* **2016**, *182*, 132–141. [CrossRef]
7. Kosera, V.S.; Cruz, T.M.; Chaves, E.S.; Tiburtius, E.R.L. Triclosan degradation by heterogeneous photocatalysis using ZnO immobilized in biopolymer as catalyst. *J. Photochem. Photobiol. A Chem.* **2017**, *344*, 184–191. [CrossRef]
8. Hasnidawani, J.N.; Azlina, H.N.; Norita, H.; Bonnia, N.N.; Ratim, S.; Ali, E.S. Synthesis of ZnO nanostructures using sol-gel method. *Procedia Chem.* **2016**, *19*, 211–216. [CrossRef]
9. Shaban, M.; Abdallah, S.; Khalek, A.A. Characterization and photocatalytic properties of cotton fibers modified with ZnO nanoparticles using sol–gel spin coating technique. *Beni-Suef Univ. J. Basic Appl. Sci.* **2016**, *5*, 277–283. [CrossRef]
10. Abas, M.T. *Physical Chemistry II*, 3rd ed.; UiTM Press: Shah Alam, Malaysia, 2009.
11. Wu, B.; Lu, S.; Xu, W.; Cui, S.; Li, J.; Han, P.F. Study on corrosion resistance and photocatalysis of cobalt superhydrophobic coating on aluminum substrate. *Surf. Coat. Technol.* **2017**, *330*, 42–52. [CrossRef]
12. Stojadinović, S.; Tadić, N.; Radić, N.; Stojadinović, B.; Grbić, B.; Vasilić, R. Synthesis and characterization of Al_2O_3/ZnO coatings formed by plasma electrolytic oxidation. *Surf. Coat. Technol.* **2015**, *276*, 573–579. [CrossRef]
13. Bansal, P.; Verma, A.; Talwar, S. Detoxification of real pharmaceutical wastewater by integrating photocatalysis and photo-Fenton in fixed-mode. *Chem. Eng. J.* **2018**, *349*, 838–848. [CrossRef]
14. Bharatvaj, J.; Preethi, V.; Kanmani, S. Hydrogen production from sulphide wastewater using Ce^{3+}–TiO_2 photocatalysis. *Int. J. Hydrogen Energy* **2018**, *43*, 3935–3945. [CrossRef]
15. Mera, A.C.; Martínez-de la Cruz, A.; Pérez-Tijerina, E.; Meléndrez, M.F.; Valdés, H. Nanostructured BiOI for air pollution control: Microwave-assisted synthesis, characterization and photocatalytic activity toward NO transformation under visible light irradiation. *Mater. Sci. Semicond. Process.* **2018**, *88*, 20–27. [CrossRef]
16. Faraldos, M.; Kropp, R.; Anderson, M.A.; Sobolev, K. Photocatalytic hydrophobic concrete coatings to combat air pollution. *Catal. Today* **2016**, *259*, 228–236. [CrossRef]
17. Ohtani, B. Photocatalysis A to Z—What we know and what we do not know in a scientific sense. *J. Photochem. Photobiol. C Photochem. Rev.* **2010**, *11*, 157–178. [CrossRef]
18. Ahmad, F.M.A.; Hassan, M.A.; Taufiq-Yap, Y.H.; Ibrahim, M.L.; Othman, M.R.; Ali, A.A.M.; Shirai, Y. Production of methyl esters from waste cooking oil using a heterogeneous biomass-based catalyst. *Renew. Energy* **2017**, *114*, 638–643. [CrossRef]

19. Ibrahim, M.L.; Nik, A.K.N.N.A.; Islam, A.; Rashid, U.; Ibrahim, S.F.; Mashuri, S.I.S.; Taufiq-Yap, Y.H. Preparation of Na_2O supported CNTs nanocatalyst for efficient biodiesel production from waste-oil. *Energy Convers. Manag.* **2020**, *205*, 112445. [CrossRef]

20. Mansir, N.; Hwa, S.T.; Lokman, I.M.; Taufiq-Yap, Y.H. Synthesis and application of waste egg shell derived CaO supported W-Mo mixed oxide catalysts for FAME production from waste cooking oil: Effect of stoichiometry. *Energy Convers. Manag.* **2017**, *151*, 216–226. [CrossRef]

21. Ahmad, F.M.A.; Hassan, M.A.; Taufiq-Yap, Y.H.; Ibrahim, M.L.; Hasan, M.Y.; Ali, A.A.M.; Othman, M.R.; Shirai, Y. Kinetic and thermodynamic of heterogeneously K_3PO_4/AC-catalysed transesterification via pseudo-first order mechanism and Eyring-Polanyi equation. *Fuel* **2018**, *232*, 653–658. [CrossRef]

22. Ibrahim, S.F.; Asikin-Mijan, N.; Ibrahim, M.L.; Abdulkareem-Alsultan, G.; Izham, S.M.; Taufiq-Yap, Y.H. Sulfonated functionalization of carbon derived corncob residue via hydrothermal synthesis route for esterification of palm fatty acid distillate. *Energy Convers. Manag.* **2020**, *210*, 112698. [CrossRef]

23. Mahlambi, M.M.; Ngila, C.J.; Mamba, B.B. Recent developments in environmental photocatalytic degradation of organic pollutants: The case of titanium dioxide nanoparticles-A review. *J. Nanomater.* **2015**, *2015*, 1–29. [CrossRef]

24. Gogate, P.R.; Pandit, A.B. A review of imperative technologies for wastewater treatment I: Oxidation technologies at ambient conditions. *Adv. Environ. Res.* **2004**, *8*, 501–551. [CrossRef]

25. Low, J.; Yu, J.; Ho, W. Graphene-based photocatalysts for CO_2 reduction to solar fuel. *J. Phys. Chem. Lett.* **2015**, *6*, 4244–4251. [CrossRef] [PubMed]

26. Zhang, Y.; Geißen, S.U.; Gal, C. Carbamazepine and diclofenac: Removal in wastewater treatment plants and occurrence in water bodies. *Chemosphere* **2008**, *73*, 1151–1161. [CrossRef] [PubMed]

27. Sirota, J.; Baiser, B.; Gotelli, N.J.; Ellison, A.M. Organic-matter loading determines regime shifts and alternative states in an aquatic ecosystem. *Proc. Natl. Acad. Sci.USA* **2013**, *110*, 7742–7747. [CrossRef]

28. Cheshme, K.A.H.; Mahjoub, A.; Bayat, R.M. Low temperature one-pot synthesis of Cu-doped ZnO/Al_2O_3 composite by a facile rout for rapid methyl orange degradation. *J. Photochem. Photobiol. B Biol.* **2017**, *175*, 37–45. [CrossRef]

29. Nasr, M.; Viter, R.; Eid, C.; Habchi, R.; Miele, P.; Bechelany, M. Optical and structural properties of Al_2O_3 doped ZnO nanotubes prepared by ALD and their photocatalytic application. *Surf. Coat. Technol.* **2018**, *343*, 24–29. [CrossRef]

30. Gao, J.; Gao, Y.; Sui, Z.; Dong, Z.; Wang, S.; Zou, D. Hydrothermal synthesis of $BiOBr/FeWO_4$ composite photocatalysts and their photocatalytic degradation of doxycycline. *J. Alloys Compd.* **2018**, *732*, 43–51. [CrossRef]

31. Al-Balushi, B.S.M.; Al-Marzouqi, F.; Al-Wahaibi, B.; Kuvarega, A.T.; Al-Kindy, S.M.Z.; Kim, Y.; Selvaraj, R. Hydrothermal synthesis of CdS sub-microspheres for photocatalytic degradation of pharmaceuticals. *Appl. Surf. Sci.* **2018**, *457*, 559–565. [CrossRef]

32. Gong, Y.; Wu, Y.; Xu, Y.; Li, L.; Li, C.; Liu, X.; Niu, L. All-solid-state Z-scheme $CdTe/TiO_2$ heterostructure photocatalysts with enhanced visible-light photocatalytic degradation of antibiotic wastewater. *Chem. Eng. J.* **2018**, *350*, 257–267. [CrossRef]

33. Wang, D.; Jia, F.; Wang, H.; Chen, F.; Fang, Y.; Dong, W.; Zeng, G.; Li, X.; Yang, Q.; Yuan, X. Simultaneously efficient adsorption and photocatalytic degradation of tetracycline by Fe-based MOFs. *J. Colloid Interface Sci.* **2018**, *519*, 273–284. [CrossRef] [PubMed]

34. Cao, Y.; Lei, X.; Chen, Q.; Kang, C.; Li, W.; Liu, B. Enhanced photocatalytic degradation of tetracycline hydrochloride by novel porous hollow cube $ZnFe_2O_4$. *J. Photochem. Photobiol. A Chem.* **2018**, *364*, 794–800. [CrossRef]

35. Osotsi, M.I.; Macharia, D.K.; Zhu, B.; Wang, Z.; Shen, X.; Liu, Z.; Zhang, L.; Chen, Z. Synthesis of $ZnWO_{4-x}$ nanorods with oxygen vacancy for efficient photocatalytic degradation of tetracycline. *Prog. Nat. Sci. Mater. Int.* **2018**, *28*, 408–415. [CrossRef]

36. Vinoth, K.J.; Karthik, R.; Chen, S.M.; Chen, K.H.; Sakthinathan, S.; Muthuraj, V.; Chiu, T.W. Design of novel 3D flower-like neodymium molybdate: An efficient and challenging catalyst for sensing and destroying pulmonary toxicity antibiotic drug nitrofurantoin. *Chem. Eng. J.* **2018**, *346*, 11–23. [CrossRef]

37. Tu, H.; Li, D.; Yi, Y.; Liu, R.; Wu, Y.; Dong, X.; Shi, X.; Deng, H. Incorporation of rectorite into porous polycaprolactone/TiO_2 nanofibrous mats for enhancing photocatalysis properties towards organic dye pollution. *Compos. Commun.* **2019**, *15*, 58–63. [CrossRef]

38. Zhou, F.; Yan, C.; Liang, T.; Sun, Q.; Wang, H. Photocatalytic degradation of Orange G using sepiolite-TiO$_2$ nanocomposites: Optimization of physicochemical parameters and kinetics studies. *Chem. Eng. Sci.* **2018**, *183*, 231–239. [CrossRef]

39. Liu, S.J.; Li, F.T.; Li, Y.L.; Hao, Y.J.; Wang, X.J.; Li, B.; Liu, R.H. Fabrication of ternary g-C$_3$N$_4$/Al$_2$O$_3$/ZnO heterojunctions based on cascade electron transfer toward molecular oxygen activation. *Appl. Catal. B Environ.* **2017**, *212*, 115–128. [CrossRef]

40. Stojadinović, S.; Vasilić, R.; Radić, N.; Tadić, N.; Stefanov, P.; Grbić, B. The formation of tungsten doped Al$_2$O$_3$/ZnO coatings on aluminum by plasma electrolytic oxidation and their application in photocatalysis. *Appl. Surf. Sci.* **2016**, *377*, 37–43. [CrossRef]

41. Moyet, M.A.; Arthur, R.B.; Lueders, E.E.; Breeding, W.P.; Patterson, H.H. The role of Copper (II) ions in Cu-BiOCl for use in the photocatalytic degradation of atrazine. *J. Environ. Chem. Eng.* **2018**, *6*, 5595–5601. [CrossRef]

42. Singh, P.; Mondal, K.; Sharma, A. Reusable electrospun mesoporous ZnO nanofiber mats for photocatalytic degradation of polycyclic aromatic hydrocarbon dyes in wastewater. *J. Colloid Interface Sci.* **2013**, *394*, 208–215. [CrossRef]

43. Elhalil, A.; Elmoubarki, R.; Farnane, M.; Machrouhi, A.; Sadiq, M.; Mahjoubi, F.Z.Z.; Qourzal, S.; Barka, N. Photocatalytic degradation of caffeine as a model pharmaceutical pollutant on Mg doped ZnO-Al$_2$O$_3$ heterostructure. *Environ. Nanotechnol. Monit. Manag.* **2018**, *10*, 63–72. [CrossRef]

44. Takanabe, K. Photocatalytic water splitting: Quantitative approaches toward photocatalyst by design. *ACS Catal.* **2017**, *7*, 8006–8022. [CrossRef]

45. Li, J.; Cushing, S.K.; Zheng, P.; Senty, T.; Meng, F.; Bristow, A.D.; Manivannan, A.; Wu, N. Solar hydrogen generation by a CdS-Au-TiO$_2$ sandwich nanorod array enhanced with au nanoparticle as electron relay and plasmonic photosensitizer. *J. Am. Chem. Soc.* **2014**, *136*, 8438–8449. [CrossRef]

46. Zhang, Z.; Wang, C.C.; Zakaria, R.; Ying, J.Y. Role of particle size in nanocrystalline TiO$_2$-based photocatalysts. *J. Phys. Chem. B* **1998**, *102*, 10871–10878. [CrossRef]

47. Ge, L.; Han, C.; Liu, J.; Li, Y. Enhanced visible light photocatalytic activity of novel polymeric g-C$_3$N$_4$ loaded with Ag nanoparticles. *Appl. Catal. A Gen.* **2011**, *409–410*, 215–222. [CrossRef]

48. Cao, J.; Yang, Z.H.; Xiong, W.P.; Zhou, Y.Y.; Peng, Y.R.; Li, X.; Zhou, C.Y.; Xu, R.; Zhang, Y.R. One-step synthesis of Co-doped UiO-66 nanoparticle with enhanced removal efficiency of tetracycline: Simultaneous adsorption and photocatalysis. *Chem. Eng. J.* **2018**, *353*, 126–137. [CrossRef]

49. Gupta, V.K.; Fakhri, A.; Azad, M.; Agarwal, S. Synthesis and characterization of Ag doped ZnS quantum dots for enhanced photocatalysis of Strychnine as a poison: Charge transfer behavior study by electrochemical impedance and time-resolved photoluminescence spectroscopy. *J. Colloid Interface Sci.* **2018**, *510*, 95–102. [CrossRef]

50. Nguyen, C.H.; Fu, C.C.; Juang, R.S. Degradation of methylene blue and methyl orange by palladium-doped TiO$_2$ photocatalysis for water reuse: Efficiency and degradation pathways. *J. Clean. Prod.* **2018**, *202*, 413–427. [CrossRef]

51. Zhang, L.; Jaroniec, M. Toward designing semiconductor-semiconductor heterojunctions for photocatalytic applications. *Appl. Surf. Sci.* **2018**, *430*, 2–17. [CrossRef]

52. Li, C.; Ma, Y.; Zheng, S.; Hu, C.; Qin, F.; Wei, L.; Zhang, C.; Duo, S.; Hu, Q. One-pot synthesis of Bi$_2$O$_3$/Bi$_2$O$_4$ p-n heterojunction for highly efficient photocatalytic removal of organic pollutants under visible light irradiation. *J. Phys. Chem. Solids* **2020**, *140*, 109376. [CrossRef]

53. Zhou, T.; Zhang, H.; Ma, X.; Zhang, X.; Zhu, Y.; Zhang, A.; Cao, Y.; Yang, P. Construction of AgI/Bi$_2$MoO$_6$/AgBi(MoO$_4$)$_2$ multi-heterostructure composite nanosheets for visible-light photocatalysis. *Mater. Today Commun.* **2020**, *23*, 100903. [CrossRef]

54. Lin, X.; Guo, X.; Shi, W.; Zhai, H.; Yan, Y.; Wang, Q. Quaternary heterostructured Ag-Bi$_2$O$_2$CO$_3$/Bi$_{3.64}$Mo$_{0.36}$O$_{6.55}$/Bi$_2$MoO$_6$ composite: Synthesis and enhanced visible-light-driven photocatalytic activity. *J. Solid State Chem.* **2015**, *229*, 68–77. [CrossRef]

55. Sakata, Y.; Tamaura, Y.; Imamura, H.; Watanabe, M. Preparation of a new type of CaSiO$_3$ with high surface area and property as a catalyst support. In *Studies in Surface Science and Catalysis*; Elsevier: Amsterdam, The Netherlands, 2006; Volume 162, pp. 331–338.

56. Zheng, Y.; Liu, J.; Cheng, B.; You, W.; Ho, W.; Tang, H. Hierarchical porous Al$_2$O$_3$@ZnO core-shell microfibres with excellent adsorption affinity for Congo red molecule. *Appl. Surf. Sci.* **2019**, *473*, 251–260. [CrossRef]

57. IUPAC. *Compendium of Chemical Terminology*; Nič, M., Jirát, J., Košata, B., Jenkins, A., McNaught, A., Eds.; IUPAC: Research Triagle Park, NC, USA, 2009.

58. Loddo, V.; Marcì, G.; Martín, C.; Palmisano, L.; Rives, V.; Sclafani, A. Preparation and characterisation of $TiO_{2(anatase)}$ supported on $TiO_{2(rutile)}$ catalysts employed for 4-nitrophenol photodegradation in aqueous medium and comparison with $TiO_{2(anatase)}$ supported on Al_2O_3. *Appl. Catal. B Environ.* **1999**, *20*, 29–45. [CrossRef]

59. Li, Y.; Yeung, K.L. Polymeric catalytic membrane for ozone treatment of DEET in water. *Catal. Today* **2019**, *331*, 53–59. [CrossRef]

60. Korolkov, I.V.; Mashentseva, A.A.; Güven, O.; Gorin, Y.G.; Kozlovskiy, A.L.; Zdorovets, M.V.; Zhidkov, I.S.; Cholach, S.O. Electron/gamma radiation-induced synthesis and catalytic activity of gold nanoparticles supported on track-etched poly(ethylene terephthalate) membranes. *Mater. Chem. Phys.* **2018**, *217*, 31–39. [CrossRef]

61. Popa, A.; Sasca, V.; Verdes, O.; Oszko, A. Preparation and catalytic properties of cobalt salts of Keggin type heteropolyacids supported on mesoporous silica. *Catal. Today* **2018**, *306*, 233–242. [CrossRef]

62. Ren, J.; Hao, P.; Sun, W.; Shi, R.; Liu, S. Ordered mesoporous silica-carbon-supported copper catalyst as an efficient and stable catalyst for catalytic oxidative carbonylation. *Chem. Eng. J.* **2017**, *328*, 673–682. [CrossRef]

63. Liu, P.; Wei, G.; He, H.; Liang, X.; Chen, H.; Xi, Y.; Zhu, J. The catalytic oxidation of formaldehyde over palygorskite-supported copper and manganese oxides: Catalytic deactivation and regeneration. *Appl. Surf. Sci.* **2019**, *464*, 287–293. [CrossRef]

64. Hu, E.; Wu, X.; Shang, S.; Tao, X.M.; Jiang, S.X.; Gan, L. Catalytic ozonation of simulated textile dyeing wastewater using mesoporous carbon aerogel supported copper oxide catalyst. *J. Clean. Prod.* **2016**, *112*, 4710–4718. [CrossRef]

65. Pawar, R.C.; Lee, C.S. Single-step sensitization of reduced graphene oxide sheets and CdS nanoparticles on ZnO nanorods as visible-light photocatalysts. *Appl. Catal. B Environ.* **2014**, *144*, 57–65. [CrossRef]

66. Zhang, D.; Zhang, L. Ultrasonic-assisted sol-gel synthesis of rugby-shaped $SrFe_2O_4$/reduced graphene oxide hybrid as versatile visible light photocatalyst. *J. Taiwan Inst. Chem. Eng.* **2016**, *69*, 156–162. [CrossRef]

67. Sacco, O.; Vaiano, V.; Matarangolo, M. ZnO supported on zeolite pellets as efficient catalytic system for the removal of caffeine by adsorption and photocatalysis. *Sep. Purif. Technol.* **2018**, *193*, 303–310. [CrossRef]

68. Sharifi, A.; Montazerghaem, L.; Naeimi, A.; Abhari, A.R.; Vafaee, M.; Ali, G.A.M.; Sadegh, H. Investigation of photocatalytic behavior of modified ZnS:Mn/MWCNTs nanocomposite for organic pollutants effective photodegradation. *J. Environ. Manag.* **2019**, *247*, 624–632. [CrossRef]

69. Zhang, G.; Jin, W.; Xu, N. Design and fabrication of ceramic catalytic membrane reactors for green chemical engineering applications. *Engineering* **2018**, *4*, 848–860. [CrossRef]

70. Neves, T.M.; Frantz, T.S.; do Schenque, E.C.C.; Gelesky, M.A.; Mortola, V.B. An investigation into an alternative photocatalyst based on CeO_2/Al_2O_3 in dye degradation. *Environ. Technol. Innov.* **2017**, *8*, 349–359. [CrossRef]

71. Boreriboon, N.; Jiang, X.; Song, C.; Prasassarakich, P. Fe-based bimetallic catalysts supported on TiO_2 for selective CO_2 hydrogenation to hydrocarbons. *J. CO_2 Util.* **2018**, *25*, 330–337. [CrossRef]

72. Hu, D.; Liu, C.; Li, L.; Lv, K.L.; Zhang, Y.H.; Li, J.L. Carbon dioxide reforming of methane over nickel catalysts supported on TiO_2 (001) nanosheets. *Int. J. Hydrogen Energy* **2018**, *3*, 21345–21354. [CrossRef]

73. Nouri, F.; Rostamizadeh, S.; Azad, M. Synthesis of a novel ZnO nanoplates supported hydrazone-based palladacycle as an effective and recyclable heterogeneous catalyst for the Mizoroki-Heck cross-coupling reaction. *Inorg. Chim. Acta* **2018**, *471*, 664–673. [CrossRef]

74. Fujita, S.I.; Mitani, H.; Zhang, C.; Li, K.; Zhao, F.; Arai, M. Pd and PdZn supported on ZnO as catalysts for the hydrogenation of cinnamaldehyde to hydrocinnamyl alcohol. *Mol. Catal.* **2017**, *442*, 12–19. [CrossRef]

75. Mitoraj, D.; Lamdab, U.; Kangwansupamonkon, W.; Pacia, M.; Macyk, W.; Wetchakun, N.; Beranek, R. Revisiting the problem of using methylene blue as a model pollutant in photocatalysis: The case of $InVO_4/BiVO_4$ composites. *J. Photochem. Photobiol. A Chem.* **2018**, *366*, 103–110. [CrossRef]

76. Li, J.; Liu, Z.; Wang, R. Support structure and reduction treatment effects on CO oxidation of SiO_2 nanospheres and CeO_2 nanorods supported ruthenium catalysts. *J. Colloid Interface Sci.* **2018**, *531*, 204–215. [CrossRef] [PubMed]

77. Larimi, A.; Khorasheh, F. Renewable hydrogen production by ethylene glycol steam reforming over Al_2O_3 supported Ni-Pt bimetallic nano-catalysts. *Renew. Energy* **2018**, *128*, 188–199. [CrossRef]

78. Shi, D.; Wang, H.; Kovarik, L.; Gao, F.; Wan, C.; Hu, J.Z.; Wang, Y. WO_x supported on γ-Al_2O_3 with different morphologies as model catalysts for alkanol dehydration. *J. Catal.* **2018**, *363*, 1–8. [CrossRef]

79. Sun, J.; Wang, Y.; Zou, H.; Guo, X.; Wang, Z. Ni catalysts supported on nanosheet and nanoplate γ-Al_2O_3 for carbon dioxide methanation. *J. Energy Chem.* **2019**, *29*, 3–7. [CrossRef]

80. Sobhani-Nasab, A.; Maddahfar, M.; Hosseinpour-Mashkani, S.M. $Ce(MoO_4)_2$ nanostructures: Synthesis, characterization, and its photocatalyst application through the ultrasonic method. *J. Mol. Liq.* **2016**, *216*, 1–5. [CrossRef]

81. Liang, B.; Han, D.; Sun, C.; Zhang, W.; Qin, Q. Synthesis of SnO/g-C_3N_4 visible light driven photocatalysts via grinding assisted ultrasonic route. *Ceram. Int.* **2018**, *44*, 7315–7318. [CrossRef]

82. Rokesh, K.; Nithya, A.; Jeganathan, K.; Jothivenkatachalam, K. A facile solid state synthesis of cone-like ZnO microstructure an efficient solar-light driven photocatalyst for Rhodamine B degradation. *Mater. Today Proc.* **2016**, *3*, 4163–4172. [CrossRef]

83. Miwa, T.; Kaneco, S.; Katsumata, H.; Suzuki, T.; Ohta, K.; Verma, S.C.; Sugihara, K. Photocatalytic hydrogen production from aqueous methanol solution with $CuO/Al_2O_3/TiO_2$ nanocomposite. *Int. J. Hydrogen Energy* **2010**, *35*, 6554–6560. [CrossRef]

84. Yang, Z.X.; Zhu, F.; Zhou, W.M.; Zhang, Y.F. Novel nanostructures of β-Ga_2O_3 synthesized by thermal evaporation. *Phys. E Low-Dimens. Syst. Nanostruct.* **2005**, *30*, 93–95. [CrossRef]

85. Zou, C.; Liang, F.; Xue, S. Synthesis and oxygen vacancy-related photocatalytic properties of ZnO nanotubes grown by thermal evaporation. *Res. Chem. Intermed.* **2015**, *41*, 5167–5176. [CrossRef]

86. Pan, Z.W. Nanobelts of semiconducting oxides. *Science* **2001**, *291*, 1947–1949. [CrossRef]

87. Li, S.; Zhao, Y.; Wang, C.; Li, D.; Gao, K. Fabrication and characterization unique ribbon-like porous $Ag/LaFeO_3$ nanobelts photocatalyst via electrospinning. *Mater. Lett.* **2016**, *170*, 122–125. [CrossRef]

88. Ahluwalia, S.; Prakash, N.T.; Prakash, R.; Pal, B. Improved degradation of methyl orange dye using bio-co-catalyst Se nanoparticles impregnated ZnS photocatalyst under UV irradiation. *Chem. Eng. J.* **2016**, *306*, 1041–1048. [CrossRef]

89. Parida, K.M.; Pradhan, A.C.; Das, J.; Sahu, N. Synthesis and characterization of nano-sized porous gamma-alumina by control precipitation method. *Mater. Chem. Phys.* **2009**, *113*, 244–248. [CrossRef]

90. Amouzegar, Z.; Naghizadeh, R.; Rezaie, H.R.; Ghahari, M.; Aminzare, M. Cubic $ZnWO_4$ nano-photocatalysts synthesized by the microwave-assisted precipitation technique. *Ceram. Int.* **2015**, *41*, 1743–1747. [CrossRef]

91. Li, B.; Wang, Y. Facile synthesis and enhanced photocatalytic performance of flower-like ZnO hierarchical microstructures. *J. Phys. Chem. C* **2010**, *114*, 890–896. [CrossRef]

92. Ayyub, P.; Multani, M.; Barma, M.; Palkar, V.R.; Vijayaraghavan, R. Size-induced structural phase transitions and hyperfine properties of microcrystalline Fe_2O_3. *J. Phys. C Solid State Phys.* **1988**, *21*, 2229–2245. [CrossRef]

93. Kumar, H.; Rani, R. Structural and optical characterization of ZnO nanoparticles synthesized by microemulsion route. *Int. Lett. Chem. Phys. Astron.* **2013**, *19*, 26–36. [CrossRef]

94. Karbassi, M.; Zarrintaj, P.; Ghafarinazari, A.; Saeb, M.R.; Mohammadi, M.R.; Yazdanpanah, A.; Rajadas, J.; Mozafari, M. Microemulsion-based synthesis of a visible-light-responsive Si-doped TiO_2 photocatalyst and its photodegradation efficiency potential. *Mater. Chem. Phys.* **2018**, *220*, 374–382. [CrossRef]

95. Li, P.; Wang, L.; Liu, H.; Li, R.; Xue, M.; Zhu, G. Facile sol-gel foaming synthesized nano foam $Bi_2Mo_3O_{12}$ as novel photocatalysts for Microcystis aeruginosa treatment. *Mater. Res. Bull.* **2018**, *107*, 8–13. [CrossRef]

96. Król, A.; Pomastowski, P.; Rafińska, K.; Railean-Plugaru, V.; Buszewski, B. Zinc oxide nanoparticles: Synthesis, antiseptic activity and toxicity mechanism. *Adv. Colloid Interface Sci.* **2017**, *249*, 37–52. [CrossRef] [PubMed]

97. Warner, J.H.; Schäffel, F.; Bachmatiuk, A.; Rümmeli, M.H. Methods for Obtaining Graphene. In *Graphene*; Elsevier: Amsterdam, The Netherlands, 2013; pp. 129–228.

98. Taylor, A.C. Advances in nanoparticle reinforcement in structural adhesives. In *Advances in Structural Adhesive Bonding*; Elsevier: Amsterdam, The Netherlands, 2010; pp. 151–182.

99. Cheaburu-Yilmaz, C.N.; Karasulu, H.Y.; Yilmaz, O. *Nanoscaled Dispersed Systems Used in Drug-Delivery Applications*; Elsevier: Philadelphia, PA, USA, 2018.

100. Tian, C.; Zhang, Q.; Wu, A.; Jiang, M.; Liang, Z.; Jiang, B.; Fu, H. Cost-effective large-scale synthesis of ZnO photocatalyst with excellent performance for dye photodegradation. *Chem. Commun.* **2012**, *48*, 2858–2860. [CrossRef] [PubMed]

101. Bhardwaj, N.; Kundu, S.C. Electrospinning: A fascinating fiber fabrication technique. *Biotechnol. Adv.* **2010**, *28*, 325–347. [CrossRef] [PubMed]

102. Sekar, A.D.; Muthukumar, H.; Chandrasekaran, N.I.; Matheswaran, M. Photocatalytic degradation of naphthalene using calcined Fe–ZnO/ PVA nanofibers. *Chemosphere* **2018**, *205*, 610–617. [CrossRef] [PubMed]

103. Rao, B.G.; Mukherjee, D.; Reddy, B.M. Novel Approaches for Preparation of Nanoparticles. In *Nanostructures for Novel Therapy*; Elsevier: Amsterdam, The Netherlands, 2017; pp. 1–36.

104. Mahdavi, R.; Ashraf Talesh, S.S. The effect of ultrasonic irradiation on the structure, morphology and photocatalytic performance of ZnO nanoparticles by sol-gel method. *Ultrason. Sonochem.* **2017**, *39*, 504–510. [CrossRef] [PubMed]

105. Fang, K.; Shi, L.; Yao, L.; Cui, L. Synthesis of novel magnetically separable $Fe_3O_4/Bi_{12}O_{17}Cl_2$ photocatalyst with boosted visible-light photocatalytic activity. *Mater. Res. Bull.* **2020**, *129*, 110888. [CrossRef]

106. Salman, S.R. Chemical Reactions Studied by Electronic Spectroscopy. In *Encyclopedia of Spectroscopy and Spectrometry*; Elsevier: Amsterdam, The Netherlands, 1999; pp. 246–252.

107. Zhang, Y.; Li, G.; Yang, X.; Yang, H.; Lu, Z.; Chen, R. Monoclinic $BiVO_4$ micro-/nanostructures: Microwave and ultrasonic wave combined synthesis and their visible-light photocatalytic activities. *J. Alloys Compd.* **2013**, *551*, 544–550. [CrossRef]

108. Bharathi, P.; Harish, S.; Archana, J.; Navaneethan, M.; Ponnusamy, S.; Muthamizhchelvan, C.; Shimomura, M.; Hayakawa, Y. Enhanced charge transfer and separation of hierarchical CuO/ZnO composites: The synergistic effect of photocatalysis for the mineralization of organic pollutant in water. *Appl. Surf. Sci.* **2019**, *484*, 884–891. [CrossRef]

109. Wang, X.; Zhou, J.; Zhao, S.; Chen, X.; Yu, Y. Synergistic effect of adsorption and visible-light photocatalysis for organic pollutant removal over $BiVO_4$/carbon sphere nanocomposites. *Appl. Surf. Sci.* **2018**, *453*, 394–404. [CrossRef]

110. Wei, X.N.; Ou, C.L.; Guan, X.X.; Peng, Z.K.; Zheng, X.C. Facile assembly of CdS-reduced graphene oxide heterojunction with enhanced elimination performance for organic pollutants in wastewater. *Appl. Surf. Sci.* **2019**, *469*, 666–673. [CrossRef]

111. Li, H.; Chen, Y.; Zhou, W.; Gao, H.; Tian, G. Tuning in $BiVO_4/Bi_4V_2O_{10}$ porous heterophase nanospheres for synergistic photocatalytic degradation of organic pollutants. *Appl. Surf. Sci.* **2019**, *470*, 631–638. [CrossRef]

112. Li, K.; Zhong, Y.; Luo, S.; Deng, W. Fabrication of powder and modular $H_3PW_{12}O_{40}/Ag_3PO_4$ composites: Novel visible-light photocatalysts for ultra-fast degradation of organic pollutants in water. *Appl. Catal. B Environ.* **2020**, *278*, 119313. [CrossRef]

113. Deng, Y.; Xiao, Y.; Zhou, Y.; Zeng, T.; Xing, M.; Zhang, J. A structural engineering-inspired CdS based composite for photocatalytic remediation of organic pollutant and hexavalent chromium. *Catal. Today* **2019**, *335*, 101–109. [CrossRef]

114. Mbiri, A.; Taffa, D.H.; Gatebe, E.; Wark, M. Zirconium doped mesoporous TiO_2 multilayer thin films: Influence of the zirconium content on the photodegradation of organic pollutants. *Catal. Today* **2019**, *328*, 71–78. [CrossRef]

115. Adhikari, S.; Sarath, C.K.; Kim, D.H.; Madras, G.; Sarkar, D. Understanding the morphological effects of WO_3 photocatalysts for the degradation of organic pollutants. *Adv. Powder Technol.* **2018**, *29*, 1591–1600. [CrossRef]

116. Bai, X.; Du, Y.; Hu, X.; He, Y.; He, C.; Liu, E.; Fan, J. Synergy removal of Cr(VI) and organic pollutants over $RP-MoS_2$/rGO photocatalyst. *Appl. Catal. B Environ.* **2018**, *239*, 204–213. [CrossRef]

117. Haque, M.M.; Muneer, M. Photodegradation of norfloxacin in aqueous suspensions of titanium dioxide. *J. Hazard. Mater.* **2007**, *145*, 51–57. [CrossRef] [PubMed]

118. Xiao, J.; Xie, Y.; Cao, H. Organic pollutants removal in wastewater by heterogeneous photocatalytic ozonation. *Chemosphere* **2015**, *121*, 1–17. [CrossRef]

119. Subash, B.; Krishnakumar, B.; Swaminathan, M.; Shanthi, M. Highly active Zr co-doped Ag-ZnO photocatalyst for the mineralization of Acid Black 1 under UV-A light illumination. *Mater. Chem. Phys.* **2013**, *141*, 114–120. [CrossRef]

120. Daneshvar, N.; Rabbani, M.; Modirshahla, N.; Behnajady, M.A. Kinetic modeling of photocatalytic degradation of Acid Red 27 in UV/TiO_2 process. *J. Photochem. Photobiol. A Chem.* **2004**, *168*, 39–45. [CrossRef]

121. Tekin, D.; Saygi, B. Photoelectrocatalytic decomposition of Acid Black 1 dye using TiO_2 nanotubes. *J. Environ. Chem. Eng.* **2013**, *1*, 1057–1061. [CrossRef]

122. Bhatia, V.; Malekshoar, G.; Dhir, A.; Ray, A.K. Enhanced photocatalytic degradation of atenolol using graphene TiO_2 composite. *J. Photochem. Photobiol. A Chem.* **2017**, *332*, 182–187. [CrossRef]

123. Tambat, S.; Umale, S.; Sontakke, S. Photocatalytic degradation of Milling Yellow dye using sol-gel synthesized CeO_2. *Mater. Res. Bull.* **2016**, *76*, 466–472. [CrossRef]

124. Ateş, S.; Baran, E.; Yazıcı, B. Fabrication of Al_2O_3 nanopores/SnO_2 and its application in photocatalytic degradation under UV irradiation. *Mater. Chem. Phys.* **2018**, *214*, 17–27. [CrossRef]

125. Taddesse, A.M.; Alemu, M.; Kebede, T. Enhanced photocatalytic activity of p-n-n heterojunctions ternary composite $Cu_2O/ZnO/Ag_3PO_4$ under visible light irradiation. *J. Environ. Chem. Eng.* **2020**, *8*, 104356. [CrossRef]

126. Zhang, J.Y.; Mei, J.Y.; Yi, S.S.; Guan, X.X. Constructing of Z-scheme 3D g-C_3N_4-ZnO@graphene aerogel heterojunctions for high-efficient adsorption and photodegradation of organic pollutants. *Appl. Surf. Sci.* **2019**, *492*, 808–817. [CrossRef]

127. Tayebee, R.; Esmaeili, E.; Maleki, B.; Khoshniat, A.; Chahkandi, M.; Mollania, N. Photodegradation of methylene blue and some emerging pharmaceutical micropollutants with an aqueous suspension of WZnO-NH_2@$H_3PW_{12}O_{40}$ nanocomposite. *J. Mol. Liq.* **2020**, *317*, 113928. [CrossRef]

128. Mahat, M.M.; Aris, A.H.M.; Jais, U.S.; Yahya, M.F.Z.R.; Ramli, R.; Bonnia, N.N.; Mamat, M.T. A preliminary study on microbiologically influenced corrosion (MIC) of mild steel by Pseudomonas aeruginosa by using infinite focus microscope (IFM). *AIP Conf. Proceed. J.* **2012**, *1455*, 117–123.

Publisher's Note: MDPI stays neutral with regard to jurisdictional claims in published maps and institutional affiliations.

MDPI

St. Alban-Anlage 66

4052 Basel

Switzerland

Tel. +41 61 683 77 34

Fax +41 61 302 89 18

www.mdpi.com

Catalysts Editorial Office

E-mail: catalysts@mdpi.com

www.mdpi.com/journal/catalysts